7th Edition

Technician's Guide
to Programmable Controllers

Terry R. Borden | Richard A. Cox

CENGAGE

Australia • Brazil • Canada • Mexico • Singapore • United Kingdom • United States

CENGAGE

Technician's Guide to Programmable Controllers, **7th Edition**
Terry Borden and Richard Cox

SVP, Higher Education Product Management:
Erin Joyner

VP, Product Management, Learning
Experiences: Thais Alencar

Product Director: Jason Fremder

Product Manager: Lauren Whalen

Product Assistant: Bridget Duffy

Learning Designer: Elizabeth Berry

Content Manager: Manoj Kumar,
Lumina Datamatics, Inc.

Digital Delivery Quality Partner:
Elizabeth Cranston

IP Analyst: Ashley Maynard

IP Project Manager: Nick Barrows

Production Service: Lumina Datamatics, Inc

Designer: Felicia Bennett

Cover and Interior Image Source: udorn_1976
/Shutterstock; Alexander Supertramp
/Shutterstock; Mr.B-king/Shutterstock

For product information and technology assistance, contact
us at **Cengage Customer & Sales Support, 1-800-354-9706
or support.cengage.com.**

For permission to use material from this text or product,
submit all requests online at **www.copyright.com.**

Library of Congress Control Number: 2021917612

ISBN: 978-0-357-62249-0

Cengage
200 Pier 4 Boulevard
Boston, MA 02210
USA

Cengage is a leading provider of customized learning solutions with
employees residing in nearly 40 different countries and sales in more
than 125 countries around the world. Find your local representative at
www.cengage.com.

To learn more about Cengage platforms and services, register or access
your online learning solution, or purchase materials for your course, visit
www.cengage.com.

Notice to the Reader

Publisher does not warrant or guarantee any of the products described herein or perform any independent analysis in
connection with any of the product information contained herein. Publisher does not assume, and expressly disclaims, any
obligation to obtain and include information other than that provided to it by the manufacturer. The reader is expressly
warned to consider and adopt all safety precautions that might be indicated by the activities described herein and to avoid all
potential hazards. By following the instructions contained herein, the reader willingly assumes all risks in connection with such
instructions. The publisher makes no representations or warranties of any kind, including but not limited to, the warranties of
fitness for particular purpose or merchantability, nor are any such representations implied with respect to the material set forth
herein, and the publisher takes no responsibility with respect to such material. The publisher shall not be liable for any special,
consequential, or exemplary damages resulting, in whole or part, from the readers' use of, or reliance upon, this material.

Printed at CLDPC, USA, 09-21

Table of Contents

Chapter 4

Chapter 5

Chapter 6

Chapter 7

Chapter 8

Chapter 12

Chapter 13

Chapter 14

Chapter 15

Chapter 19

Chapter 20

Chapter 21

Chapter 22

PREFACE

Programmable logic controllers, first introduced in 1969, have become an unqualified success. They can be found in almost every industrial and manufacturing facility worldwide. This computer-based device has become the industry standard, replacing the hard-wired electromechanical devices and circuits that had controlled the process machines and driven equipment of industry in the past.

Programmable logic controllers, or **PLCs** as they are referred to, vary in size and sophistication. When PLCs were first introduced, they typically used a dedicated programming device for entering and monitoring the PLC program. The programming device could only be used for programming a specific brand of PLC. These dedicated programmers, while user friendly, were very expensive and could not be used for anything except programming a PLC. Today, personal laptop computers have replaced the dedicated programming devices of the past.

Many electricians and/or technicians seem apprehensive about PLCs and their application in industry. One of the purposes of this text is to explain PLC basics using a plain, easy-to-understand approach so that electricians and technicians with no PLC experience will be more comfortable with their first exposure to PLCs.

Half the battle of understanding any PLC is to first understand the terminology of the PLC world. This text covers terminology, as well as explaining the input/output section, processor unit, memory organization, program instructions, communications, and much more.

A chapter has been included to explain not only ladder diagrams but also Relay Ladder Logic, which is the programming language used in the majority of programmable controllers. Additional chapters are included to cover Function Block, Structured Text, and Sequential Function Chart programming.

Examples of basic programming techniques used with typical PLCs are discussed and illustrated, as are the commonly used commands and functions. There are a variety of PLCs on the market today, and it would be impossible to write a book that explains how they all work and are programmed. Instead, this book is intended to discuss PLCs in a somewhat general, or generic, sense, and to cover the basic concepts of operation that are common to all. All of the examples used in this text are based on the Allen-Bradley Logix 5000 family of PLCs.

New to This Edition: Completely updated with all new color figures and photos, this edition features step-by-step programming examples using the latest Rockwell Automation, Inc. software and hardware. It also features expanded sections on analog configuration and troubleshooting. The introductory chapter on communication networks has been updated to include the latest industrial protocols. Programming instructions along with examples have been included throughout this edition including array and PID instructions.

Although this text only scratches the surface of available instructions and other advanced programming capabilities of the PLCs on the market today, the reader will gain a basic understanding of the most commonly used instructions and how they work. A chapter, "PLC Programming Examples," will provide the reader with several examples of PLC programming code/logic utilizing many of the instructions covered in previous chapters. The examples are intended to help the reader gain a better understanding of the various PLC instructions and how they can be combined to provide simple control logic solutions.

As with any new skill, a firm base of understanding is required before an electrician or technician can become proficient. After completing the text, the reader will possess a good foundation upon which additional PLC skills and understanding can be built.

The best teacher, of course, is experience, and the only way to really fully understand any given PLC is to work with that PLC and its associated software. If a PLC is not available, the next best thing is a workshop or seminar sponsored by a local PLC distributor. If a workshop or seminar is not available, obtain as much literature and other information as possible from a local electrical distributor, PLC representative, or the Internet.

PLC manufacturers and third-party vendors are continually adding new hardware that can communicate across different platforms. The day of proprietary hardware and communications is all but gone. With the rapid advancements in PLCs and software, the technician without an electronics background need not feel intimidated. The manufacturers are doing everything possible to make the PLC easy to install, program, troubleshoot, and maintain.

Technician's Guide to Programmable Controllers is organized to take the student from "What is a PLC?" to learning the basic hardware, software, communications, and programming languages used with most PLCs on the market today. Each chapter is designed to build on the knowledge gained from previous chapters. Students that complete the text will have a solid foundation and understanding of what PLCs are, how they are used, and the programming language required to make them perform the automation tasks required in today's industrial world.

INSTRUCTOR RESOURCES

Additional instructor resources for this product are available online. Instructor assets include an Instructor's Manual, PowerPoint® slides, and a test bank powered by Cognero®. Sign up or sign in at **www.cengage.com** to search for and access this product and its online resources.

NEW USERS

If you're new to Cengage.com and do not have a password, contact your sales representative.

About the Authors

Terry R. Borden is owner of Adept Consulting in Ione, Washington, and recently retired from Seattle City Light as Operations Manager of a 1,000-megawatt hydroelectric project in northeast Washington. Mr. Borden is a former member of the Electrical Maintenance and Automation Department at Spokane Community College in Spokane, Washington. He holds a degree in Industrial Automation and Robotics. Mr. Borden has worked as a control engineer and was a partner in Applied Solutions, LLC.

Richard A. Cox is the executive director of COXCO Training and Consulting in Spokane, Washington, and is a retired member of the Electrical/Robotics Department at Spokane Community College. He holds a Bachelor of Science degree from the University of New York, a Master of Science degree from Eastern Washington University, and is also a retired member of the International Brotherhood of Electrical Workers, Local 73.

The authors wish to acknowledge the cooperation and assistance of Rockwell Automation, Inc. whose product information, photographs, and software are used throughout the text to illustrate PLC concepts and components. Rockwell Automation, Inc. is a world leader in the PLC field, and offers a full line of programmable logic controllers and software for industrial automation. It is not a question of which PLC is best, but rather which PLC best fits your needs and individual applications.

Chapter 1

What Is a Programmable Logic Controller (PLC)?

Learning Objectives

After completing this chapter, you should have the knowledge to:

- Describe several advantages of a programmable logic controller (PLC) over hardwired relay systems.
- Identify the four major components of a typical PLC.
- Describe the function of the four major components of a typical PLC.
- Define the acronyms PLC, CPU, PC, I/O, ADC, and DAC.
- Define the term *discrete*.
- Define the term *analog*.

Introduction to Programmable Logic Controllers

A programmable logic controller is a solid-state system originally designed to perform the logic functions previously accomplished by components such as electromechanical relays, drum switches, mechanical timers/counters, etc., for the control and operation of manufacturing process equipment and machinery.

The electromechanical relay (control relays, pneumatic timer relays, etc.) had served well for many generations, often under adverse conditions. The ever-increasing sophistication and complexity of modern processing equipment required faster-acting, more reliable control functions than electromechanical relays and/or timing devices could offer. Relays had to be hardwired to perform a specific function, and when the system requirements changed, the relay wiring had to be changed or modified. In extreme cases, such as in the auto industry, complete control panels had to be replaced since it was not economically feasible to rewire the old panels with each model changeover.

It was, in fact, the requirements of the auto industry and other highly specialized, high-speed manufacturing processes that created a demand for smaller, faster-acting, and more reliable control devices. The electrical/electronics industry responded with modular-designed, solid-state electronic devices. These early devices, while offering solid-state reliability, lower power consumption, expandability, and elimination of much of the hardwiring, also brought with them a new language. The language consisted of AND, OR, NOT gates, J-K flip flops, and so on.

What happened to simple relay logic and ladder diagrams? That was the question plant engineers and maintenance electricians/technicians asked the solid-state device manufacturers. The reluctance of the end user to learn a new language and the advent of the microprocessor gave the industry what is now known as the **programmable logic controller (PLC)**. The first PLC was invented in 1969 by Richard (Dick) E. Morley, the founder of the Modicon Corporation.

Internally, there are still AND gates, OR gates, and so forth in the processor, but the design engineers have preprogrammed the PLC so that programs can be entered using Relay Ladder Logic. While Relay Ladder Logic may not have the mystique of other computer languages such as Python, Java, and C++, it is however a high-level, real-world, graphic programming language that is understood by most electricians and technicians. Relay Ladder Logic programming is the most common programming language used today, but other programming languages such as Sequential Function Chart, Structured Text, and Function Block languages can also be found. A brief description of each will be covered in later chapters.

The International Electrotechnical Commission (IEC) IEC61131 defines a programmable controller as:

> A digitally operating electronic system, designed for use in an industrial environment, which uses a programmable memory for internal storage of user-oriented instructions for implementing specific functions such as logic, sequencing, timing, counting and arithmetic, to control, through digital or analogue inputs and outputs, various types of machines or processes. Both the PLC and its associated peripherals are designed so that they can be easily integrated into an industrial control system and easily used in all their intended functions.

What does a PLC consist of, and how is it different from a computer system? The PLC consists of a programming device (computer), processor unit, power supply, memory, and an input/output (**I/O**) interface such as the system illustrated in Figure 1–1. And while there are similarities, there are also some major differences.

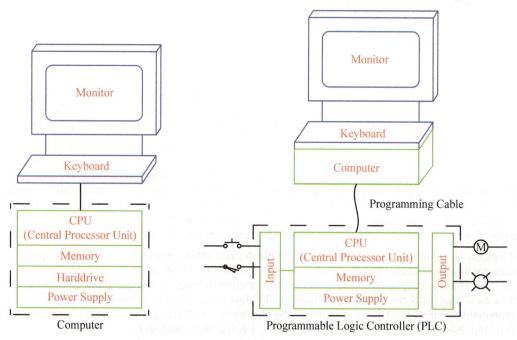

Figure 1–1 Comparison of a Computer System and a PLC

Note: An interface occurs when two systems come together and interact or communicate. In the case of the PLC, the communication or interaction is between the inputs (limit switches, push buttons, sensors, and the like), outputs (coils, solenoids, lights, and so forth), and the processor (CPU). This interface happens when any I/O voltage (AC or DC) or current signal is changed to or from a low-voltage DC signal that the processor uses internally for the decision-making process.

PLCs are designed to be operated and maintained by plant engineers and maintenance personnel with limited knowledge of computers. Like the computer, which has an internal memory for its operation and storage of programs and data, the PLC also has memory for storing the user program, or Logic, as well as memory for storing data. But unlike the computer, the PLC is typically programmed in Relay Ladder Logic, *not* one of the computer languages and is designed for controlling the operation of a process machine or driven equipment.

The PLC is also designed to operate in an industrial environment with wide ranges of ambient temperature, vibration, and humidity, and is not usually affected by the electrical noise that is inherent in most industrial locations.

Note: Electrical noise is discussed in Chapter 3.

Maybe one of the biggest, or at least the most significant difference between the PLC and a computer, is that PLCs have been designed for installation and maintenance by plant electricians who are not required to be highly skilled electronics or computer technicians. Troubleshooting is simplified in most PLCs because they include fault indicators, blown-fuse indicators, I/O status indicators, and written fault information that can be displayed on the PLC and/or programmer.

Although the PLC and the personal computer (PC) are different in many ways, the PC is often used for programming and monitoring the PLC. Using PCs in conjunction with PLCs will be discussed in later chapters.

The Main Components of a Programmable Logic Controller System

A typical PLC can be divided into four main components. These components consist of the **processor unit**, **power supply**, input/output section (I/O interface), and the programming device (**programmer**).

The processor unit or central processing unit (**CPU**) houses the processor, which is the decision-maker or "brain" of the system. The brain is a microprocessor-based system that replaces control relays, counters, timers, sequencers, and so forth, and is designed so that the user can enter the desired program in Relay Ladder Logic. The processor then makes all the decisions necessary to carry out the user program, based on the status of the inputs and outputs for control of a machine or process. It can also perform arithmetic functions, data manipulation, and communications between the local I/O section, remotely located I/O sections, and/or other networked PLC systems. Figure 1–2 shows the Allen-Bradley 1756-L85E PLC processor unit.

The power supply is necessary to convert the available power source, 120 or 240 volts AC, 24, 48, or 125 volts DC, to the low-voltage DC required for the logic circuits of the processor, and for the internal power required for the I/O modules. The power supply can be a separate unit as, shown in Figure 1–3, one of modular design that plugs into the processor chassis or rack, or, depending on the manufacturer, one that is an integral part of the processor.

Note: The power supply does not supply power for the actual input or output devices themselves; it only provides the power needed for the internal electronic circuitry of the input and output modules. Power for the input and output devices, if required, must be provided from a separate source.

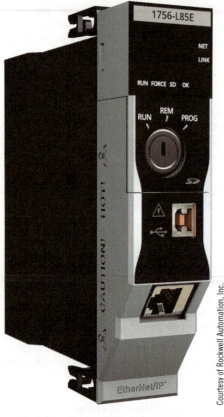

Courtesy of Rockwell Automation, Inc.

Figure 1–2 Allen-Bradley 1756-L85E PLC Processor Unit

Courtesy of Rockwell Automation, Inc.

Figure 1–3 Allen-Bradley PLC Power Supply

The size or power rating of the power supply is based on the number and type of I/O modules that are to be installed. Power supplies are normally available with output power ratings of 30 to 75 watts.

Note: Consider future needs and the possibility of expansion when initially sizing the power supply. It is cheaper in the long run to install a larger power supply initially than to try to add additional capacity at a later date.

The I/O section consists of I/O modules that are either fixed or modular. The number of inputs and outputs necessary is dictated by the requirements of the equipment that is to be controlled by the PLC. Figure 1–4 shows a chassis or rack with modular I/O modules installed. I/O modules are "plugged in" and added as needed.

I/O modules are where the real-world devices are connected. The real-world **input** (I) **devices** can be push buttons, limit switches, analog sensors, pressure switches, selector switches, etc., while the real-world **output** (O) **devices** can be hardwired motor starter coils, solenoid valves, indicator lights, positioning valves, and the like. The term *real world* is used to distinguish actual devices that exist and must be physically wired to the I/O modules as compared to internal functions of the PLC system that duplicate the function of relays, timers, counters, and so on, even though none physically exists. This may seem a bit strange and hard to understand at this point, but the distinction between what

Figure 1–4 PLC Chassis with I/O Modules Installed

the processor can do internally—which eliminates the need for all the previously used control relays, timers, counters, and so forth—will be graphically shown and readily understandable later in the text.

Real-world I/O devices are of two types: discrete and analog. Discrete I/O devices are either *ON* or *OFF*, open or closed, while analog devices have a range of possible values. Examples of discrete devices are limit switches, push buttons, motor starter coils, and indicator lamps. Examples of analog devices are pressure sensors, temperature probes, panel meters, variable speed drive signals, and modulating valves. When reference is made to an I/O device, the terms ***discrete input*** *device*, ***discrete output*** *device*, ***analog input*** *device*, and ***analog output*** *device* are commonly used to describe the type of device.

A reference was made earlier in this chapter to the I/O section as an interface. Although not a common reference, it is an accurate one. The I/O section contains the circuitry necessary to convert input voltages from *discrete input devices* to low-level DC voltages (typically 5V) that the processor uses internally to represent the status or condition (*ON* or *OFF*). Similarly, the I/O section changes low-level DC signals from the processor to AC or DC voltages required to operate the *discrete output devices*. The I/O section also converts varying voltage or current signals from *analog input devices* into corresponding decimal values by way of an Analog-to-Digital converter (ADC). This same process, but reversed via a Digital-to-Analog converter (DAC), is used by the I/O section to convert decimal values into corresponding voltage or current signals necessary to operate *analog output devices*. The field signals from both digital and analog devices are normally isolated from the low-level logic circuitry of the processor by means of optical coupling. This is a brief overview of the I/O section and its function. How I/O devices are wired to I/O modules, optical coupling, and more information about the module circuitry itself is covered in Chapter 3.

The programming device is used to enter the desired program or sequence of operation into the PLC memory. The program is entered using relay ladder logic, or one of the other PLC programming languages, and it is this program that determines the sequence of operation and ultimate control of the process equipment or driven machinery. The programming device is typically a PC.

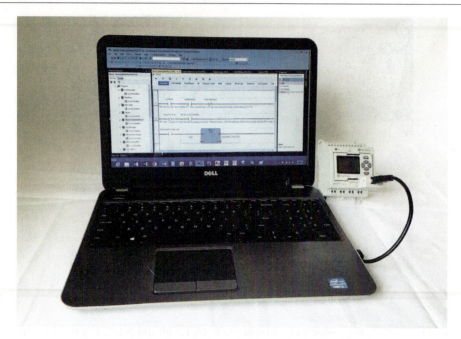

Figure 1–5 Laptop Computer Connected to an Allen-Bradley Micro PLC

A **PC** is used to program nearly all the PLCs on the market today. The PLC programming software that is installed on the PC and a communications cable is sometimes all that is required to program a PLC. At other times special hardware keys and/or communication cards are required to be installed on the PC for it to work successfully as a programming device. The PC provides the benefit of a large viewing screen that allows more of the program to be viewed at one time and makes troubleshooting and memory access much easier. It also provides program storage, as well as runs all the various software packages we have come to depend on today, such as spreadsheets, word processing, etc. Figure 1–5 shows a laptop PC that, with the appropriate software, is used to program and monitor a PLC.

Chapter Summary

Programmable logic controllers (PLCs) have made it possible to precisely control large process machines and driven equipment with less physical wiring and lower installation costs than is required with standard electromechanical relays, pneumatic timers, drum switches, and so on. The programmability allows for fast and easy changes in the Relay Ladder Logic to meet the changing needs of the process or driven equipment without the need for expensive and time-consuming rewiring. Designed to be "technician friendly," the modern PLC is easier to program and can be used by plant engineers and maintenance technicians who have little or no electronic or computer background.

Key Terms

programmable logic controller (PLC)

processor unit

programmer

input device

discrete input

analog input

I/O

power supply

CPU

output device

discrete output

analog output

Review Questions

1. List the four main components of a programmable logic controller.
2. Define the term *interface*.
3. Define the term *real world*.
4. Define the term *discrete*.
5. What do the following acronyms stand for?

   CPU   ADC   PC

   I/O   DAC   PLC
6. Define the term *analog*.
7. What is typically used to program and monitor PLCs?
8. Relay Ladder Logic is a high-level graphic computer language.

   T   F
9. What is the major advantage of a PLC system over the traditional hardwired control system?

Chapter 2

Numbering Systems

Learning Objectives

After completing this chapter, you should have the knowledge to:

- Describe decimal, binary, octal, hexadecimal, and binary coded decimal (BCD) numbering systems.
- Convert from one numbering system to another.
- Express negative numbers in 2s complement.
- Add signed numbers.
- Convert a negative binary display to its decimal equivalent.
- Complete a subtraction and addition problem using 2s complement.

Electricians, technicians, or other personnel who are required to program, modify, or maintain a PLC must have a "working" knowledge of the different numbering systems that are used. For example, the input/output addresses may use the octal numbering system; the timer and counter addresses may use the decimal numbering system; accumulated and preset values of the timers and counters may use the binary numbering system; and the hexadecimal system may be used for setting I/O module configurations. The numbering system used in each area discussed varies with the different PLC manufacturers, but it is obvious that to fully understand and program a PLC, an understanding of the various numbering systems is necessary.

Decimal System

Electricians and technicians use the decimal numbering system every day, therefore it is a system they are comfortable with. This system uses 10 unique numbers, or digits, which are 0–9. A numbering system that uses 10 digits is said to have a base of 10. The value of the decimal number depends on the digit(s) used and each digit's place value. Each position can be represented as a power of 10, starting with 10^0, as shown in Figure 2–1. In the decimal system, the first position to the left of the decimal point is called the units place, and any digit from 0–9 can be used. The next position to the left of the units place is the tens place; next is the hundreds place, the

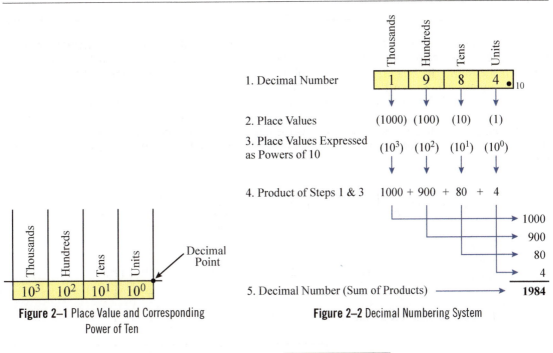

Figure 2–1 Place Value and Corresponding Power of Ten

Figure 2–2 Decimal Numbering System

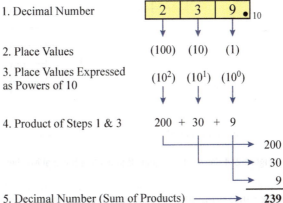

Figure 2–3 Decimal Numbering System

thousands place, and so on, with each place extending the capability of the decimal system by 10, or a power of 10.

Note: Any number that uses an exponent of 0, such as 10^0, has a place value of 1. Exponent 10^0 equals 1.

A specific decimal number can be expressed by adding the place values, as shown in Figure 2–2.

Mathematically, each place value is expressed as a digit number times a power of the base, or 10, in the decimal numbering system. Another example is shown in Figure 2–3 using the decimal number 239.

Binary System

The **binary** system uses only two digits: 1 and 0. Since only two digits are used, this system has a base of 2. Like the decimal system—and all numbering systems for that matter—each digit has a certain place value. The first place to the left of the starting point, or binary point, is the units or 1s location (base 2^0). The next place, to the left of the units place, is the 2s place, or base 2^1, as shown in Figure 2–4. The next place value is the 4s place, or base 2^2, then the 8s place, or base 2^3, and so forth. A binary number is always indicated by placing a 2 in subscript to the right of the unit's digit. Figure 2–4 illustrates how a binary number is converted to a decimal equivalent number. Note the subscripted 2 at the lower right-hand corner of the binary number line that indicates a base 2, or binary number.

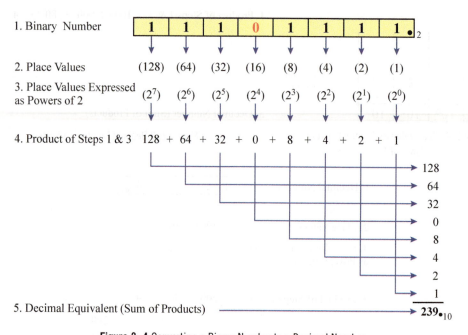

Figure 2–4 Converting a Binary Number to a Decimal Number

To convert a decimal number into a binary number, or to any numbering system for that matter, use the following procedure, as shown in Figure 2–5. Divide the decimal number by the base you wish to convert to, in this case 2. The remainder is the 1s value (see Step 1). Now divide the quotient from the first division again; the remainder becomes the value that is placed in the 2s location (see Step 2). The quotient of each preceding division is then divided by the base 2 until the base can no longer be divided (see Step 8), and the remainder (1) becomes the last digit in the binary number.

It is important to arrange the remainders correctly when making the decimal-to-binary conversion. The first digit placed in the 1s position is called the *least* significant digit, whereas the last digit is called the *most* significant digit. The last digit placed has the highest place value (128s) which is why it is called the most significant digit. This reference to least and most significant digits is common, and refers to the relative position of any given digit within a number.

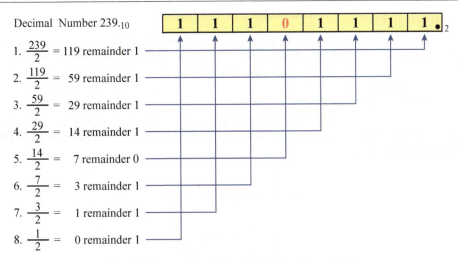

Figure 2–5 Converting a Decimal Number to a Binary Number

The following steps summarize this decimal-to-binary conversion.

Step 1. The decimal number is divided by 2 (base of the binary numbering system). The quotient is listed (119) as well as the remainder (1).

Step 2. Divide the quotient of Step 1 (119) by base 2, and list the new quotient (59) and the remainder (1).

Step 3. Divide the quotient of Step 2 (59) by base 2, and list the new quotient (29) and remainder (1).

Step 4. Divide the quotient of Step 3 (29) by 2, and list the new quotient (14) and the remainder (1).

Step 5. Divide the quotient of Step 4 (14) by 2, and list the new quotient (7) and remainder (0).

Step 6. Divide the quotient of Step 5 (7) by 2, and list the new quotient (3) and remainder (1).

Step 7. Divide the quotient of Step 6 (3) by 2, and list the new quotient (1) and remainder (1).

Step 8. Divide the quotient of Step 7 (1) by 2, and list the new quotient (0) and remainder (1).

Note: When using a calculator to do the division, the value to the right of the decimal must be multiplied by the base to get the actual remainder. For example, when 239 is divided by 2 (Step 1) on a calculator, the answer is 119.5. To find the actual remainder, the 0.5 is multiplied by 2, the base, to find the remainder 1. This procedure is true for any numbering system. The base times the value to the right of the decimal point equals the actual remainder.

The binary numbering system is used to store information in the processor memory in the form of **bits**.

2s Complement

Virtually all programmable controllers, computers, and other electronic calculating equipment perform counting functions using the binary system. For those PLCs that are programmed to perform arithmetic functions, a method of representing both positive (+) and negative (−) numbers must be used. The most common method is **2s complement**. The 2s complement is simply a convention for binary representation of negative decimal numbers.

Before going any further with a discussion of 2s complement, a review of adding binary numbers may be helpful. In decimal addition, numbers are added according to an addition table. A partial addition table is shown in Figure 2–6.

	0	1	2	3	4	5
0	0	1	2	3	4	
1	1	2	3	4	5	
2	2	3	4	5	6	
3	3	4	5			
4	4	5				

Figure 2–6 Decimal Addition System

To use the table, the first number to be added is located on the vertical line, and the second number on the horizontal line. The sum, or total, is found where the two imaginary lines intersect. For example, 3 + 2 = 5 (as shown in Figure 2–7).

	0	1	2	3	4	5
0	0	1	2	3	4	
1	1	2	3	4	5	
2	2	3	4	5	6	
3	3	4	5			
4	4	5				

Figure 2–7 Adding 2 and 3

For binary addition, a similar addition table is constructed. The table is small because the binary system only has two digits (1 and 0) (Figure 2–8).

	0	1
0	0	1
1	1	10

Figure 2–8 Binary Addition Table

To use the table, the first number (digit) to be added is located on the vertical line, the second digit is located on the horizontal line. The sum, or total, is found where the two imaginary lines intersect. Figure 2–9 shows an example of adding 1 + 0 = 1.

Figure 2–9 Adding Binary 1 and 0

Notice that if 1 and 1 are added, the table shows 1 0, not 2, as might be expected. 1 0 is the binary representation of 2 (Figure 2–10).

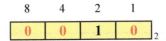

Figure 2–10 Binary Representation of 2

Figure 2–11 shows how binary numbers 1011_2 and 110_2 are added.

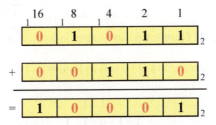

Figure 2–11 Adding Binary Numbers

In the 1s column 1 + 0 = 1.
In the 2s column 1 + 1 = 0, with a carryover of 1.
In the 4s column 1 + 0 + 1 = 0, with a carryover of 1.
In the 8s column 1 + 1 + 0 = 0, with a carryover of 1.
In the 16s column 1 + 0 + 0 = 1.
The sum (total) of 1011_2 and 110_2 is, therefore, 10001_2.

To verify our results we can convert the binary numbers to decimal equivalent numbers and add them, as shown in Figure 2–12.

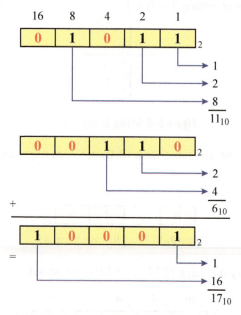

Figure 2–12 Converting Binary Numbers to Decimal Equivalents

Another example of adding binary numbers is shown in Figure 2–13, where 11011_2 and 11_2 are added.

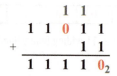

Figure 2–13 Addition of Binary Numbers

In the 1s column $1 + 1 = 0$, with a carryover of 1.
In the 2s column $1 + 1 + 1 = 1$, with a carryover of 1.

| **Note:** $1 + 1 + 1 = 3$. The binary equivalent of 3_{10} is 11_2.

In the 4s column $1 + 0 = 1$.
In the 8s column $1 + 0 = 1$.
In the 16s column $1 + 0 = 1$.
The sum of 11011_2 and 11_2 is 11110_2.

To verify this method, convert the binary numbers to decimal numbers, and add them, as shown in Figure 2–14.

$$\begin{array}{r} \mathbf{1\ 1\ 0\ 1\ 1} \quad = 27_{10} \\ +\ \underline{\mathbf{1\ 1}} \quad = \ \ 3_{10} \\ \mathbf{1\ 1\ 1\ 1\ 0_2} \quad = 30_{10} \end{array}$$

Figure 2–14 Comparing Binary and Decimal Addition

To represent negative numbers using the binary numbering system, one bit is designated as a signed bit. If the designated bit is a 0 (zero), the number is positive, and if the bit is a 1, the number is negative.

Using a 4-bit word length, and using bit 4 as the designated signed bit, 0001_2 represents +1 decimal (see Figure 2–15).

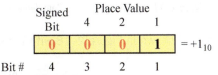

Figure 2–15 4-Bit Word with a Signed Bit

The table in Figure 2–16 shows all of the possible numbers for a 4-bit word using 2s complement.

Binary Number	Decimal
0111	+7
0110	+6
0101	+5
0100	+4
0011	+3
0010	+2
0001	+1
0000	0
1111	−1
1110	−2
1101	−3
1100	−4
1011	−5
1010	−6
1001	−7
1000	−8

Figure 2–16 2s Complement Numbers for a 4-Bit Word

Notice that the negative numbers go to –8 while the positive numbers only go to +7. In this case, the signed bit is used for its place value, which is 8. The same holds true for 8-, 16-, and 32-bit words. The maximum negative number is always one number *higher* than the maximum positive number.

To display a negative binary number requires that the same value positive number be complemented (all 1s changed to 0s and all 0s changed to 1s) and a value of 1 added. The result is the 2s complement of the number. Figure 2–17 shows the steps to express –5 in 2s complement using a 4-bit word.

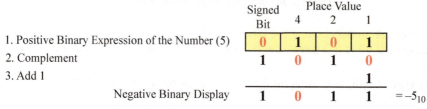

Figure 2–17 Expressing –5 in 2s Complement

Another example of 2s complement is shown in Figure 2–18 with the steps required to express –7 in 2s complement.

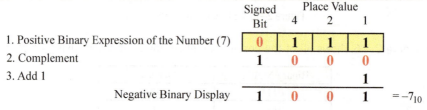

Figure 2–18 Expressing –7 in 2s Complement

To convert a negative binary number to the decimal equivalent, the negative binary display is complemented, 1 is added, the binary sum is converted to decimal, and the negative sign (–) is added. Figure 2–19 shows what steps are necessary to determine the negative value of 1110_2.

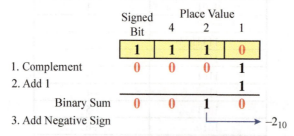

Figure 2–19 2s Complement to Decimal Equivalent

An easy way to convert a negative binary number to its equivalent negative decimal number is to subtract the place value of the signed bit from the value of the binary digits. In Figure 2–19, the binary sum 1110 is equal to -2_{10}, which is the decimal number –2. In this example, a 4-bit word is used with bit 4 being the signed bit. The fourth bit normally would have a place value of 8. In this example, the value 8 is subtracted from the value of the binary number 110_2 which is equal to 6, $6 - 8 = -2$.

Another example using this method of converting negative binary numbers to their decimal equivalent is shown in Figure 2–20, using an 8-bit word with bit 8 being the signed bit.

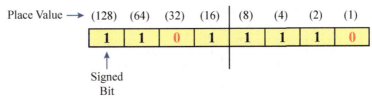

Figure 2–20 8-Bit Word 2s Complement

In this example, the signed bit would have a value of 128 (2^7), whereas the numeric value of the other bits would be 94, as shown in Figure 2–21.

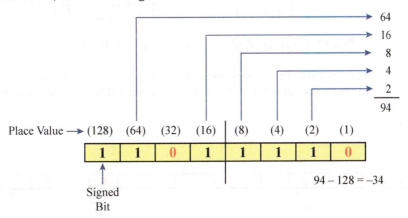

Figure 2–21 8-Bit Word (2s Complement) with Bits Added

Subtracting the place value of the signed bit (128) from the numeric value of the other bits (94) gives us the decimal number: $94 - 128 = -34$. To verify this answer, complement the original binary number 1101 1110 to get 0010 0001. Then add 1.

$$\begin{array}{r} 0010\ 0001 \\ 1 \\ \hline 0010\ 0010 \end{array}$$

The answer is $2^5 + 2^1$ or $32 + 2 = 34$; adding the negative sign gives us a final answer of –34, the same answer we got when we subtracted the place value of the signed bit (128) from the numeric value of the binary number 94.

If the PLC system uses 2s complement for arithmetic, the highest positive number that a 16-bit word can represent is +32,767, as shown in Figure 2–22. The highest positive number that a 32-bit word can represent is +2,147,483,647.

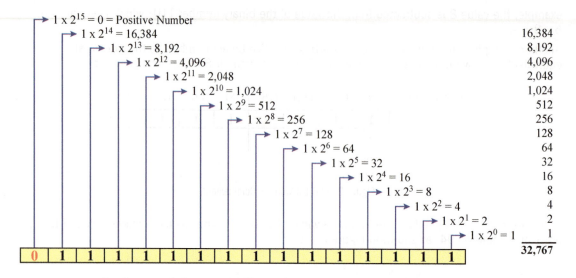

Figure 2–22 Maximum Positive Value of 2s Complement 16-Bit Word

The largest negative number that can be represented by a 16-bit word is –32,768, as shown in Figure 2–23. The largest negative number that can be represented by a 32-bit word is –2,147,483,648.

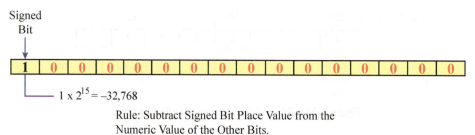

Rule: Subtract Signed Bit Place Value from the Numeric Value of the Other Bits.

Signed Bit Place Value = 32,768
0 – 32,768 = –32,768

Figure 2–23 Maximum Negative Value of 2s Complement 16-Bit Word

Another method of converting positive numbers to 2s complemented negative numbers is as follows: starting at the least significant bit and working to the left, copy each bit up to and including the first 1 bit, and then complement or change each remaining bit. Figure 2–24 shows this alternate method of expressing –2 in 2s complement using a 4-bit word.

1. Original Positive Number (+2)	0	0	1	0
2. Copy Up to First (1) Bit			1	0
3. Complement the Remaining Bits	1	1	1	0
2s Complement –2	1	1	1	0

Figure 2–24 Alternate Method of 2s Complement

A further example is shown in Figure 2–25 for 2s complementing the value 24 using an 8-bit word.

0	0	0	1	1	0	0	0

1. Original Positive Number (+24)

				1	0	0	0

2. Copy Up to First (1) Bit

1	1	1	0	1	0	0	0

3. Complement the Remaining Bits

1	1	1	0	1	0	0	0

2s Complement –24

Figure 2–25 2s Complement of –24 Decimal

By using 2s complement, negative and positive values can now be added. The two steps for adding -7_{10} and $+5_{10}$ using 2s complement with a 4-bit word are shown in Figure 2–26.

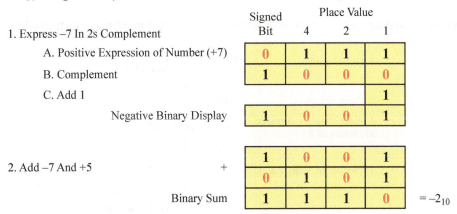

1. Express –7 In 2s Complement

    A. Positive Expression of Number (+7)

    B. Complement

    C. Add 1

                Negative Binary Display

2. Add –7 And +5

                Binary Sum

Figure 2–26 Adding Positive and Negative Numbers

| **Note:** When adding signed binary numbers, any carryover from the signed bit column is discarded.

Once addition of signed numbers is possible, the other arithmetic functions (subtraction, multiplication, and division) are also possible, because they are achieved by successive addition on a PLC.

Example: Subtracting the number 20 from 26 is accomplished by complementing 20 to obtain –20, and then performing addition.

Subtracting 20 from 26 by complementing 20 and performing addition using an 8-bit word is shown in Figure 2–27.

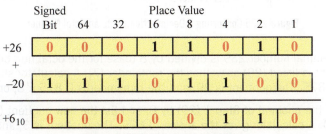

Figure 2–27 Subtraction by Addition

Octal System

The **octal** system, or base 8, is made up of 8 digits, numbers 0–7. The first digit to the *left* of the octal point is the units place, or 1s, and has a base or power of 8^0. The next place is eights (8s) or base 8^1. The next place is sixty-fours (64s) or base 8^2, followed by five hundred twelves (512s) or base 8^3, and four thousand ninety-sixes or base 8^4, and so on. An octal number will always be expressed by placing an eight in subscript to the right of the unit's digit, as shown in Figure 2–28.

$$357._8$$

Figure 2–28 Octal Number

The method of converting an octal number to a decimal equivalent number is illustrated in Figure 2–29.

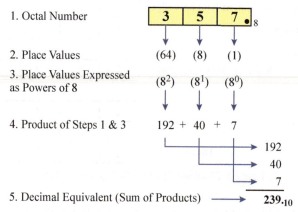

1. Octal Number

2. Place Values

3. Place Values Expressed as Powers of 8

4. Product of Steps 1 & 3

5. Decimal Equivalent (Sum of Products)

Figure 2–29 Converting an Octal Number to a Decimal Number

The decimal number 239 is converted to an octal number in Figure 2–30.

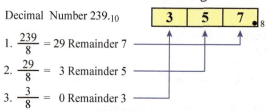

Decimal Number $239._{10}$

1. $\dfrac{239}{8}$ = 29 Remainder 7

2. $\dfrac{29}{8}$ = 3 Remainder 5

3. $\dfrac{3}{8}$ = 0 Remainder 3

Figure 2–30 Converting a Decimal Number to an Octal Number

Step 1.   The decimal number 239 is divided by 8 (base for the octal numbering system). The quotient is listed (29) as well as the remainder (7). A calculator shows the answer as 29.875. The quotient is 29, and the remainder is 0.875 × 8, or 7.

Step 2.   Divide the quotient of Step 1 (29) by 8, and list the new quotient (3) and the remainder (5). A calculator gives the answer 3.625. The quotient is 3, and the remainder is 0.625 × 8, or 5.

Step 3.   Divide the quotient from Step 2 (3) by 8, and list the new quotient (0) and remainder (3). The quotient 3 divided by 8 equals 0.375. The new quotient is 0, and the remainder is 0.375 × 8, or 3.

The decimal number 239 is the same as the octal number 357.

Since the largest single number that can be expressed using the octal numbering system is seven (7), each octal digit can be represented by using only three (3) binary bits (base 2). Figure 2–31 illustrates how to convert an octal number to a binary number. The figure shows three sets of binary bits and the place value of each bit. For the *least* significant digit (7), a one (1) must be placed in the 1s place, the 2s place, and the 4s place to equal 7. For the middle digit (5), a 1 is placed in the 1s place and the 4s place, while a 0 is placed in the 2s place. This combination equals 5. For the *most* significant digit (3), a 1 is placed in the 1s place and the 2s place, and a 0 is placed in the 4s place. This combination adds up to 3.

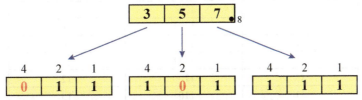

Figure 2–31 Conversion of Octal Number to Binary

Some older PLCs used the octal numbering system for I/O addressing. The terminals of the input and output modules would be labeled 00 through 07 and 10 through 17, rather than 0 through 15 as would be the case with decimal numbering (which is used in many PLCs today). When using the octal numbering system, words are labeled 000–007, 010–017, 020–027, and so forth, whereas the bits are labeled 00–07 and 10–17. Figure 2–32 shows a memory word with the internal bits addressed using the octal numbering system.

17	16	15	14	13	12	11	10	07	06	05	04	03	02	01	00
0	0	0	0	0	0	0	0	0	0	0	0	0	0	0	0

WORD 010

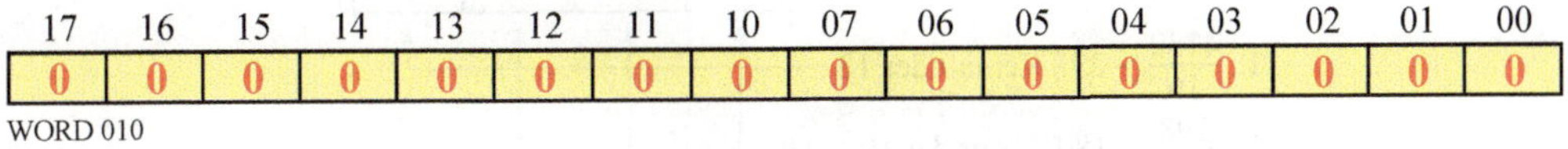

Figure 2–32 Word and Bit Labeling Using the Octal Numbering System

Hexadecimal System

The **hexadecimal** system, often referred to as HEX, consists of a number system with base 16.

It seems logical that the numbers used in base 16 would be 0–15. However, only numbers 0–9 are used, and the letters A–F represent numbers 10–15, respectively. The place values from the hexadecimal point are 1s—16^0, 16s—16^1, 256s—16^2, 4096s—16^3, and so on.

Each hexadecimal digit is represented by four binary digits. The binary equivalents are shown in the table in Figure 2–33.

Hexadecimal	Binary Number	Decimal
0	0000	0
1	0001	1
2	0010	2
3	0011	3
4	0100	4
5	0101	5
6	0110	6
7	0111	7
8	1000	8
9	1001	9
A	1010	10
B	1011	11
C	1100	12
D	1101	13
E	1110	14
F	1111	15

Figure 2–33 Hexadecimal Equivalents for Binary and Decimal

The decimal number 4,780 is converted to hexadecimal, as illustrated in Figure 2–34.

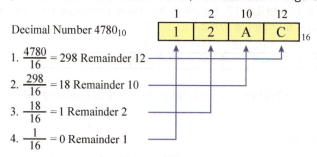

Decimal Number 4780_{10}

1. $\dfrac{4780}{16} = 298$ Remainder 12

2. $\dfrac{298}{16} = 18$ Remainder 10

3. $\dfrac{18}{16} = 1$ Remainder 2

4. $\dfrac{1}{16} = 0$ Remainder 1

Figure 2–34 Converting a Decimal Number to a Hexadecimal Number

Step 1. The decimal number is divided by 16 (base for the hexadecimal numbering system). The quotient is listed (298) as well as the remainder (12). A calculator provides the answer 298.75. The quotient is 298, and the remainder is 0.75 × 16, or 12.

Step 2. Divide the quotient of Step 1 (298) by 16, and list the new quotient (18) and the remainder (10). The answer is 18.625. The quotient is 18, and the remainder is 0.625 × 16, or 10.

Step 3. Divide the quotient from Step 2 (18) by 16, and list the new quotient (1) and the remainder (2). Eighteen divided by 16 equals 1.125. The quotient is 1, and the remainder is 0.125 × 16, or 2.

Step 4. Divide the quotient from Step 3 (1) by 16, and list the new quotient (0) and the remainder (1). One divided by 16 equals 0.0625. The quotient is 0, and the remainder is 0.0625 × 16, or 1.

Converting a hexadecimal number to a decimal number is illustrated in Figure 2–35.

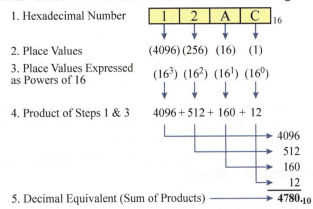

Figure 2–35 Converting a Hexadecimal Number to a Decimal Number

| **Note:** Remember that A is equivalent to 10, and C is equivalent to 12.

The binary equivalent of the hexadecimal number 12AC is shown in Figure 2–36.

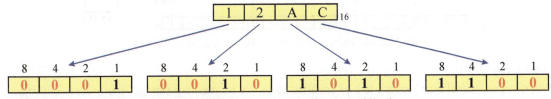

Figure 2–36 Binary Equivalent of a Hexadecimal Number

Since the largest number that can be displayed using the hexadecimal numbering system is 15, or F (as shown in the table in Figure 2–33), only four binary bits are needed to display each hexadecimal digit. The conversion to binary simply places the 1s in the correct binary locations to duplicate the hexadecimal digit (1 through F), as illustrated. The C has a value of 12, so 1s are placed in the 8s and 4s locations, while zeros (0) are placed in the 2s and 1s locations for a total binary value of 12. The same procedure is followed for the remaining digits A (10), 2, and 1.

The HEX system is used when large numbers need to be processed. The hexadecimal system is also used by some PLCs for entering output instructions into a sequencer.

Figure 2–37 shows the conversion of a 16-bit binary number to its hexadecimal equivalent.

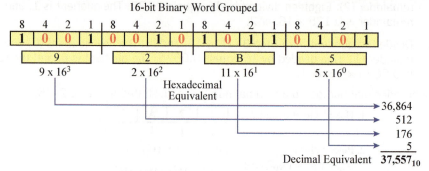

Figure 2–37 16-Bit Binary to Hexadecimal

The first step in converting 16-bit binary to hexadecimal is to group the 16-bit binary word into groups of four (conversion to BCD). Each group of four digits is converted to its hexadecimal equivalent. In Figure 2–37, the hexadecimal number is $92B5_{16}$.

The conversion of $92B5_{16}$ to decimal is $37,557_{10}$. Figure 2–38 shows the conversion of the original 16-bit binary number to its decimal equivalent.

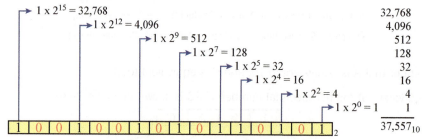

Figure 2–38 Converting a 16-Bit Digital Number to a Decimal Number

BCD System

When large decimal numbers are to be converted to binary for memory storage, the process becomes somewhat cumbersome. To solve this problem and speed conversion, the Binary Coded Decimal (**BCD**) system was devised. In the BCD system, four binary digits (base 2) are used to represent each decimal digit. To distinguish the BCD numbering system from a binary system, the designation BCD is subscripted and placed to the lower right of the units place. Converting a BCD number to a decimal equivalent is shown in Figure 2–39.

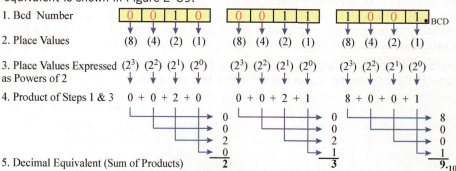

Figure 2–39 Converting a BCD Number to a Decimal Number

When using a BCD numbering system, three decimal numbers may be displayed using 12 bits (3 groups of 4), or 16 bits (4 groups of 4) may be used to represent four decimal numbers or digits.

When only three decimal digits are to be represented, using 12 bits, they are further identified as *most significant digit* (MSD), *middle digit* (MD), and *least significant digit* (LSD) (Figure 2–40).

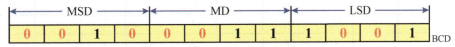

Figure 2–40 Significant Digits

When using the BCD system, the largest decimal number that can be displayed by any four binary digits is 9. The table in Figure 2–41 shows the four binary digit equivalents for each decimal number 0–9.

Place Value				Decimal Equivalent
2^3 (8)	2^2 (4)	2^1 (2)	2^0 (1)	
0	0	0	0	0
0	0	0	1	1
0	0	1	0	2
0	0	1	1	3
0	1	0	0	4
0	1	0	1	5
0	1	1	0	6
0	1	1	1	7
1	0	0	0	8
1	0	0	1	9

Figure 2–41 Binary to Decimal Equivalents

Using Numbering Systems

The alphanumeric keys of many programming terminals generate standard ASCII characters and control codes. **ASCII** is an acronym for American Standard Code for Information Interchange. The ASCII code uses different combinations of 7-bit binary (base 2) information for communication of data. The data may be communicated to a printer, barcode reader, or be shown on the display of the programmer and/or computer.

Note: ASCII information is often expressed in hexadecimal (base 16). Figure 2–42 shows the 128 standard ASCII control codes and character set with both the binary and hexadecimal numbering systems.

		Most Significant Bit (**MSB**)							
BINARY →		000	001	010	011	100	101	110	111
HEX →		0	1	2	3	4	5	6	7
0000	0	NUL	DLE	SP	0	@	P	\	p
0001	1	SOH	DC1	!	1	A	Q	a	q
0010	2	STX	DC2	"	2	B	R	b	r
0011	3	ETX	DC3	#	3	C	S	c	s
0100	4	EOT	DC4	$	4	D	T	d	t
0101	5	ENQ	NAK	%	5	E	U	e	u
0110	6	ACK	SYN	&	6	F	V	f	v
0111	7	BEL	ETB	'	7	G	W	g	w
1000	8	BS	CAN	(	8	H	X	h	x
1001	9	HT	EM	)	9	I	Y	i	y
1010	A	LF	SUB	*	:	J	Z	j	z
1011	B	VT	ESC	+	;	K	[	k	{
1100	C	FF	FS	,	<	L	\	l	|
1101	D	CR	GS	-	=	M	]	m	}
1110	E	SO	RS	.	>	N	^	n	~
1111	F	SI	US	/	?	O	—	o	DEL

*Least Significant Bit (**LSB**)*

Figure 2–42 Standard ASCII Control Code and Character Set

The digital or hexadecimal number is determined by first locating the vertical column where the code or character is located, and then the horizontal row.

Example: The letter A is in column 4, row 1. The binary number that transmits the letter A is 100 0001. The hexadecimal number is 41. The symbol # is 010 0011 in binary and 23 in HEX.

An eighth bit is often used by programmers to provide error-checking of information that is transmitted. This eighth bit is called the **parity bit**.

For *even* parity, the parity bit (the eighth bit) is added to the seven bits that represent the ASCII codes and characters so that the number of 1s will always add up to an even number.

Example: The binary number for the # symbol is 010 0011. The 1s add up to three, an odd number. By adding an eighth bit and making it a 1, the total of 1s is now 4, or even, as shown in Figure 2–43.

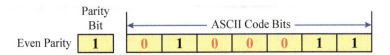

Figure 2–43 Parity Bit Set to 1 for Even Parity

The letter A, which is the binary number 100 0001, has two 1s and is already even. In this case, the parity bit would be a 0, as shown in Figure 2–44.

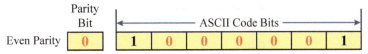

Figure 2–44 Parity Bit Set to 0 for Even Parity

The ASCII control code BS (backspace) is binary number 000 1000. For even parity, a 1 is added for the parity bit, as shown in Figure 2–45.

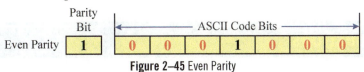

Figure 2–45 Even Parity

By checking each character or control code that is sent for an even number of 1s, transmission errors can be detected when an odd number of 1s is found.

For systems that operate on *odd* parity, the parity bit is used to make the total of 1s add up to an odd number.

Example: The number 5 has a binary number of 011 0101. The 1s add up to 4. The parity bit is set to 1, making the 1s total 5, or an odd number. Figure 2–46 illustrates this concept.

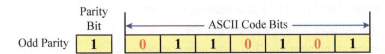

Figure 2–46 Parity Bit Set to 1 for Odd Parity

For systems that do not use a parity bit for error-checking, the 8 bit is always a zero (0).

Chapter Summary

There are several numbering systems that are used to store information in the form of binary digits (bits) into the memory system of a processor. The specific numbering system or the combination of numbering systems used depends on the hardware requirements of the specific PLC manufacturer. The important thing to remember, however, is that no matter which numbering system or systems are used, the information is still stored as 1s and 0s.

For programmable controllers to perform arithmetic functions, a way must be found to represent both positive and negative numbers. One of the most common methods used is called 2s complement. Using 2s complement, negative and positive numbers can be added, subtracted, divided, and multiplied. In reality, however, all arithmetic functions are accomplished by successive addition.

Key Terms

binary	2s complement
octal	BCD
hexadecimal	bits
most significant digit	ASCII
least significant digit	middle digit
parity bit	

Review Questions

1. When information is stored using only 1s and 0s, it is called a _____ system.
2. A *bit* is an acronym for _____.
3. The decimal numbering system uses 10 digits, or a base of 10. List the base for each of the following numbering systems.
   a. binary   base _____
   b. hexadecimal   base _____
   c. octal   base _____
4. Convert *binary* number 11011011 to a *decimal* number.
5. Convert *decimal* number 359 to a *binary* number.
6. Convert *hexadecimal* number 14CD to a *decimal* number.
7. Convert *decimal* number 3247 to a *hexadecimal* number.
8. Convert *decimal* number 232 to an *octal* number.
9. How do we prevent binary numbers 10 and 11 from being confused as decimal numbers?
10. Convert the following *binary* values to *decimal*.
    a. 10011000
    b. 01100101
    c. 10011001
    d. 00010101

11. Convert the following *BCD* values to *decimal*.
   a.   1001 1000
   b.   0110 0101
   c.   1001 1001
   d.   0001 0101

12. The BCD value 1001 0011 0101 is *not*
   a.   935 decimal
   b.   0011 1010 0111 binary
   c.   647 octal
   d.   3A7 hexadecimal

13. The hexadecimal value 2CB is *not*
   a.   715 decimal
   b.   1313 octal
   c.   0010 1100 1011 binary
   d.   0111 0001 0011 BCD

14. Express the following signed decimal numbers in 2s complement. Use 8-bit words. Show all work.
   a.   (–)7
   b.   (–)4
   c.   (–)3

15. Convert the following decimal numbers to 2s complement and add. Use 8-bit words. Show all work.
   a.   (+)4
        (–)7
   b.   (–)10
        (+)22
   c.   (+)22
        (+)33

16. Convert the following negative binary values (8-bit word) to decimal. Show all work.
   a.   10010011
   b.   11001100
   c.   11100111

17. Convert the following decimal numbers to 2s complement and subtract. Use 8-bit words. Show all work.

    a.   (+)30

         (+)21

    b.   (+)48

         (+)32

    c.   (+)67

         (+)116

Chapter 3

Understanding the Input/Output (I/O) Section

Learning Objectives

After completing this chapter, you should have the knowledge to:

- Describe the I/O section of a programmable logic controller.
- Describe how basic AC and DC I/O modules work.
- Define *optical isolation*.
- Describe why *optical isolation* is used.
- Describe the proper wiring connections for I/O devices and their corresponding modules.
- Explain why a hardwired emergency-stop function is desirable.
- Define the term *interposing*.
- Describe what I/O shielding does.
- List environmental concerns when installing PLCs.

I/O Section

The input/output section (**I/O section**) is the major reason that PLCs are so versatile when used with process machines or driven equipment. The I/O section has the ability to change virtually any type of voltage or current signal into a logic-level signal (typically 5V DC) that is compatible with the PLC processor. The I/O section automatically makes the conversions necessary for the processor to interpret input signals and to activate output devices, even when the I/O devices are of various voltage and current levels.

A DC input module, for example, can be used with a 24V DC proximity switch to turn on a 240V AC motor starter coil that is connected to an AC output module. The conversion and interfacing is all accomplished automatically in the I/O section of the PLC, and it is the ease with which the interfacing is accomplished that has made the PLC such a viable tool in industrial and process control.

The input modules of the I/O section provide the status (*ON* or *OFF*) of push buttons, limit switches, proximity switches, and the like, to the processor so decisions can be made to control the machine or process in the proper sequence. Outputs, such as motor starter coils, indicator lights, and solenoids are interfaced to the processor through the output section of the I/O. Once a decision has been made by the processor, a signal is sent to the output section to control the flow of current to the output device. In general, the status of the inputs are relayed to the processor and, based on the logic of the program that has been written, a decision is made to turn the outputs to *ON* or *OFF*. All of the different types and levels of signals (voltages and currents) used in the control process are **interfaced** in the I/O section.

The I/O section generally can be divided into two categories: fixed I/O and modular I/O.

Fixed I/O

PLCs with fixed I/O typically come in a complete unit that contains the processor, I/O section, and power supply. The I/O section contains a fixed number of inputs and outputs. For example, the Allen-Bradley Micro850 PLC shown in Figure 3–1 has a combination of digital inputs (14) and outputs (10) in a self-contained unit. Like most small PLCs, it can be DIN-rail or panel mounted.

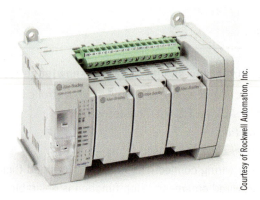

Figure 3–1 Allen-Bradley Micro850 PLC with Embedded I/O

If more I/O capability is required or different voltages are needed, expansion units with various I/O configuration can be added to many of the micro and small PLCs on the market today. Figure 3–2 shows an Allen-Bradley 5370 CompactLogix PLC with 16 embedded digital inputs and outputs with the capacity for six optional expansion modules and up to an additional 80 I/O points.

Figure 3–2 Allen-Bradley CompactLogix 5370 PLC with Embedded I/O

Micro and small PLCs with fixed I/O typically have a discrete input and output section. As discussed in Chapter 1, discrete-type I/O signals are *ON* or *OFF* and do not vary in level. For example, when a 120V limit switch contact closes or is *ON,* the signal to the input section will be 120V, and the signal will be 0V when the limit switch contact is open (*OFF*). Many manufacturers offer models with a combination of digital and analog inputs and outputs.

While these PLCs are small in size, they are big on features. Most include full-feature instruction sets, embedded communications ports, real-time clocks, and much more. One should consult the specific dealer for a full list of features.

As the cost of these compact units has decreased, their use has increased. The costs are so competitive that any control processes that uses only a small number of relays and/or timers can now be accomplished using a small micro type PLC. The use of a small PLC not only saves money, but also gives added reliability and flexibility.

Modular I/O

Modular I/O, as the name implies, is modular in nature, more flexible than fixed I/O units, and provides added versatility when it comes to the type and number of I/O devices that can be connected to the system. The various types of I/O modules that make up the I/O section are housed, or installed, in an I/O chassis or rack. Chassis-based I/O modules are specifically designed for a particular PLC controller. You can install the I/O modules locally in the same chassis as the controller, or in some cases remotely through the use of I/O communication networks. In some smaller PLCs, the I/O modules do not actually mount in a physical chassis, rather they connect to each other and the processor to create a rackless design.

The I/O chassis or rack is a framework or housing into which modules are inserted. Figure 3–3a shows a 10-slot I/O chassis and Figure 3–3b a 17-slot chassis. Figure 3–3c shows a chassis with the I/O modules installed to the right of the power supply.

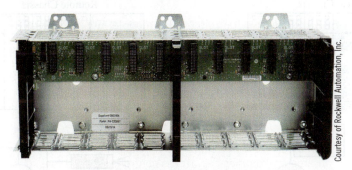

Courtesy of Rockwell Automation, Inc.

Figure 3–3a Allen-Bradley 1756-A10 I/O Chassis

Courtesy of Rockwell Automation, Inc.

Figure 3–3b Allen-Bradley 1756-A17 I/O Chassis

Figure 3–3c I/O Chassis with I/O Modules & Power Supply Installed

Chassis come in many sizes and typically allow for 4 to 17 modules to be inserted. Chassis that contain I/O modules and the PLC controller are referred to as the *local chassis* or local I/O. Chassis that contain I/O modules, remote communication modules, and are mounted separately or away from the PLC controller, are referred to as a *remote chassis* or remote I/O. An advantage of remote I/O is that you can distribute the I/O closer to the field devices (sensors and actuators), reducing wiring costs. The number of remote I/O chassis that a processor can control varies with each manufacturer. The communication between the remote chassis(s) and the PLC controller is accomplished using several different types of I/O communication methods. These methods include coaxial cable, twin axial cable, shielded-twisted pair, or fiber optic cable. If distance or electrical noise are considerations, the fiber optic communication method may be the best option. Figure 3–4 shows a local and remote I/O chassis.

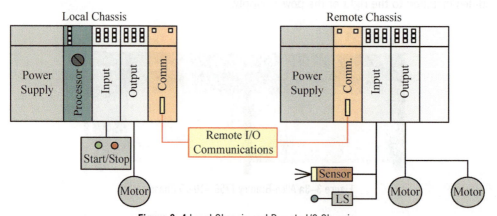

Figure 3–4 Local Chassis and Remote I/O Chassis

Most modern chassis or racks today do not require any additional configurations other than to be mounted and modules installed. On the other hand, older PLC systems required that jumpers or switches had to be set or configured.

I/O modules can be separated into three basic groups: discrete or digital input/output modules, analog input/output modules, and specialized modules.

Discrete I/O Modules

Discrete I/O modules are types of modules that only accept digital or *ON-* and *OFF-*type signals. These modules only recognize these two states or conditions, *ON* or *OFF*. If a discrete device, such as a limit switch, is connected to a digital input module, the module determines the state, or position, of the limit switch, and communicates that status to the processor. If the limit switch contact is open (*OFF*), the module indicates to the processor that the limit switch is *OFF*. This *OFF* condition is stored in the processor memory as a zero (0). Had the limit switch contact been in a closed position, the module would have sent a signal to the processor indicating that the limit switch was *ON*, or closed. The *ON* condition would have been stored in the processor memory as a one (1). All information stored in the processor memory about the status or condition of discrete I/O devices is always in ones and zeros. Understanding this basic concept is critical in learning to program and interpret PLC Ladder Logic.

Discrete modules are the most common type used in a majority of PLC applications and can be divided into two groups: input and output.

Discrete Input Module

The **discrete input module** communicates the status of the various real-world input devices connected to the module (*ON* or *OFF*) to the processor.

Note: As discussed in Chapter 1, the term *real world* is used to indicate that an actual device is involved. As you will learn later in the text, the PLC has the ability to provide timing and counting functions for a machine, even though the timers and counters exist only within the processor, and are not wired into the circuit as with real-world, or actual, devices.

Once the real-world input device is connected, an open or closed electrical circuit exists, depending on the position (open or closed) of the device. The status of the real-world input device is then converted to a logic-level DC electrical signal by the input module and sent to the processor.

Discrete input modules come in a wide range of voltages for various applications. Some of the more common voltages are 120V AC, 240V AC, 24V DC, and 130V DC. Some manufacturers give their modules an AC/DC rating to increase their flexibility and reduce required inventory. It is important to note, however, that while the module may be used with either AC or DC input voltages, the voltages *cannot* be intermixed on the same module.

Input modules can be purchased with a wide range of input terminals or points, which determine the number of individual field devices that can be connected to the module. Common sizes, depending on the manufacturer, are 8, 16, and 32 points. Sixteen- and 32-point modules are often referred to as high-density modules since they are physically the same size as an 8-point module. High-density modules usually provide lower cost per point, or input device, but are also more difficult to wire. The increased difficulty in wiring is caused by the closer proximity of the wiring terminals and the increased number of wires in the wiring harness.

AC Discrete Input Module

Figure 3–5 shows a simplified diagram of one of the input circuits of a typical AC discrete input module. Resistors and capacitors are used to filter and limit the voltage signal. The filter requires that the AC signal be not only of a specific value but also be present for a specific amount of time before the module views it as a real signal and communicates the results to the processor. A valid signal is relayed through an **optically coupled** circuit, across the backplane of the I/O chassis to the processor.

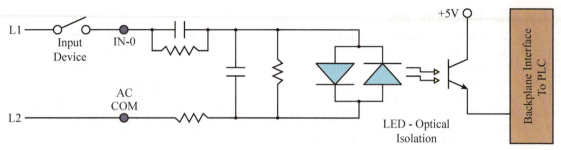

Figure 3–5 Simplified AC Input Module Circuit

The optically coupled circuit uses a light-emitting diode (LED) to turn *ON*, or forward bias, a photo transistor to complete the electrical circuit to the processor. When the LED is turned *ON* to indicate that the actual input device has closed its contact and is providing a voltage to the input terminal, the light from the LED is picked up by the photo transistor, which makes the transistor conduct or switch on, completing a 5V DC logic circuit, and the status of the input is communicated to the processor. This form of optical coupling is also referred to as **optical isolation**. By employing optical coupling or isolation, there is no actual electrical connection between the input device and the processor. This eliminates any possibility of the input line voltage, i.e., 120 or 240V AC, from coming in contact with and damaging the low-voltage DC section of the processor or I/O module. Optical isolation also protects the processor and I/O module from electrical noise, voltage transients, or spikes. In summary, optical isolation prevents any unwanted voltage from the I/O section from reaching the logic section of the processor or I/O module.

Individual status lights are provided on the front of I/O modules for each input terminal (Figure 3–6). The status light is *ON* when the input device is providing a voltage to the input terminal (contact closed) and is *OFF* when a voltage is absent (contact open). With the status lights showing the actual state of the various input devices connected to the input module, they make a valuable troubleshooting aid. The electrician or technician need only look at the input status lights on the input module to determine the state, or status, of any input device.

Figure 3–6 AC Input Module with Input Indicator Lights

A typical I/O module consists of two parts: a printed circuit board and a removable terminal block (RTB) assembly. The printed circuit board plugs into a slot, or connector, in the I/O chassis, or in the case of a rackless system, the adjacent module and contains the **solid-state** electronic circuits that interface the I/O devices with the processor. The RTB assembly then attaches to the front edge of the printed circuit board, which may or may not have a protective cover, depending on the manufacturer. Figure 3–7 shows a typical 16-point AC input module *without* the RTB attached.

Courtesy of Rockwell Automation, Inc.

Figure 3–7 16-Point AC Input Module

Figure 3–8 shows how the input module is installed in the I/O chassis.

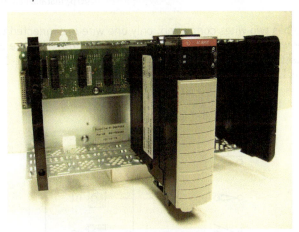

Figure 3–8 Installing I/O Module in I/O Chassis

After the input modules have been installed in the I/O chassis or rack, they are ready to have one side of each input device connected to their RTB (Figure 3–9). Do to the density and size of the terminal blocks, you may find it easier to wire the removable terminal blocks before installing them on the module. In some cases, as an alternative to buying RTBs and connecting the wires yourself, you can buy a wiring system that connects to the I/O modules through prewired and pretested cables.

RTBs allow I/O modules to be replaced without unwiring them. Figure 3–10 shows the RTB installed on the I/O module.

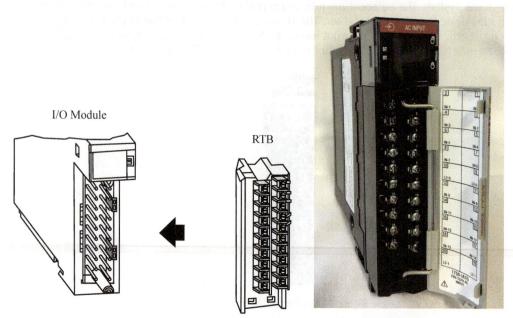

I/O Module

RTB

Figure 3–9 Removable Terminal Block (RTB)

Figure 3–10 Removable Terminal Block
(RTB) Installed on I/O Module

While each input device has two wires connected, only one wire is connected directly to the input module. The other wire of each input device is connected to L1 (Figure 3–11a). The same connection scheme is used for 8-, 16-, or 32-point input modules. Figure 3–11b shows the wiring connections for a typical 120V AC 16-point input module.

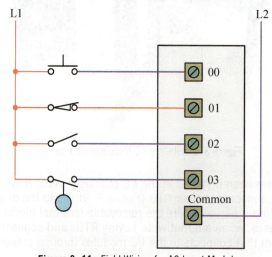

Figure 3–11a Field Wiring for AC Input Module

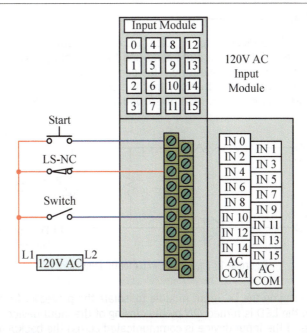

Figure 3–11b Typical Connections for a 120V AC 16-Point Input Module

The wires from the individual devices are referred to as **field wiring**, since the wires are external to the PLC and are connected in the field. On larger PLC systems, the field wiring that is brought into the I/O chassis or rack can consist of hundreds of wires. The basic rule is that one side of each input device is wired to a hot conductor (L1 for AC or + for DC), and the other side of the device is wired to an input terminal on the RTB of the input module. The input module has a common connection for the neutral, or grounded potential (L2), for AC modules, and the negative (–) for DC modules. Consult the manufacturer's literature that comes with each input module to ensure that the correct wiring connections are made.

Figure 3–12 shows two simplified circuits. In the first, or traditional circuit, the input device (single-pole switch) is connected to, and controls, the light. In the PLC circuit, the input device is connected directly to the input module instead of the light. The module converts the AC input signal to 5V DC, and communicates the status of the single-pole switch to the processor which, in turn, controls a light that is connected to an output module.

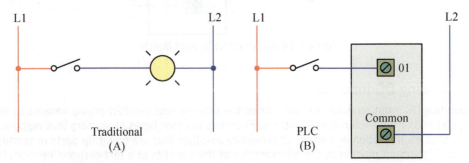

Figure 3–12 Traditional Wiring for Single Pole Switch Compared to PLC Wiring

DC Discrete Input Module

Figure 3–13 shows a simplified diagram for a typical DC input module. Resistors are used to lower, or drop, the incoming voltage. Filtering circuits condition the low-voltage signal and add a very small amount of additional delay in the response time. This period of delay is less than the delay used in the AC input module, but it is also used to verify that the signal received is a valid signal of a proper duration, and not a signal caused by electrical noise, voltage transients, or the like.

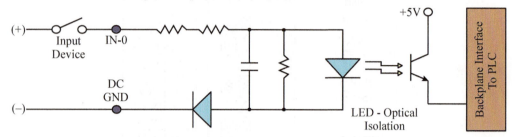

Figure 3–13 Simplified Circuit for DC Input Module

Optical isolation is also used on the DC input module to isolate the processor from the higher voltage of the input devices. When the LED is turned *ON* by the closing of the input device, the photo transistor conducts, and the status of the input device is communicated across the backplane of the I/O chassis or rack to the processor. Figure 3–14 shows how a typical DC sinking input module is wired. DC sourcing input modules are also available as well as Transistor-Transistor Logic (TTL) input modules.

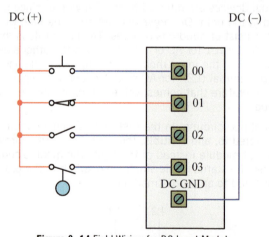

Figure 3–14 Field Wiring for DC Input Module

Fast-Responding DC Input Modules

Fast-responding DC input modules are used when the process requires fast-acting sensors to respond to high-speed and/or high-volume applications. Encoders or other types of sensors that respond with many pulses for each rotation of a shaft, or proximity switches that are counting parts or products produced by high-speed machines, are all examples of the benefits of a fast-responding module. The internal operation of the module is the same as the discrete DC input module, with the exception of the filtering circuit. In addition, modern fast digital input modules have software configurable features such as pulse capture, per point time-stamping and change of state, and configurable filter times. Fast-responding DC input modules should only be used in applications where very fast input detection is required.

Diagnostic Digital Input Modules

Diagnostic digital input modules are specialized input modules that provide features like diagnostic change of state, open wire detection, and field power loss detection. Because of their specialized nature, they are only available in 8-point 120V AC and 16-point 24V DC modules. In order to use the open wire detection feature, the field device must provide a minimum leakage current to function properly. On contact input devices, a leakage resistor must be placed across the contacts. This leakage current must exist when the input contact is open. Figure 3–15 shows a simplified schematic and field wiring diagram for an Allen-Bradley 1756-IA8D Diagnostic Input Module.

Discrete input modules come in a variety of types and styles that fit most applications requiring an *ON-* or *OFF*-type of signal. Allen-Bradley, for example, manufactures nearly 20 different types of discrete input modules. Selecting the proper module for any given application is relatively easy when the voltage level, voltage type, and response time are known.

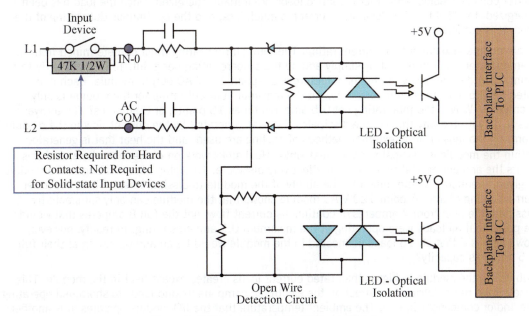

Figure 3–15 Diagnostic Digital Input Module Circuit

Discrete Output Modules

The purpose of a **discrete output module** is to control the current flow to real-world devices such as motor starter coils, pilot lights, control relays, and solenoid valves. As was true for the discrete input module, the discrete output module also works on a digital, or *ON* and *OFF*, basis. When the processor has made a decision to turn a specific device *ON*, the processor places a 1 in the memory location assigned to that output device, and later in the process, the information is communicated by way of the backplane of the I/O chassis to the output module, and the required real-world device will be turned *ON*. Similarly, when the processor determines that a device needs to be turned *OFF*, a 0 is placed in the device's memory location and the device will be turned *OFF*. The output module acts like a remote control switch that is controlled by the processor for turning output devices *ON* and *OFF*.

Output modules are generally classified as AC and DC. Both types offer a wide range of voltages, typical for the input modules discussed earlier. Read the information sheets carefully to determine that the module selected is appropriate for the loads (output devices) that are to be controlled.

Output modules are sized by the number of output devices that can be connected to them. The numbers of terminals, or points, for connecting output devices are typically 8, 16, and 32. Output modules have a current rating for each terminal or connection point, as well as an overall rating for the module. The individual terminal rating indicates the maximum current, or load, that can be controlled. This rating is a continuous duty rating. A surge rating is usually provided, indicating the maximum current that can be controlled and the length of time. The time rating may be given in milliseconds or in cycles. A typical current rating for a 120V AC output module would be: 0.5 amperes maximum continuous duty, with a surge current rating of 5 amperes for 43 msec.

The surge current rating is necessary for the inrush, or "pull-in," current that is present when motor starter coils, solenoids, and other inductive loads are initially energized. Once the load has been energized, the "hold-in," or "seal-in," current is much less, and the continuous duty rating of the module is sufficient.

Each module also has a total current rating. The total current rating must be determined from the manufacturer's literature, not by simply adding the ampere rating per point. To further clarify this point, consider the rating of one manufacturer's 16-point 120V AC output module. Each point is rated for 0.5 amperes continuous duty, yet the maximum current rating for the module is only 4 amperes. Why is the total rating not 0.5 amperes times 16 points, or 8 amperes? The answer has to do with the way the module dissipates the heat generated by the current flow in the module. Normally, no fans or other external methods of cooling are used, and the heat that is generated within the module is dissipated using **heat sinks**. Heat sinks work on convection alone, and this limits the amount of heat that they can effectively dissipate. The total current rating that a module is given, therefore, is determined by the ability of the module to dissipate heat. Thus, the lower current rating of the 16-point 120 V AC module shows that the module can only satisfactorily dissipate the heat from 4 amperes of continuous current flow, not the full 8 amperes that would be possible if all loads were operating at the maximum 0.5 amperes rating. In reality, however, how likely is it that all 16 loads connected to the module would be on and operating at their full 0.5-amperes capacity?

Subjecting the module to higher-than-rated current loads creates excess heat in the module. This excess heat has a detrimental effect on the electronic components and leads to shortened operating life and/or component failure. The ambient temperature that the I/O module operates in is another factor that must be considered. PLCs are normally designed to operate in a temperature range of 32°–140°F or 0°–60°C. Operating in temperatures above these limits affects the module's ability to dissipate heat and can lead to early component failure.

With each manufacturer, their output module ratings differ, it is necessary to carefully review the manufacturer's literature and specifications when selecting output modules.

AC Output Module

The internal circuitry for one point of a typical AC output module is shown in Figure 3–16. The AC output module usually consists of a **Triac** (shown in the figure). However, some manufacturers use a Silicon-Controlled Rectifier (**SCR**) instead of a Triac. When the processor determines that the output is to be turned *ON*, a signal is sent across the backplane of the I/O chassis and the LED is turned *ON*. The light from the LED causes the photo transistor to conduct, which provides current for the gate of the Triac. This portion of the output module optically isolates the logic section of the processor from the line voltage of the output devices.

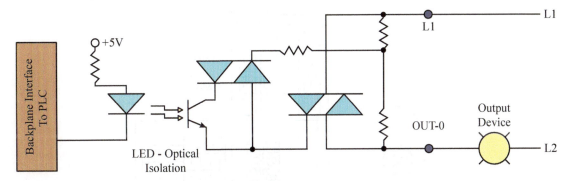

Figure 3–16 Typical AC Output Module Circuit

The Triac is used as an electronic switch to turn output devices *ON* and *OFF*. The Triac itself is the equivalent of two silicon-controlled rectifiers in inverse parallel connection with a common gate. The gate controls the switching state (*ON* or *OFF*) of the device. Once a signal is applied and the break-over voltage point is reached on the gate (normally 1 to 3V), the Triac freely conducts in either direction, completing the path for current flow to the output device.

A Triac is a solid block of crystalline material and is more sensitive to applied voltages and currents than standard relay contacts. Triacs are also limited to a maximum peak applied voltage, and, if this value is exceeded, a "dielectric-type" breakdown can result, causing a permanent short-circuit condition. A snubber circuit that consists of a resistor in series with a metal oxide varistor (MOV) is used to protect the Triac from damage resulting from electrical noise and voltage spikes. The metal oxide varistor is designed to break down, or conduct, at certain voltage levels. In a 120V AC circuit, the peak voltage is approximately 170V. The MOV would typically be set to break down at 190V. When the MOV conducts, it clips the excess voltage and thus prevents damage to the module.

Triacs constructed of a solid "block of material" have some characteristics that are not found with standard relay contacts. The Triac, rather than having *ON* and *OFF* states, actually has low and high resistance levels, respectively. In the *OFF* state (high resistance), a small leakage current still flows through the Triac. This leakage current, which is usually only a few milliamperes or less, normally causes no problems. When low-resistive pilot lights are connected to AC output modules, a faint glow of the filament may be detectable, even when the module is *OFF*. Similarly, the coils of some small control relays or solenoids may produce a detectable hum due to the Triac leakage current, even though the Triac is technically *OFF*. This small leakage current also causes false readings in some applications. Troubleshooting techniques for Triacs are covered in Chapter 21.

While Triacs are capable of carrying surge currents higher than their continuous current ratings, such surges must be of short duration (1/2 to 1 cycle) and not repetitive. Exceeding the manufacturer's listed surge values or the maximum continuous current rating, usually referred to as maximum RMS on-state current, results in a permanent short circuit.

After an output module is installed in the I/O chassis or rack, the actual real-world output devices are connected. Figure 3–17 shows the proper termination for an 8-circuit AC output module. Each output device has two wires; one wire from each output device is wired to the L2 potential. The other wire from each device is wired to one terminal of the output module, as shown in the figure. L1 is then connected to a common terminal on the module to supply the other potential for the output devices.

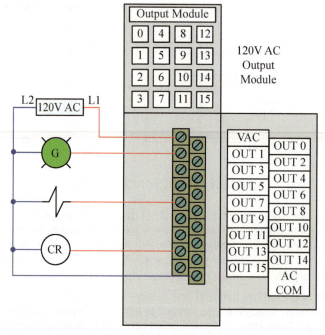

Figure 3–17 Field Wiring for AC Output Module

Figure 3–18 shows two simplified circuits, each including an output device. In the first circuit, the output is wired to the L2 potential, and the single-pole switch is used to control the L1 potential to complete the circuit and turn *ON* the light. In the PLC circuit, the output is again wired to L2, but the switch is replaced by the output module, which is wired to the L1 potential. Simply stated, the output module may be viewed as an electronic switch that is controlled by the PLC's processor.

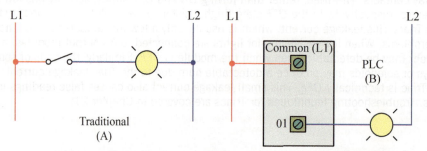

Figure 3–18 Traditional Wiring for an Output Device Compared to PLC Wiring

Output Fuses

In order to prevent damage to output modules, it is important not to exceed the current rating or to exceed its inrush capability. It is also important that output modules be protected from short circuits and ground faults. To provide protection for overcurrent, short circuit, and ground faults, output modules are always fused. The number of fuses used varies with each manufacturer. Some modules have an individual fuse for each output terminal, or point, others have one fuse for each four outputs or groups, while still others use one fuse to protect all points on their output module. Some output modules have internal electronic fusing to prevent too much current flow.

Most output modules come with "blown-fuse" indicators to show that a fuse has blown. Some modules have a "blown-fuse" indicator for each output terminal. If a fuse is blown on output terminal 6 for example, an indicator lamp will be lit to indicate the location of the blown fuse. Other modules have only one "blown-fuse" indicator for the whole module. This one indicator lamp only indicates that a fuse has blown; it does not indicate its location, and it is up to the electrician or technician to determine the blown fuse(s). When individual indicators are used for each output terminal, troubleshooting is greatly simplified. Figure 3–19a shows an Allen-Bradley 16-point AC output module and Figure 3–19b shows a close-up view of the indicator lights, including blown-fuse indicators.

Figure 3–19a Allen-Bradley
16-Point AC Output Module

Figure 3–19b AC Output Module Indicator Lights

As you may expect, access to the fuses varies with the manufacturer. Some modules must be removed from the I/O chassis and a cover removed before the fuses can be changed, while other modules provide direct access to the fuses on the front edge of the module.

It is important that when a blown fuse is removed, its replacement fuse be of the same voltage rating, amperage rating, interrupting rating, response time, and physical size as recommended by the manufacturer.

To speed troubleshooting and fuse replacement, many plants now add additional fuses to each output circuit. Terminal blocks are available that have built-in fuse holders and individual blown-fuse indicators. A separate fuse can then be wired in series with each output device, external to the output module. The fused terminal blocks are then mounted in a convenient location for easy access and troubleshooting.

Status Lights

Status lights are provided for each output point to indicate when that point has been turned *ON*. For troubleshooting purposes, it is important to understand how these status lights are wired. If the power for the lights comes from the processor side of the module, an illuminated light indicates that the processor has sent a signal to turn *ON* the output to that particular point. It is *not* an indication that current is flowing to the output device. If the status light is powered from the actual output power, then it is an indication that the output is turned on.

In addition to the individual output status lights, almost all I/O modules have a module status light. This status light provides an indication of the module's condition. It can provide such information as to whether the I/O is being actively controlled or not by the processor, the module has detected an internal problem, or the module has lost communications with the processor.

The status lights are a powerful troubleshooting tool, but it is important to understand exactly what the status lights indicate, and not to read more information into them than they are able to provide.

Module Keying

The various I/O modules look alike. This is true not only for Allen-Bradley modules, but also for other manufacturers products as well. In all instances, the manufacturers label and/or color code the fronts of all modules to distinguish between the different types (AC input, DC output, Analog, and so on). Most manufacturers have designed each module so it can be **keyed** either electronically or physically. Older PLC I/O modules used keying bands that were physically installed on the chassis backplane connector so only a specific type of module could be installed in that slot. Today, electronic keying is widely used. Electronic keying automatically compares the expected module to the physical module before I/O communications can begin. If enabled, electronic keying helps to prevent communication to a module that does not match the type and revision expected. Electronic keying is enabled in the PLC software at the time the module is configured.

DC Output Modules

DC output modules are basically the same in operation as AC output modules. The difference is the use of a power transistor instead of a Triac for the control of output current. The power transistor, typically a Metal-Oxide Semiconductor Field-Effect Transistor (MOSFET), has a quicker switching capability than the Triac; therefore, the response time for DC modules is faster than for AC modules. The MOV used in the AC output module is replaced with a diode to provide protection from electrical noise and spikes. A typical DC output circuit is shown in Figure 3–20.

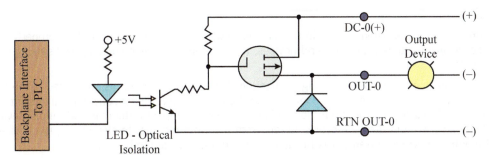

Figure 3–20 Simplified DC Output Module Circuit

DC output modules are available in ranges from 10 to 130V DC depending on the manufacturer. It is important to check to make sure that the module is appropriate for the voltage and current level of the output devices that are intended to be connected to the module. DC modules will be fused like their AC counterparts and also have blown-fuse indicators, as well as status lights. Figure 3–21 shows how the output devices are wired to a DC output module.

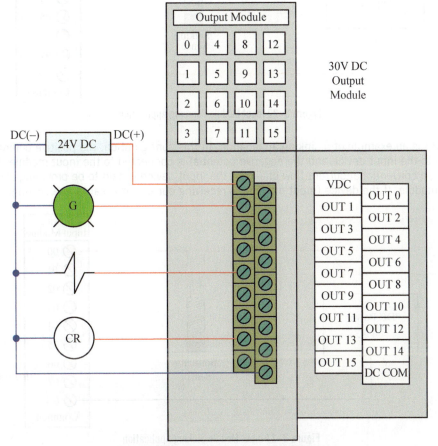

Figure 3–21 Field Wiring for a DC Output Module

Sourcing and Sinking

The terms *sourcing* and *sinking* refer to the manner in which DC devices are wired. To properly interface DC I/O modules with real-world devices, the difference between sourcing and sinking must be understood.

Figure 3–22 is an example of a sourcing application. The positive potential is connected to the input module and the negative potential is connected to the input device. Using conventional current flow (+ to −), it is said that the input module is the source of supply for the real-world input device. Stated simply, the input device receives current from the input module. In electronics, if a device (input module) provides current, or is the source of current, it is said to be sourcing.

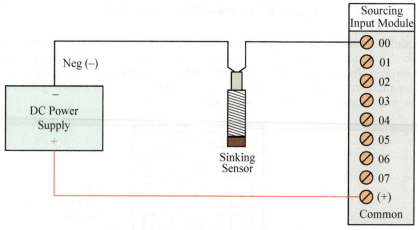

Figure 3–22 Input Module Sourcing Application

Figure 3–23 is an example of a sinking application. In this configuration, the positive potential is connected to the input device and the negative potential is connected to the input module. In this case, using a conventional current flow of + to −, the input device is said to be providing current to the input module. If the device (input module) is receiving current, it is said to be sinking.

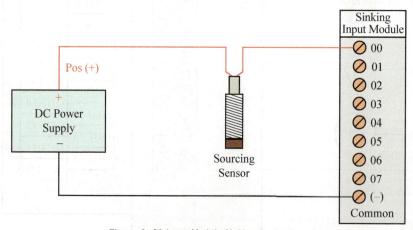

Figure 3–23 Input Module Sinking Application

When an output module is connected positive, as shown in Figure 3–24, and provides current to the real world (output device), the module is referred to as sourcing.

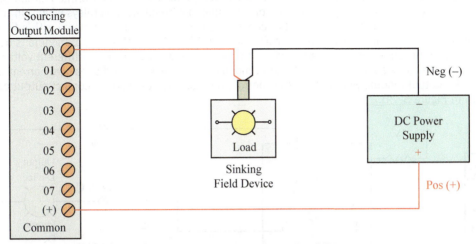

Figure 3–24 Output Module Source Capability

Figure 3–25 shows an output module wired negative, and the output device wired positive. When the output device (real world) provides the current to the module, the module is referred to as sinking.

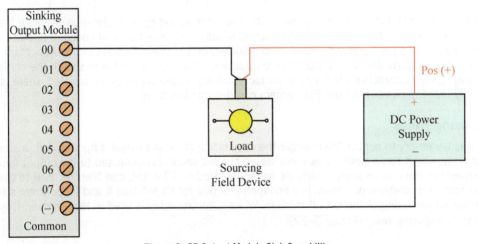

Figure 3–25 Output Module Sink Capability

There is always a great deal of confusion surrounding sourcing and sinking as it applies to I/O modules and devices. As a general rule, sinking modules are used with output modules when interfacing with electronic equipment (TTL or complementary metal-oxide-semiconductor (CMOS) compatible), while sourcing modules are used for such DC loads as solenoids. The safest way to make sure that connections to DC devices are correct is to select a module that works for the application, based on the manufacturer's specifications and wiring diagrams.

Contact Output Modules

Contact output modules have small electromechanical relays mounted on their printed circuit boards. A signal from the processor energizes the coil of the small electromechanical relays which, in turn, opens or closes a set of contacts. Each set of contacts is isolated and can be ordered with a Form-A normally open (N.O.) contact or Form-C (1-N.O. and 1-N.C.) contacts. This type of module is used when extra current ratings are required or when it is desirable to isolate loads of different voltages or phases, from the same source. A contact output module is also used when the leakage current of a standard AC output module would affect the control process. A typical Form-C contact output circuit is shown in Figure 3–26.

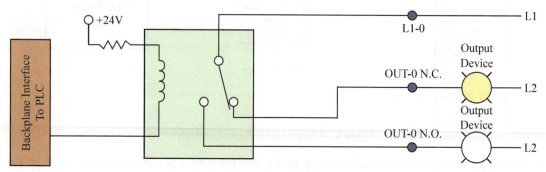

Figure 3–26 Form-C Contact Output Module

For example, a contact module can be used with a variable speed drive application. Assume that the drive has been programmed for multiple speed settings. To select a given speed, a circuit must be completed at two points on a terminal strip mounted on the drive. One set of contacts is used for each speed setting. As the drive supplies its own power, all that is needed is to switch the control signal through the contacts on the output contact module. As the contacts are isolated, there is no possibility that the power from the PLC system can damage the drive.

Interposing Relay

When it is necessary to control loads larger than the rating of an individual output circuit, a standard control relay, which has a small inrush and sealed current value, is connected to the output module. The contacts of the control relay, which are generally rated at 10 amps, can then be used to control a larger load. This method of control is a common practice for NEMA size 4 and large motor starters, depending on the rating of the output module. When a control relay is used in this manner, it is called an **interposing relay** (Figure 3–27).

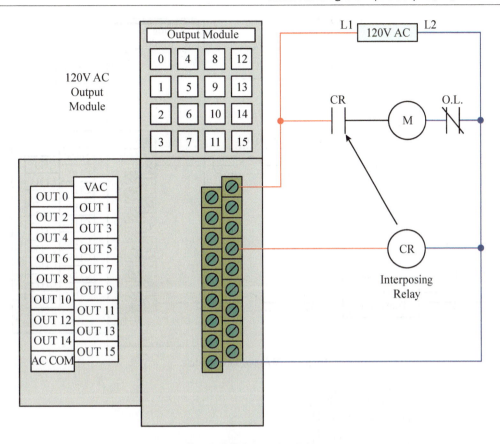

Figure 3–27 Interposing Relay

Transistor-Transistor Logic (TTL) I/O Modules

TTL input modules are designed to be compatible with other solid-state controls, sensing instruments, many types of photoelectric sensors, and some 5V DC level control devices. TTL output modules are used for interfacing with discrete or integrated circuit (IC) TTL devices, LED displays, and various other 5V DC devices.

Analog I/O Modules

Analog input modules are used to convert analog signals (i.e., 4–20 mA, 0–10V DC) that sense such process variables as temperature, pressure, speed, and position to binary values for use by the PLC logic as required. The conversion from analog to digital is accomplished using an analog-to-digital converter (ADC). The analog output module changes binary values generated in the PLC logic into analog signals using a digital-to-analog converter (DAC). These analog output signals can be used for controlling variable speed controllers, valve positioners, displays, etc.

Figure 3–28 shows a typical analog input module and Figure 3–29 a typical analog output module. Analog signals will be covered in greater detail in Chapter 17.

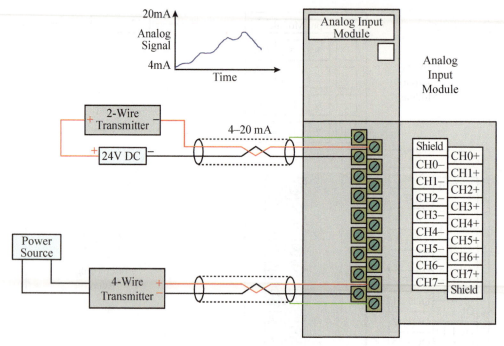

Figure 3–28 Analog Input Module

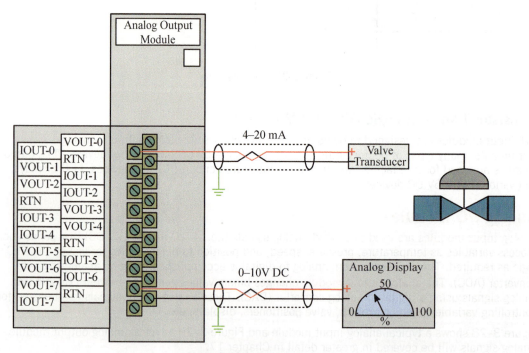

Figure 3–29 Analog Output Module

| Note: PLC manufacturers are occasionally introducing new and special application modules. A few modules have been discussed in this chapter for a basic understanding only. The PLC manufacturer(s) should be contacted for a full and complete list(s) of modules that are available.

Safety Circuit

Consideration needs to be given to the use of emergency-stop functions that are independent of the PLC unless the PLC is so designed for such functions. In most cases and as standard practice with conventional PLC systems, independent, **hardwired** emergency-stop (E-Stop) circuits are included in the design, as shown in Figure 3–30. Where a programmable logic controller is controlling machinery, there are codes and standards that do address the requirements for safety circuits such as emergency stops. For example, the NFPA 79 Electrical Standard specifies safety provisions for industrial machine and equipment and specifically states that "each machine shall be equipped with a Category 0 stop..." and goes on to state that the "final removal of power to the machine actuators shall be ensured and shall be by means of electromechanical components. Where relays are used to accomplish a Category 0 emergency stop function, they shall be non-retentive relays. Exception: Drives, or solid state output devices, designed for safety related functions shall be allowed to be the final switching element, when designed according to relevant safety standards." Safety risk assessments are normally conducted to determine the required safety devices and circuits for a given machine or system during the design phase.

While programmable logic controllers of today are rugged and dependable, where safety is concerned do not depend on solid-state devices and circuitry of the PLC, or PLC logic unless it has been designed specifically for that function. The NFPA 79 recommendation recognizes the importance of hardwired emergency-stop circuits like the one shown in Figure 3–30 to remove power to the output devices.

When hardwired emergency stop circuits are used to remove power to the I/O circuits, the L1 or positive potential is typically wired to one of the PLC inputs so that the PLC programmed logic can take appropriate action for a safe restart of the system. It is not good practice to have outputs come on immediately when the emergency stop relay is reenergized. If it is not practical or possible to directly wire the L1 or positive potential to a PLC input, or in the case where only the PLC output circuits are de-energized, a second contact on the emergency stop relay can be used for this function. Figure 3–31 provides several examples.

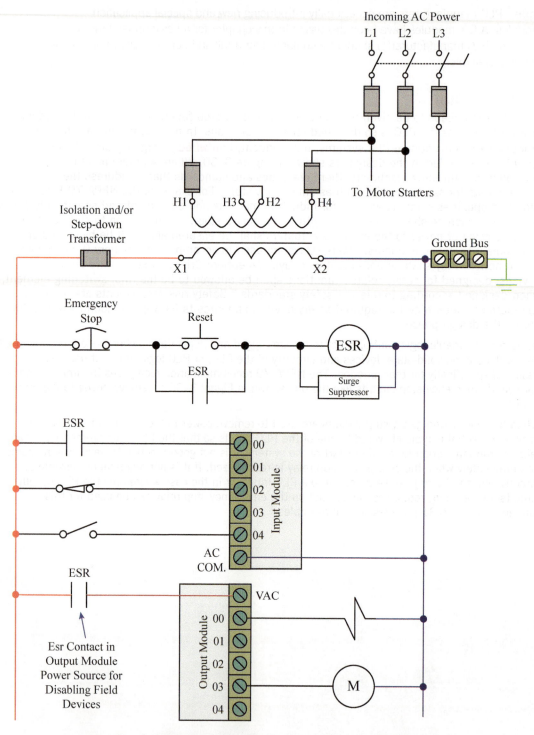

Figure 3–30 Power Distribution with Emergency Stop Relay (ESR) for a Grounded AC System

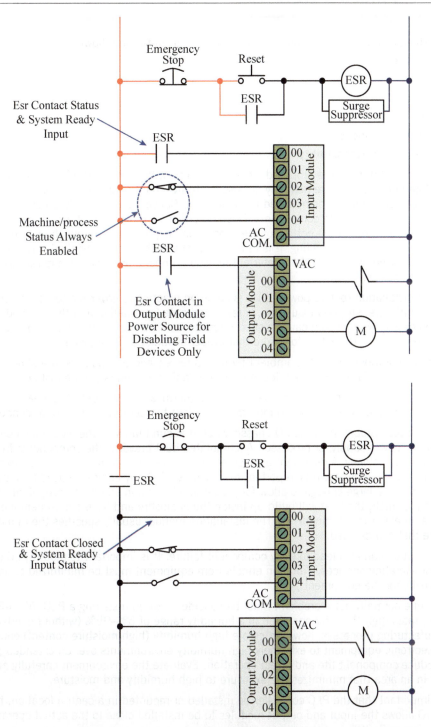

Figure 3–31 Emergency Stop Relay (ESR) Input Examples

Chassis Installation

Before installing a chassis (rack), consideration must be given to the following:

- temperature
- dust
- vibration
- humidity
- field wiring distances
- troubleshooting accessibility

The ambient temperature of the proposed location should not be lower than 32°F or higher than 140°F (0°C and 60°C). Fans are normally not used with I/O chassis, and all cooling of the electrical or electronic components is accomplished by convection. Convection cooling is accomplished when warm air caused by heat in the components rises and creates a movement of air. This movement of air draws cool air in through the bottom of the chassis and expels warm air out through the top. To maintain efficient convection cooling, it is important that the chassis be installed correctly and not used as a shelf for notebooks or other material that would impede or block the natural flow of heat up through the chassis.

During initial installation (before power-up) it is common practice to cover the top of the chassis to prevent any scrap wire, stripped insulation, screws, and nuts from falling into the I/O modules, power supply, or processor, which could cause a short circuit or other electrical failure. The protective cover must be removed after installation to assure that proper cooling can take place.

Under adverse conditions, when the ambient temperatures exceed the manufacturers' recommended maximums, enclosure fans, ventilation louvers, or air conditioning units can be installed.

When the temperature is expected to go below 32°F, a thermostatically controlled heater is sometimes used inside the enclosure to maintain sufficient temperature and prevent condensation.

Dust can also cause problems in the I/O chassis when it accumulates on the electronic components of the modules, power supply, or processor. Accumulated dust prevents the components from dissipating heat effectively. A dust-tight enclosure with a cover and gasket can be used to prevent problems that dust can create. It is important to remember that any enclosure used to house PLC components must be large enough to allow for proper air circulation and heat dissipation. If the enclosure is too small, the heat will build up inside the enclosure and have a detrimental effect on the electrical or electronic components. The installation manual usually specifies the minimum size of enclosure that can be used.

Excessive vibration can also lead to early component failure. It is important to mount PLC equipment on solid, non-vibrating surfaces. Vibration effects from equipment must be minimized to assure proper longevity for the equipment.

Humidity, while normally not a problem, must be considered when installing a PLC. Allen-Bradley, for example, rates their PLCs for operation in a humidity range of 5%–95% (without condensation). Some manufacturing processes, however, create high humidity (high moisture content) conditions. Exposing electronic equipment to extremely high humidity environments over an extended period of time can reduce component life and affect operation. Evaluate the environment carefully and mount equipment in an area that minimizes the exposure to high humidity and moisture.

While it is important that the PLC controller be installed or mounted in a central location, the use of remote I/O allows the input and output modules to be installed close to the actual operating equipment (Figure 3–32).

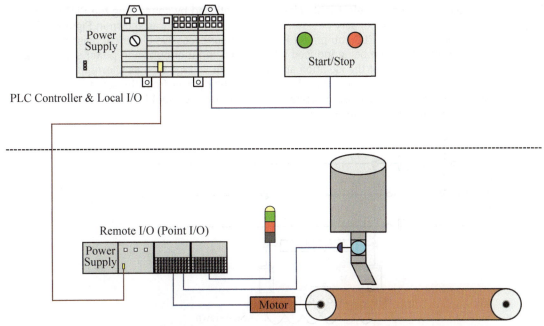

Figure 3–32 Remote I/O Close to Operating Equipment

By mounting the I/O close to the actual equipment, the amount of conduit, cable, and other associated wiring and labor costs will be decreased. The only wiring needed for communication back to the processor will be a shielded-twisted pair, twin axial cable, fiber optic cable, or the like. By having the input and output modules located close to the process or driven equipment, troubleshooting is also easier and more efficient. As discussed earlier in this chapter, each input and output module has status lights that indicate whether an input or output device is *ON* or *OFF*. Having this capability close to the actual equipment shortens troubleshooting time.

Before mounting the I/O chassis and other associated equipment, give careful consideration to location and accessibility. If access is restricted or the equipment is difficult to reach, troubleshooting, repair, and maintenance will be more difficult and time consuming.

When mounting the I/O racks or chassis, ensure there is good electrical connection between each chassis and the enclosure back-panel through each mounting fastener. Remove paint from around fasteners to ensure good electrical connection.

Electrical Noise (Surge Suppression)

Electrical noise is generated whenever inductive loads such as relays, solenoids, motor starters, and motors are operated by "hard contacts" such as push buttons, selector switches, and relay contacts. The noise, or high transient voltages (spikes), is caused by the collapsing magnetic field when the inductive device is switched *OFF*. The level of the voltage spike can be very high and is capable of causing erratic operation of the PLC processor and/or output module, or can cause permanent damage to the module. The interference caused by these voltage spikes and the accompanying electrical noise is often called Electromagnetic Interference (EMI). There are several steps that can be taken to reduce or eliminate the effects of EMI. Two of the most common are isolation and suppression.

Power distribution isolation of the electrical noise is accomplished by connecting the PLC power supply and I/O circuits to the secondary of a step-down transformer as was shown in Figure 3–30. In many applications where the PLC equipment is located near excessive electrical noise and/or to prevent output transients from being induced into the PLC input circuits and power supplies, a second transformer is installed of the isolation and/or constant voltage type as shown in Figure 3–33. In this arrangement, the PLC output circuits should derive their power from the same source as the isolation transformer and the PLC power supplies and input circuits should derive their power from the secondary side of the isolation transformer.

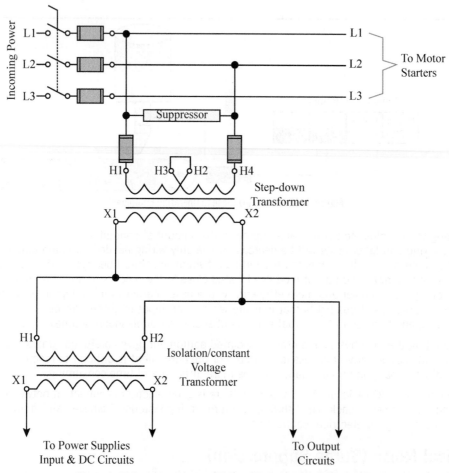

Figure 3–33 Reducing Electrical Noise with an Isolation Transformer

A second method in reducing EMI is to install surge-suppression at its source. Surge-suppression devices can consist of an RC circuit (resistor/capacitor), an MOV, or an RC combination for AC inductive loads and a diode for DC coils. The collapsing magnetic field of the inductive load is, in a sense, dissipated by the suppression device and reduces the effects of EMI. Bear in mind that inductive loads switched by solid-state output modules or devices alone do not require surge-suppression. However, where hard contacts are in series or parallel with a solid-state AC output module that is controlling an inductive load, surge-suppression is required to protect the output module circuits and suppress transient EMI. Figure 3–34 shows examples of where to use surge-suppression.

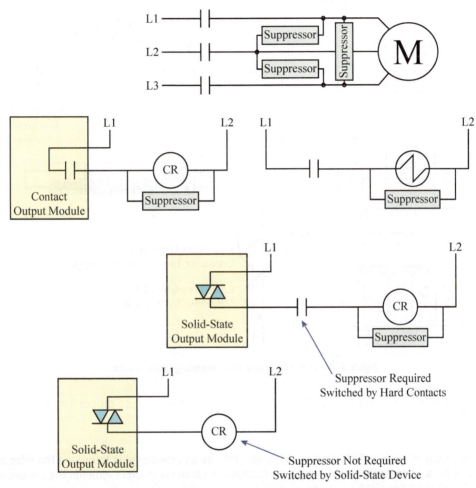

Figure 3–34 Typical Suppression Device Placement

The type of surge suppressor to use depends on the size and type of load. An equipment representative or local electrical distributor should be consulted for help with selection and application.

Grounding

With solid-state control systems, proper grounding helps eliminate the effects of EMI as well as for safety. Figure 3–35 shows a typical installation using equipment-grounding conductors connected to the PLC chassis(s). The equipment-grounding conductor is attached to the metal frame of each chassis. It is recommended that the chassis equipment-grounding conductor be either 1-inch copper braid or 8 AWG minimum stranded copper wire, or per the manufacturer's installation instructions. Where a PLC chassis mounts on a DIN rail and the DIN rail provides the ground connection to the chassis, it is important to connect an equipment-grounding conductor directly to a DIN rail mounting bolt and to the enclosure ground bus.

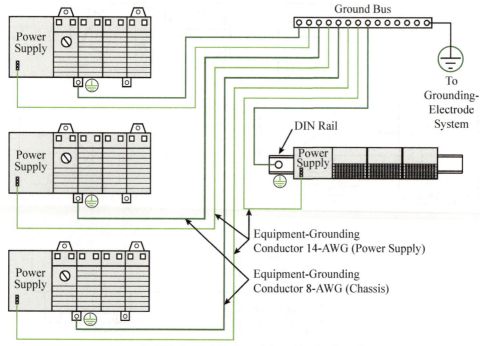

Figure 3–35 Typical PLC Equipment Grounding Configuration

| **Note:** Check local codes and manufacturers' recommendations to ensure proper installation.

I/O Shielding

Certain I/O modules such as TTL, analog, RTD, etc. require shielded cables to reduce the effects of electrical noise. The cable shield, which surrounds the cable conductors, shields the conductors from electrical noise coupling.

When installing shielded cable, it is important that the shield only be grounded at one end. If the shield is grounded at both ends, a ground loop is created, which can introduce ground currents that may result in faulty input signals and/or operation of the PLC processor or I/O modules. Unless specifically stated otherwise, the shield should be grounded at the I/O chassis end, not at the device end.

Try to avoid breaking shields when installing in the field. If you must break the shields at junction boxes for example, do not strip the shields back any further than is necessary and connect the shields of the two cable segments to ensure continuity.

Figure 3–36 shows two methods of grounding the shield of analog shielded cables. In the first example, the shield is connected directly to a chassis or module ground as some manufacturers recommend. The second example uses shield ground clamps.

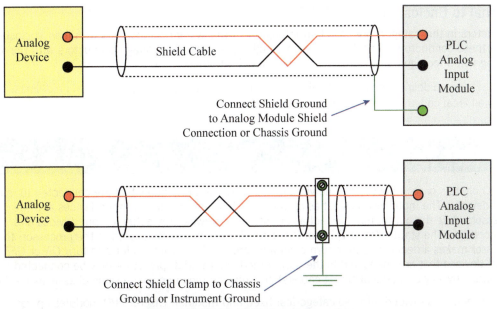

Figure 3–36 Shielded Cable Ground Connections

Categorizing Conductors and Routing

To further guard against electrical noise on analog signals and other low-voltage digital signals, follow these general rules and route the conductors per the guidelines except where it is not possible or practical.

Control and AC Power

Conductors in this category are considered high-power conductors in a PLC system and are more tolerant of electrical noise. For example, AC power to the PLC power supplies and I/O modules, high-power digital conductors to I/O modules that are rated for high power and high noise immunity. Conductors and cables in this category can be routed in the same cable tray or raceway with machine power conductors of up to 600 VAC.

Signal and Communication

Conductors in this category are considered low-power conductors that are less tolerant to electrical noise than AC power conductors, i.e., analog cables, low-power digital AC and DC conductors to I/O modules, and communication cables. Conductors and cables in this category should be routed in a raceway separate from AC power conductors and properly shielded as applicable. If routed in a contiguous metallic raceway or conduit, route at least 3 inches away from AC power conductors that are less than 20 amps and 6 inches when more than 20 amps. When AC power conductors are carrying more than 100 kVA, increase that distance to 12 inches. If not routed in a contiguous metallic raceway, double those distances. Where conductors and cables must cross AC power conductors, it should be done at a right angle (90°) if at all possible.

Internal to Enclosure

Conductors in this category are ones that interconnect the system components within an enclosure. Route these conductors and cables away from high-power and high-voltage conductors with the same spacing listed for signal and low-voltage conductors if at all possible.

Note: These guidelines are for controlling noise immunity only and are just general guidelines. Follow all local codes for safety requirements.

Chapter Summary

The I/O chassis or rack houses the individual input and output modules that are connected to real-world devices. The input modules act as an interface between the actual input devices and the processor, while the output modules act as an interface between the actual output devices and the processor. The status (*ON* or *OFF*) of the input devices is communicated to the processor; the processor makes a decision, and in turn communicates to the output modules to turn *ON* or *OFF* the output devices that are connected to the output module. The PLC processor may be connected to the I/O chassis by way of interconnecting cable(s) or it may be mounted in the same chassis as the I/O.

The I/O section is divided into two categories: fixed and modular. Discrete I/O modules operate on digital, or *ON* and *OFF* signals, whereas analog I/O operates on a variety of signal levels and types. I/O modules are available in a variety of voltages and normally can control 8, 16, or 32 individual I/O devices. Optical coupling, or isolation, is used to protect the low-voltage (5V DC) side of the PLC processor and I/O module from the line-voltage input and output signals that can be as high as 240V.

AC output modules typically use Triacs for switching *ON* and *OFF* the actual output devices. When in the high-resistive, or *OFF* state, Triacs have a small leakage current that flows through the output device. When Triacs fail, they normally fail in the *ON* condition. Fuses used for the protection of output modules are carefully selected by the manufacturer for current and time characteristics, and only fuses recommended by the manufacturer should be used to prevent possible damage to the equipment.

PLC troubleshooting is simplified by the addition of status lights on the I/O modules. The lights indicate which inputs and outputs are *ON* or *OFF*. Indicator lights also indicate if an output module has a blown fuse. To prevent an incorrect module from being installed in a given chassis slot, the modules are often keyed.

A wide variety of input and output modules are available that fit almost any application. Care must be taken to ensure that the module has the correct voltage, current, and time characteristics. The various PLC manufacturers continue to introduce new modules to meet the changing requirements of automated equipment and processes.

Proper installation of PLC equipment requires that the environment (dust, heat, humidity, and vibration) be considered, as well as the physical location for access and troubleshooting. Reduction and/or elimination of electrical noise, voltage spikes, voltage variation, and the like is necessary to ensure proper operation of the system.

Key Terms

I/O section	interfaced
discrete input module	discrete output module
optically coupled	optical isolation
solid-state	field wiring
heat sinks	Triac
SCR	keyed
interposing relay	analog input module
analog output module	hardwired
electrical noise	

Review Questions

1. Describe briefly the purpose of the I/O section.
2. State two reasons for employing optical isolation.
3. Draw an AC input module with four input devices, show all necessary electrical connections, and identify potentials L1 and L2.
4. Draw an AC output module with four output devices, show all necessary electrical connections, and identify potentials L1 and L2.
5. Triacs are susceptible to "dielectric-type" breakdown if the maximum peak voltage level is exceeded.

   T   F
6. Briefly describe why a hardwired emergency-stop circuit is recommended for PLC installations.
7. Briefly describe the function of an interposing relay.
8. I/O modules are keyed to prevent unauthorized personnel from removing them from the I/O rack.

   T   F
9. Which of the following are *not* normally sources of electrical noise?

   a.   solenoid

   b.   relay

   c.   indicator lamp

   d.   motor starter

   e.   motor

   f.   overload heaters

10. To ensure the maximum benefit of shielding, the shield of a shielded cable must be terminated and grounded at both ends.

    T   F

11. E-Stop refers to

    a.   extra stop

    b.   emergency-stop

    c.   every stop

    d.   elevator stop

    e.   energy stop

12. Electromagnetic interference (EMI) can be reduced with the proper grounding of equipment.

    T   F

13. Solid-state output devices tend to

    a.   never fail

    b.   fail in the *open* or *OFF* condition

    c.   fail in the *shorted* or *ON* condition

    d.   not be affected by overload

14. List three environmental considerations when installing PLC equipment.

Chapter 4

Processor Unit

Learning Objectives

After completing this chapter, you should have the knowledge to:

- Describe the function of the processor.
- Describe a typical synchronous program scan.
- Describe a typical asynchronous program scan.
- Identify the two distinct types of memory.
- Describe the function of the *watchdog timer*.
- Identify various memory designs.
- Explain the terms *on-line* and *off-line* programming.

The PLC **processor** houses the microprocessor, memory module(s), and the communications circuitry necessary for the processor to operate and communicate with the I/O and other peripheral equipment. The DC power required for the processor is provided either by a power supply that is an integral part of the processor unit, or by a separate power supply unit. The processor, or "brain," of the PLC is the decision-maker that controls the operation of the equipment to which it is connected. The processor controls the operation of the output devices that are connected to the output modules based on the status of the input devices and the program logic that has been entered into its memory (Figure 4–1). The processor is often referred to as the central processing unit (CPU) or simply the controller.

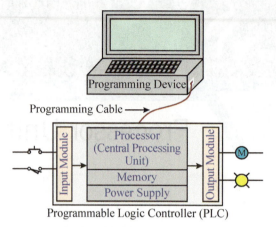

Figure 4–1 Basic PLC Configuration

Processors are available that control as few as 8, or more than 100,000, real-world inputs and/or outputs. The size of the processor unit to be used is dependent on the size of the process(es) or driven equipment to be controlled. The larger the number of I/O devices that are required for the process, the more powerful the processor must be to properly control the number of I/O that will be connected to it. One PLC can control more than one machine or process line and is limited only by the I/O required, physical distance, and memory capacity of the PLC used.

Note: It is difficult to discuss processor unit configuration due to the differences in PLC hardware from the various manufacturers. The discussion that follows is general and is not intended to cover all the PLCs on the market today. It should also be noted that when pictures are used to illustrate a given configuration or concept, one of a particular manufacturer's models is illustrated; however, the manufacturer also has other models larger and/or smaller, and in different configurations.

The Processor

The processor may be a self-contained unit, or may be modular in design, and plug directly into the I/O chassis (rack) as shown in Figure 4–2. Whatever the configuration, the processor consists of micro-processor(s), memory chips, circuits necessary to store and retrieve information from the memory, and communication circuits required for the processor to interface with programmers and other external devices. The memory and communication circuits can be modules separate from the microprocessor module. The actual hardware configuration will depend on the PLC design.

Courtesy of Rockwell Automation, Inc.

Figure 4–2 ControlLogix XT with a Logix 5563 Processor Installed

The microprocessor is the device that

1. Monitors the state or status (*ON* or *OFF*) of the input devices.

2. Systematically solves the logic of the user program.

3. Controls the state of the output devices (*ON* or *OFF*).

4. Communicates with other devices (operator interface terminals, PCs, other PLCs, etc.).

5. Manages memory and updates internal functions (timers, counters, PID, etc.), and internal memory registers.

The execution or completion of these tasks is referred to as the processor **scan**. When the PLC is powered up or turned *ON,* the processor runs an internal self-diagnostic, or self-check, prior to initiating its first scan. If any part of the processor system is not functioning, such as a faulty memory, improper communication with the I/O section, or failure in a remote rack, the processor fault light or other indicator will come on alerting the technician or electrician to the faulted condition. If a programming workstation is connected to the processor, a written explanation or fault code will be displayed on the screen.

Once the processor has passed the self-diagnostic check, it is ready to go to work. Figure 4–3 illustrates a typical four-step PLC scan when the inputs and outputs are updated **synchronous** to the program scan in a non-preemptive multitasking controller. In the first step of the scan, the processor determines the status of the input devices. It does so by looking at the memory locations that have been designated for all the input devices. Remember, as stated earlier in the text, the actual status (*ON* or *OFF*) of any input device is stored in a memory location as either a 1 or a 0. A 1 indicates that a device is *ON* or closed, while a 0 indicates that the input device is *OFF* or open. Based on the 1s and 0s, the processor determines the actual condition of all the input devices.

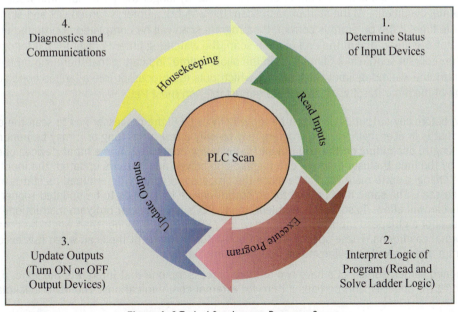

Figure 4–3 Typical Synchronous Processor Scan

The second step in the processor scan is to interpret the logic of the program that has been written and stored in the processor memory. Based on the program requirement, the processor will turn the required output devices *ON* or *OFF,* which is the third step in the processor scan. This third step is referred to as **updating** the outputs. This updating process occurs once during each scan. The fourth step of the scan is often referred to as housekeeping. During this part of the scan, the processor will perform any necessary memory management (housekeeping). Memory management will include resetting the internal watchdog timer, checking for memory errors, updating internal registers, etc. In some processors during this part of the scan, the processor will also communicate with any connected devices (communications). Many PLCs today have two or more processors, one that interprets the logic, and a second that handles the I/O updates, and a third that handles external communications.

The scan is continuous and the four-step process is repeated over and over every few milliseconds or in some PLCs microseconds. To summarize, the four steps of the scan are:

1. Determine the status of the input devices (*ON* or *OFF*).

2. Read and solve the logic of the program (ladder logic).

3. Update the output devices (turn *ON* or *OFF*).

4. Perform communications and housekeeping tasks.

The time it takes to complete one scan will vary from a fraction of a millisecond to 20 or more milliseconds. Scan time is dependent on many factors such as size and number of program tasks, program instruction types, and memory used. Some PLC manufacturers allow you to specify the percentage of processors' time that is devoted to communications, task execution, and background functions, which also impacts overall scan times.

Note: As part of the processor's internal self-diagnostic system, a **watchdog timer** is used. The watchdog timer is preset to an amount of time that is slightly longer than the scan time would be under normal conditions. At the start of each scan, the watchdog timer is turned ON and starts to accumulate time. If the program is correct, the program scan will be completed prior to the time set on the watchdog timer, and at the end of each scan, the watchdog timer is reset to 0. If for some reason the program scan is not completed in the allotted time, indicating that there is a problem with the program, the watchdog timer will time out, which puts the processor into a faulted condition. The range of the timer is software selectable (adjustable) on most PLCs.

Normally, before any output devices can be turned *ON* or *OFF*, the processor has to scan the entire program that is in user memory. The program may be only a few rungs long or it may be hundreds of rungs in length, depending on the equipment that is being controlled. Some input devices operate so fast that by the time the user program can be read and solved and outputs updated, the input device may have changed state more than once since the processor originally determined its status at the start of the scan. The same may be true for an output device that needs to be updated sooner than a normal scan will allow. To solve this problem, many PLCs have special program instructions that allow critical or high-speed input and output devices to be updated sooner than would be possible under normal conditions. These special instructions actually interrupt the scan when it is executing the program and allows I/O devices to be updated immediately.

As mentioned earlier, some PLCs have two or more processors, one that interprets the logic, a second that handles I/O updates, and perhaps a third for external communications. The I/O or backplane processor operates independently and therefore I/O status is being updated during the execution of the PLC program(s). This is also true with pre-emptive multitasking controllers. This means the I/O is updated **asynchronous** to the execution of the logic, which improves overall scan and I/O update times but introduces additional programming considerations which will be covered in later chapters. Figure 4–4 illustrates a typical asynchronous PLC scan when the processor is a pre-emptive multitasking controller or where there is a separate I/O processor. As you can see, the only difference is how and when the I/O is updated.

Note: Additional information, and a more detailed discussion on how the processor scans the user logic, is covered in Chapter 9.

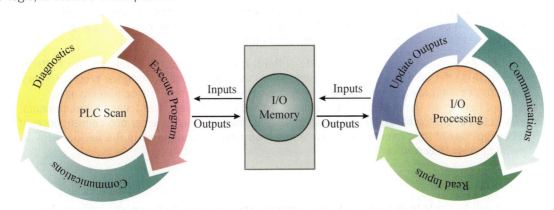

Figure 4–4 Typical Asynchronous Processor Scan

The memory section of the processor consists of thousands of locations where information is stored. In the broadest sense, the memory is divided into two classifications: user and storage. The **user memory** is for the storage of the user program that contains the logic, or instructions that control the driven equipment or process. The **storage memory** is used to store information such as input/output status, timer or counter preset and accumulated values, and internal control relays, etc., which is necessary for the processor to control the equipment or process. The actual memory structure will be covered in Chapter 5, while the purpose or use of each memory is covered later in this chapter.

Memory chips used in the processor can be separated into two distinct groups: volatile and nonvolatile. A **volatile memory** is one that loses its stored information when power is removed or shortly thereafter. A loss of power erases any information stored or programmed on a volatile memory chip. A **nonvolatile memory** has the ability to retain stored information when power is removed, accidentally or intentionally. To protect a volatile memory, backup batteries are included in the processor or power supply. The batteries may be rechargeable nickel cadmium or non-rechargeable lithium types.

Caution: Extra care must be exercised when disposing of batteries, since they are classified as hazardous waste. Special care must be taken with lithium batteries because they may explode when exposed to, or dropped into, water.

When batteries are included, they may be located in the processor, or in the power supply, depending on the PLC. Wherever they are located, there is a battery indicator light(s) to indicate the condition, or state of charge, of the batteries. Common indicator lights are *BAT OK* and *BAT LOW*. A simpler system uses one light to indicate that the battery condition is normal. When the light goes out, it is a warning that the batteries need to be replaced.

When the battery indicator light comes on (or goes out), indicating that the batteries need to be replaced, the memory is still protected for a minimum of two weeks or more. Depending on the size of the memory and the type of batteries used, in many cases the memory remains protected for one year or more with fully charged batteries. In reality, rarely is the power interrupted or off for more than a few hours.

The type of batteries used and the number required will vary with each manufacturer. Because lithium batteries are not rechargeable and must be replaced periodically, care must be taken to always replace the batteries with the type specified, paying special attention to the orientation of each battery in the battery holder to ensure that proper polarity is maintained. Some batteries are available with leads that simply plug into a connector on the PLC.

Caution: As a general rule, a copy is made of the current project prior to changing the batteries. This copy is referred to as a "backup" copy, and is used to replace the original program if for some reason the program in memory is lost. The batteries in some PLCs can be changed without turning off the power. Changing batteries is one of the few maintenance requirements of a PLC. Failure to change the batteries in a timely manner may have serious consequences if a backup copy of the project is not maintained. Common sense dictates that a backup copy be made of every PLC project.

Memory Types

No attempt will be made to explain solid-state memory types in more than a generalized way for basic understanding. Detailed explanations of solid-state memory types are available online or in the electronics section of most libraries.

The most common type of volatile memory is Random Access Memory (**RAM**). Information can be written into, or read from, a RAM chip, and it is often referred to as read/write memory. Information stored in memory can be retrieved or read, while "write" indicates that the user can program or write information into the memory. Random access refers to the ability of any location (address) in the memory to be accessed or used. RAM memory is used for both the user memory and storage memory in many PLCs. Since RAM is volatile, it must have battery backup to retain or protect the stored information.

Nonvolatile memories are memories that retain their information or program when power is lost, and do not require battery backup. A common type of nonvolatile memory is Read Only Memory (**ROM**). "Read only" indicates that the information stored in memory can only be read, and cannot be changed. Information in ROM is placed there by the manufacturer for the internal use and operation of the PLC, and the manufacturer does not want the information changed or altered.
Other types of nonvolatile memory are EEPROM and FLASH.

Electrically Erasable Programmable Read Only Memory (**EEPROM**) is a memory chip that can be programmed and can be erased by the proper signal being applied to the erase pin on the memory chip. EEPROM is used primarily as a nonvolatile backup for the user program in RAM. If the user program in RAM is lost or erased, a copy of the program stored on an EEPROM chip can be downloaded into RAM. It is common on some PLCs for the processor to load the program from the EEPROM chip into RAM memory each time the processor is powered up or after a power failure.

FLASH memory is a nonvolatile memory chip that can be electrically erased and reprogrammed. It is a specific type of EEPROM that is erased and programmed in large blocks. The CompactFlash memory card and Secure Digital (SD) card are the most common nonvolatile memory cards used in PLCs today. Figure 4–5 shows an SD memory card being installed into a PLC controller.

Figure 4–5 SD Memory Card Install

Regardless of the memory type, the user memory containing the program logic can be protected from changes by a key switch located on the front of the PLC controller. With the key switch in the "Run" position, the program logic in the processor will be executed but cannot be changed or overwritten (program download). The key switch can also be used to place the controller in program mode preventing it from running the program logic. Figure 4–6 shows a typical PLC controller with a key switch.

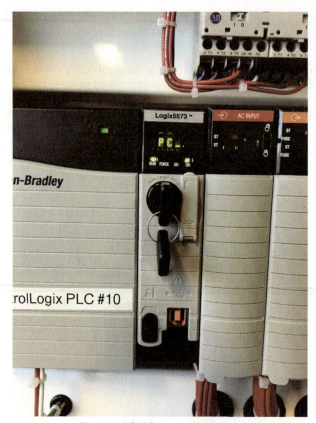

Figure 4–6 PLC Processor Key Switch

Another popular method of restricting access to the program is to use passwords. Passwords restrict access to the program to only those personnel who know the correct password and how to enter it using the programming device. Passwords are often referred to as "software" locks, whereas key switches are referred to as "hardware" locks.

Memory Size

PLCs are available with memory sizes ranging from as little as 4K for small systems up to 40MB (megabytes) for larger systems. Memory size can be expressed in K values: 2K, 4K, 16K, and so on. K, or kilo, which usually stands for 1,000, actually represents 1,024 in computerese. The difference between a standard K (1,000) and the 1,024K value used with PLC processors and computers is due to the way the words were counted. One of the counting or numbering systems used with PLCs is the binary system. The binary system has a base 2, as contrasted to the decimal system we use every day that is base 10. Base 10 represents the numbers 0 through 9, which is 10 digits. The binary numbering system with a base 2 only has 2 digits. The digits are 1 and 0. As with the decimal system that has place values (tens, hundreds, thousands), the binary system also has place values. These are 1, 2, 4, 8, 16, and so on, and each place value is equal to twice the value of the previous number. Base 2^0 represents the number 1; base 2^1 represents 2; base 2^2 represents the number 4

$(2 \times 2 = 4)$; base 2^3 represents the number 8 $(2 \times 2 \times 2 = 8)$; and so on. Counting in this fashion, 2^{10} would equal 1,024. While 1,024 is actually larger than the 1,000 that K actually represents. This also explains the reason for the odd memory sizes of individual memory chips: 256 (1×2^8) and 512 (1×2^9). A memory chip of 256 words would be ¼K and 512 words would be ½K. A PLC with a total memory of 64K would actually have 65,536 words of memory (64 times 1,024). Words, word structure, and numbering systems were covered in Chapter 2.

While it was common for older PLCs to measure their memory capacity in words, it is important to know the number of bits in each word. A PLC that uses 8-bit words would have half the memory capacity of a PLC that uses 16-bit words. For example, the PLC that uses 8-bit words has 65,536 bits of storage with an 8K word capacity ($8 \times 8 \times 1024 = 65,536$), whereas a PLC using 16-bit words has 131,072 bits of storage with the same 8K memory ($16 \times 8 \times 1024 = 131,072$). It is important to know the word size of any given PLC before memory size can be accurately compared.

Note: PCs and most PLC manufacturers today size memory in bytes, not words. A byte is 8 bits, or half of a 16-bit word. A 32-bit word would have 4 bytes. Memory sized in bytes is normally expressed in megabytes (MB) or gigabytes (GB). For example, a 32-bit PLC controller with 20MB memory would have 5,000,000 words of user memory.

The actual size of the memory required depends on the application. In the event that future expansion is planned, there are two options: buy a PLC with more memory than is presently necessary to allow for future expansion, or buy a PLC that meets present needs and add memory (upgrade) when the need arises. Depending on the manufacturer, adding memory may be as simple as replacement of the memory chip, or in most cases replacing the entire processor with one that has more memory.

Memory Structure

As indicated earlier, the processor memory is divided into two general classifications: user memory and storage memory.

User memory contains the instructions programmed by the user. The instructions are entered by a programming device.

Storage memory is where the status (*ON* or *OFF*) of all input and output devices is stored. Numeric values of timers and counters (preset and accumulated), numeric values for arithmetic instructions, and the status of internal relays also are stored in this memory area.

While the information presented in this section applies generally to all PLCs, more specific information on memory structure can only be obtained by reviewing the specifications and literature of the individual manufacturers. In subsequent chapters, the memory structure of specific PLCs will be discussed and illustrated, but the text does not cover all the PLCs on the market today.

Programming Devices

A programming device is needed to enter, modify, and monitor the PLC program(s), or to check the status of the controller. Once a PLC project has been entered (downloaded) into the controller, the programming device is no longer needed for the PLC to execute the user program(s) and may be disconnected.

In the early days of PLCs, the programming device was a dedicated device purchased from the PLC manufacturer for use with their PLCs only. Today, PCs are primarily used as PLC programming devices. Some small PLCs have a built-end LCD display and keypad for viewing and entering data. Figure 4–7 shows a personal laptop computer connected to a micro PLC. Figure 4–8 shows an Allen-Bradley Micro810 PLC with LCD and keypad.

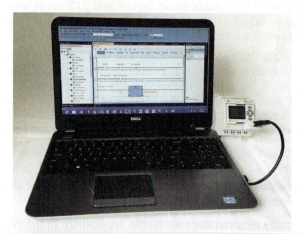

Figure 4–7 Personal Computer Connected to Micro PLC

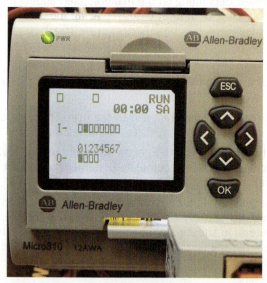

Figure 4–8 Allen-Bradley Micro810 LCD Display & Keypad

Computer Programmers

With software available for all major brands of PLCs, the small personal laptop computer makes for an excellent programming device because of its portability. The color monitor of PCs can vividly show circuit elements to indicate their status, as well as displaying multiple rungs of program logic. The PC has the added ability to interface the PLC software with other software programs for "cut and paste" program development and editing. Documentation of the PLC project is also easily accomplished using a PC. The documentation may be in the form of labeling each circuit element, or writing rung comments. Added graphic capabilities are also part of the programming software as shown in Figure 4–9.

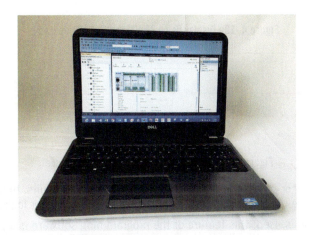

Figure 4–9 PLC Software Graphics

When a project is first developed and edited, it is normally done in the program mode or *off-line* mode. Off-line indicates that the project is being developed or viewed on the programming device only, not while connected to the PLC controller. The project is not operational until it is downloaded into the processor memory, and the controller is placed in the *RUN* mode. Once the project has been developed in the off-line mode and checked, it is downloaded into the controller for final testing and further verification and/or modification. Changes that are made after the project has been loaded into the controller's memory is called *on-line* edits or on-line programming.

Caution: Making changes to the program while the program is running and the driven equipment is operational (on-line programming) must only be done by trained personnel who not only understand the PLC program logic, but also thoroughly understand the driven equipment and/or process.

As mentioned earlier, the monitor of a PC shows multiple rungs of program logic, as well as highlighting the circuit elements to indicate status. When the PLC project has been downloaded into the user memory of the PLC controller, placed in the *RUN* mode, and the program logic activated, the computer monitor gives a visual display of the circuit or logic conditions.

Actual logic (circuit) condition is shown on the computer display in a specific color or intensity on circuit elements that are considered to be logically true or have a true input state. You can also think of it in terms of passing current or having power flow. Figure 4–10a illustrates how a logic circuit appears before the *START* button is pushed for software that intensifies and color codes contacts, interconnecting lines, and coils. Figure 4–10b shows the computer display after the *START* button is pushed and the holding contacts close.

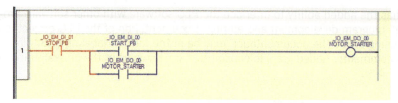

Figure 4–10a Display Before START Button Is Depressed

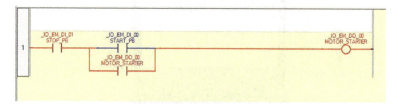

Figure 4–10b Display After START Button Is Released

The terms *passing current* or *power flow* are holdovers from hardwired circuits. In fact, there is no power flow, or current flow, as we normally think of it; rather, it is logic continuity or logically true statements.

No matter which method is used, this feature of the programming software is a powerful troubleshooting aid. By viewing the display on the computer monitor, the electrician or technician can determine which contacts are closed and which outputs are turned on.

Note: An output coil that is intensified only indicates that the output module circuit is ON. It does not guarantee that the actual output device is ON. However, if the output device and associated wiring are complete, the output device will be ON anytime the output module circuit is ON.

To provide a dependable backup of the PLC project in case the controller memory fails or is inadvertently cleared or altered, a flash drive or hard drive is used to store a copy of the PLC project. In fact, a major advantage of using a PC for programming is the ability of the PC to store a copy of the PLC project internally or on an external memory storage device. If for any reason the project is lost, the restoration of the PLC project is simple. Merely download a copy of the PLC project from the computer to the PLC controller.

Chapter Summary

The processor contains the circuitry necessary to monitor the status (*ON* or *OFF*) of all inputs and control the condition (*ON* or *OFF*) of all outputs. It also has the ability to solve and execute the individual program steps in the user program, has a memory for storing the user program and other numeric information, and has the ability to retrieve and use any and all information stored in memory. The memory used in PLCs is of two distinct groups: volatile and nonvolatile. Volatile memory requires a battery backup to prevent the program from being lost due to a power failure; nonvolatile memory retains the program when power is lost or turned off. Programs can be stored on various types of memory chips, as well as on the hard drive of a computer.

The programming device, or programmer, is used to enter, modify, and monitor the user project. Contacts and coils are either color coded or intensified to indicate power flow or logic continuity. Programming the PLC is not difficult, but time must be spent to become familiar with the specific PLC and its programming software.

Key Terms

processor

synchronous

watchdog timer

user memory

volatile memory

RAM

EEPROM

off-line

scan

updating

asynchronous

storage memory

nonvolatile memory

ROM

FLASH

on-line

Review Questions

1. The processor is often referred to as the _____ of the programmable controller.

2. Briefly define *volatile memory*.

3. Briefly define *nonvolatile memory*.

4. 1K of memory is actually
   a.   1,000 words
   b.   1,010 words
   c.   1,024 words
   d.   1,042 words

5. What is *off-line* programming?

6. The most common type of volatile memory is
   a.   FLASH
   b.   ROM
   c.   EEPROM
   d.   RAM

7. Which of the following are types of nonvolatile memory?
   a.   EEPROM
   b.   RAM
   c.   SD
   d.   FLASH

8. What is commonly used as a programming device?

9. What is meant by synchronous scan?

10. What is meant by asynchronous scan?

11. What is a *watchdog timer*?

12. What special precautions should be taken with lithium batteries?

13. What is *on-line* programming?

14. When a PLC is first turned *ON*, it will run a self-diagnostic or self-check test.

    T   F

15. Describe the four steps of a typical synchronous PLC processor scan.

16. The actual scan time, or time it takes the PLC to complete a four-step scan, decreases as the number of program words increases.

    T   F

Chapter 5

Memory Organization

Learning Objectives

After completing this chapter, you should have the knowledge to:

- Identify the two broad categories of memory.
- Describe the function of the two broad categories of memory.
- Identify the types of information stored in each category of memory.
- Define the term *byte*.
- Define the acronym *bits*.
- Describe the types of program tasks.
- Describe what a *tag* is.
- Identify the different data types.
- Identify the structure of an I/O tag address.
- Describe the difference between tasks, programs, and routines.

Memory Words and Word Locations

For the programmable controller to function properly and control a process or driven equipment, it must be able to perform the user program repeatedly and accurately. The system must also be able to perform its control function with great speed, which is achieved by processing all information in binary signals. The key to the speed with which binary information can be processed is that there are only two states, each of which is distinctly different. Binary signals fall into one of two states: 1 and 0. The 1 and 0 can represent *ON* or *OFF*, true or false, voltage or no voltage, high or low, or any other two conditions depending on the system. There is no in-between state or condition, and when information is processed, the decision is either *yes* or *no*. There is no *maybe*, *almost*, or any other alternative.

As indicated in Chapter 4, the processor **memory** consists of hundreds, thousands, or millions of locations that are referred to as words. Each word is capable of storing binary data in the form of binary digits, or **bits** (**BI**nary digi**TS**). A binary digit, like a binary signal, can only be a 1 or a 0.

The number of bits that a word can store will depend on the system or PLC. Words can be made up of 64 bits, 32 bits, 16 bits, or 8 bits. Figure 5–1 shows a 16-bit word.

| 0 | 0 | 1 | 1 | 0 | 1 | 0 | 0 | 0 | 1 | 0 | 1 | 1 | 0 | 0 | 1 |

←————————————————— 16-Bit Word —————————————————→

Figure 5–1 16-Bit Word

If a memory size is 256 words, then it can actually store 4,096 bits of information using 16-bit words (256 words × 16 bits per word) or 8,192 bits using a 32-bit word (256 words × 32 bits per word). When comparing memory sizes of different PLC systems, it is important to know how memory is sized. Many older PLCs sized their memory on the number of words of memory, whereas today most PLC memory is sized in bytes, same as your PC. There are four units of measuring memory. The bit (binary digit) is the smallest unit of measurement having only two states, 1 or 0. Next, is a nibble which is a grouping of 4 bits, followed by a **byte** which is a grouping of 8 bits and the smallest unit which can represent a data item or a character. Last is the **word**, which is a group of a fixed number of bits that is processed as a unit and varies from PLC to PLC or computer to computer. The length of a PLC word is called word size or word length. It may be as small as 8 bits (byte) or may be as long as 64 bits. The most common word size in use today by most PLC manufacturers is 16, 32, and 64 bit words. Table 5–1 shows the breakdown of binary memory units.

In order to locate information in memory, a memory address is used. A memory address is a unique identifier used by a PLC or computer for data tracking. This binary address is used by the processor to track the location of each memory byte or word. As already mentioned, PLCs primarily store data in words and each word is given a unique identifier or address (tag name). Addressing words in the memory serves the same function as the address used for homes and apartments. Word 100, for example, represents a specific word location in memory, just like 100 N. Lincoln represents the address of an apartment building. The bits in word 100 are found by referencing a given bit number, just like the occupant of the apartment complex is found by the given apartment number. The type of unique identifier for a given memory address (word) is determined by the PLC system being used.

Table 5–1 Binary Memory Units

Bit (Binary Digit)	A binary digit is logical 0 and 1 representing a state (passive or active) of a component in a memory circuit. A binary digit (bit) in computer memory is just like a wall switch, there are only two states, *ON* or *OFF*, and they both are memory devices. In its simplest form, computer memory is made up of millions of small electrical switches (bits) that the computer either turns *ON* or *OFF* that represent a logical 1 or 0.
Nibble	A grouping of 4 bits
Byte	A grouping of 8 bits
Word	A computer or PLC word that is made up of a fixed number (group) of bits. Common word sizes are 8, 16, 32, and 64.
1 Kilobyte	Equals 1,024 bytes or 8,192 bits
1 Megabyte	Equals 1,024 kilobytes or 8,388,608 bits
1 Gigabyte	Equals 1,024 megabytes or 8,589,934,592 bits

It may be numbers, characters, or a combination of both. The point to remember is that it will be unique and specific to a given location in memory whether that location is a word, byte, or bit.

Since a bit can only be a 1 or 0 (*ON* or *OFF*), how is the status of bits within a word determined? Words that store the status of individual bits for input devices are set to 1 (*ON*) or 0 (*OFF*), depending on the status (*ON* or *OFF*) of the input devices that the bit locations represent. Other bits are set to 1 or cleared to 0 by the processor in response to the logic of the user program or special instructions, which, in turn, control the status (*ON* or *OFF*) of other bits that represent output devices.

A simple example of how this works is illustrated in Figure 5–2.

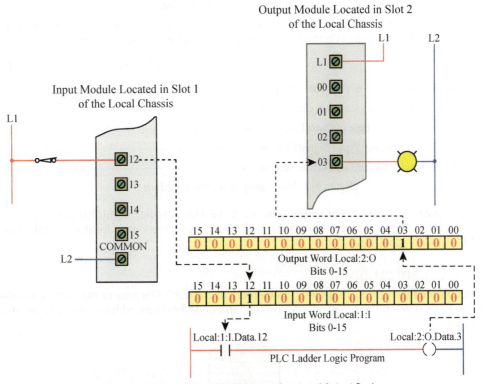

Figure 5–2 Relationship of Bit Address to Input and Output Devices

Note: Although the example in Figure 5–2 uses ControlLogix memory identifiers (numbers), the concepts illustrated are common to all PLCs.

Assume that when a given limit switch is closed, the closure will turn an indicator lamp *ON*. The limit switch is connected to an input module in the I/O chassis, while the indicator lamp is connected to an output module. In Figure 5–2, the limit switch is connected to terminal 12 on the input module and is given an address of Local:1:I.Data.12. This indicates that bit 12 of the input module located in slot 1 in the local chassis stores the status (*ON*-[1] or *OFF*-[0]) of the limit switch. The indicator lamp is connected to terminal 03 of an output module and is given an address of Local:2:O.Data.3. This address indicates that bit 3 of the output module located in slot 2 in the local chassis controls the status (*ON*-1 or *OFF*-0) of the lamp.

By programming a simple circuit into the user memory of the processor as shown at the bottom of Figure 5–2, the processor controls the indicator lamp using the logic of the user program. The logic states that if the contact connected to input module address Local:1:I.Data.12 closes, the lamp connected to output module address Local:2:O.Data.3 should light, or turn *ON*. When power is applied to the processor, the processor starts its scan and looks at bit 12 of word address Local:1:I. Data to see if the bit is set to 1 or 0. If the limit switch is open, the bit will be set to 0, or *OFF*. If the limit switch is closed, as indicated in Figure 5–2, the input module sends a signal to the processor, and bit 12 of word address Local:1:I.Data will be set to 1, or *ON*.

The next part of the scan solves the user program. The logic of the ladder diagram, or user program, indicates that when the contact connected to input module address Local:1:I.Data.12 is closed, or *ON,* the indicator lamp connected to output module address Local:2:O.Data.3 should be turned *ON.*

The address Local:1:I.Data.12 tells us that the limit switch is an input device, and is wired to terminal 12 of the input module located in slot 1 of the local chassis.

Figure 5–3 illustrates the significance of each letter/digit or group of digits used for addressing the I/O memory in the Allen-Bradley ControlLogix family of programmable logic controllers.

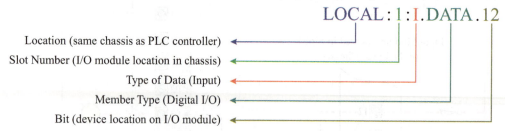

LOCAL : 1 : I . DATA . 12

Location (same chassis as PLC controller)
Slot Number (I/O module location in chassis)
Type of Data (Input)
Member Type (Digital I/O)
Bit (device location on I/O module)

Figure 5–3 Allen-Bradley Logix I/O Address Format

While the I/O address system discussed is specific to the Allen-Bradley ControlLogix PLC, most PLC manufacturers use an addressing scheme that identifies memory word locations relative to the physical I/O module locations.

Memory Organization

Depending on the type of PLC controller, the memory of the controller may or may not be divided. Many older PLCs divided their memory into two general classifications, storage and user, as shown in Figure 5–4.

Storage Memory

User Memory

Figure 5–4 Two Broad Categories of Memory

Storage Memory

Storage memory is that portion of memory that will store information on the status of input and output devices, preset and accumulated values of timers and counters, internal relay equivalents, numerical values for arithmetic functions, and so on. The entire storage memory is called data tables, register tables, data memory, or other names, depending on the PLC manufacturer. A register is defined as an area for storing information (logic or numeric). Although the names or titles that are given to sections or subsections of the storage memory vary, the principles involved do not.

For example, the section of the memory that stores the status of the real-world input devices may be referred to as an input image table, input register, input data, or I/O memory section. No matter what name is used, the information is stored in the same way. The status (*ON* or *OFF*) of each input device is stored as either a 1 or a 0 (*ON* or *OFF*) in one bit of a memory word. When the processor is executing the user program (ladder logic), it scans the input device status stored in the storage memory to determine which inputs are *ON* or *OFF*.

The section of storage memory set aside for output status may be referred to as the output image table, output register, output data, or I/O memory. Again, the name does not change the function of this section of the storage memory, or the method by which information is placed in memory for control of the actual output devices. As the processor executes the user program, it sends binary data (1s or 0s) to the output section of memory to control the output devices. Each output device is represented by one bit of a memory word.

Numeric information for timer or counter preset and accumulated values, arithmetic functions, sequencer functions, data manipulation, etc., uses a part of the storage memory that is called data registers or data memory. Information is entered and stored in this part of memory using the binary, BCD, or hexadecimal numbering systems (the various numbering systems were covered in Chapter 2). The numbering system(s) used depends on the PLC hardware and system requirements. The storage of numeric information requires that several bits of one word be used to represent numbers. In a practical sense, any word used to store numerical information is not available for additional storage, even if all the bits of the word are not used. For example, the value 10 may only require 1 byte of memory storage, but all bits within the word structure (16, 32, etc.) are reserved or used.

Internal relays will replace the numerous control relays used in most hardwired control circuits. The concept and use of internal, or dummy, relays is covered later in the text.

User Memory

The **user memory**, or logic memory as it is sometimes called, is where the programmed ladder logic or program source code is stored. Within the user memory, words are set aside as holding registers. Holding registers typically store information generated and used by the processor when it is solving the user program. Holding registers that are set aside to store intermediate values or other short-term bits of information are sometimes referred to as *scratch areas* or *scratch pads*.

In many PLCs today, the controllers do not divide their memory. They store all data elements in one common memory area and the controller allocates memory as needed. The controller memory is simply just referred to as user memory. When purchasing a PLC controller, you specify the model number with the amount of user memory and I/O needed for your system.

Allen-Bradley LOGIX MEMORY

The Allen-Bradley Logix PLC controllers have memory that is either in one contiguous section, or separated into isolated sections. Figure 5–5 shows an Allen-Bradley controller with one contiguous section of memory. This type of controller allocates memory as needed. The second type of PLC controller has a processor memory separated into three isolated sections or segments (Figures 5–6). The logic and data memory area stores the program source code (ladder logic) and tag data. The I/O memory area stores the I/O data, force tables, message buffers, and produced/consumed tags. The project documentation memory area stores comment descriptions, alarm log, and extended tag properties. Tags will be covered later in this section. As you can see, this is somewhat different from the "Storage" and "User" memory just discussed.

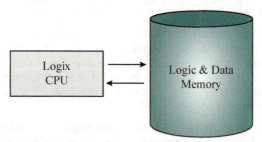

Figure 5–5 Allen-Bradley Logic Controller with Contiguous Memory

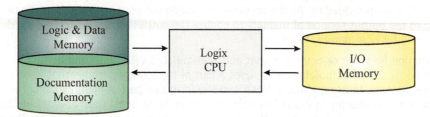

A. Single CPU with Isolated Memory Sections

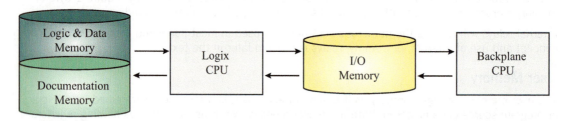

B. Dual CPUs with Isolated Memory Sections

Figure 5–6 Allen-Bradley Logic Controller with Segmented Memory

The Logix CPU in Figure 5–6(a) executes the user's program code and messages. The backplane CPU, on the other hand, communicates with the I/O modules and sends/receives data from the backplane. The backplane CPU also operates independently from the Logix processor. As you recall from Chapter 4, the I/O is typically updated at the end of the program scan in many older PLCs that had a single non-preemptive multitasking processor. In the Logix family of PLCs, the inputs are updated asynchronously in relation to the program or task execution. This means input data is being updated in the controller's memory at any time during the program scan. In traditional PLCs, the controller polled the input modules to obtain the status of the inputs. In the Logix system and in many other PLCs on the market today, the input modules send their data either upon a change of state (COS) or at the requested packet interval (RPI) rate based on how the input module was configured. On the other hand, output data is sent by the controller to the output modules at the rate specified in the output modules RPI configuration, or in the case of output modules in the local chassis (only), at the end of every task execution. Regardless of the number of processors or how the I/O modules are configured, I/O data is being updated asynchronous to the program scan. Understanding when and how the status of I/O data is being updated is very important when designing the PLC logic and configuring I/O modules.

When working with the Logix family of PLCs and many others, you will often hear the term **producer/consumer**. The method of communicating to the I/O modules just described uses the producer/consumer model. In the case of input modules, the input module is the producer of input data and the PLC controller is the consumer of the data. Conversely, in an output module, the PLC controller is the producer of the data and the output module is the consumer.

Most PLCs typically have register-based memory as discussed earlier in this chapter. Logix memory, on the other hand, is a tag-based memory that is common to users that have a computer background. In the Logix processor, the user determines the memory structures and can adjust these to match the application. A user can create tag names and define the data type as needed. There is no predefined memory layout as was typical with many older PLCs.

For example: When the user adds an I/O module (see Figure 5–7) in the Logix PLC software, the Logix processor automatically creates and configures the necessary I/O memory (tags) and data types. Each I/O tag name created follows this format as was previously discussed:

Figure 5–7 Logix I/O Configuration

Location:SlotNumber:Type.MemberName.SubMemberName.Bit

- Location – Identifies the network location of the I/O module, Local or the Adapter Name. If the module is located in the local chassis with the processor, then "Local" is placed in the location field of the tag. If the I/O module is located remotely from the processor chassis, then the remote adapter or bridge name is placed in the location field.

- SlotNumber – Identifies the slot number of the I/O module in its chassis.

- Type – Type of data being addressed. This can be I (input), O (output), C (configuration), or S (status).

- MemberName – Is the specific data from the I/O module; depends on what type of data the module can store. For a digital module, a "Data" member usually stores the input or output bit values. For an analog module, a Channel member (CH#) usually stores the data for a channel.

- SubMemberName (optional) – Specific data related to a MemberName.

- Bit (optional) – Specific point on the I/O module; depends on the size of the I/O module.

The good news is that the Logix processor automatically creates the correct I/O memory tags for the modules that you install and configure. These I/O memory tags are automatically assigned (scoped) as controller tags and all program routines can access controller scoped tags.

If you notice in Figure 5–7 there are two I/O modules that have been configured: a 1756-IA16 input module and a 1756-OA16 output module. The number in [ ] is the chassis slot number location of the module. When the I/O tags are created for each module, they are created in the controller tag area of memory. Figure 5–8 shows the tags that were created for the two I/O modules. Notice at the top of the tag edit screen (right pane) under "Tag Name" that tags were created for each module. Just as previously described, the tag name begins with "Local" because each module is located in the local chassis with the processor. Next is the slot number of the module and then the data type. Two data types are shown for the 1756-IA16 input module and three are shown for the 1756-OA16 output module. Can you guess the data types for each module? The "C" indicates a configuration type tag, and each module has one. The "I" indicates an input data type tag and the "O" indicates an output data type tag.

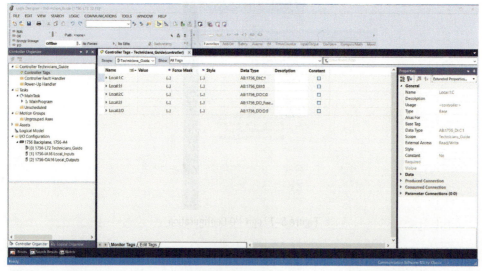

Figure 5–8 Logix I/O Tags

Note: It is worth mentioning that the Logix family of controllers and their associated I/O modules have many features that older traditional PLCs do not. Some of those features include diagnostic information, time stamping, fuse blown indication, etc. This book makes no attempt to cover all the features and capabilities. It is only intended to give the reader a basic understanding of the different memory and addressing methods used today.

If you expand the "Local:1:I" tag in Figure 5–8 by clicking on the ">" sign to the left of the tag, you will see all of the input data tags associated with the 1756-IA input module located in slot 1 of the local chassis (see Figure 5–9). Notice that there are two input tags. One is a fault input tag and the other is a data input tag. The fault tag would contain fault information associated with that module and the data tag would contain the actual input status of the inputs wired to that module.

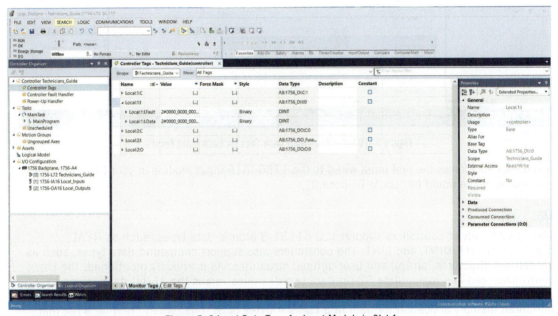

Figure 5–9 Input Data Tags for Input Module in Slot 1

If we again expand the "Local:1:I.Data" tag we will see all of the individual input data tags (bits) for the 1756-IA input module (Figure 5–10). It is worth noting that the Logix family of controllers are 32-bit (word) controllers and store all data in a minimum of 4 bytes or 32 bits of data. That's why there are more than 16 input data tags shown. Only the first 16 are actually used by this input module because it is a 16-point input module.

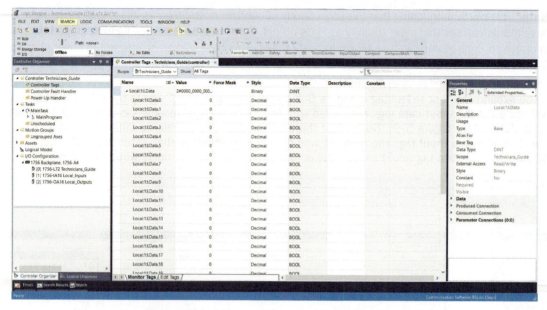

Figure 5–10 Individual Input Data Tags for Each Input Point

If you were to address the first input wired to the 1756-IA16 input module in your PLC logic, the address tag name would be "Local:1:I.Data.0".

Data Types

The Logix family of controllers support IEC 61131-3 atomic data types, such as REAL, BOOL, SINT, INT, DINT, and LINT. The controllers also support compound data types, such as predefined structures, arrays, and user-defined structures. As previously mentioned, the Logix controllers read and manipulate 32-bit data (word) values versus the 16-bit data values found with many traditional PLC controllers. Since the processor manipulates 32-bit data values, the minimum memory allocation for data in a tag is 4 bytes or 32 bits. When creating a tag that stores data that are less than 4 bytes, the processor uses what is needed and the remainder becomes unused memory, as was the case with 1756-IA16 input module data tags discussed in Figure 5–10.

The data type of a tag defines the amount and function of bits and bytes (words) of memory assigned to the tag. The predefined data types (DINT, BOOL, etc.) are used to store the following types of data:

- BOOL – A memory location for a single bit where 1 = ON and 0 = OFF.

- INT – A memory location for storing an integer value between −32,768 and +32,767.

- DINT – A memory location for storing a base integer number in the range of −2,147,483,648 to +2,147,483,647. DINT stands for double integer or double word (two 16-bit words).

- SINT – A memory location for storing a short integer (8 bits) number in the range of −128 to +127.

- REAL – A memory location for a 32-bit value that contains a mantissa or an exponent (raised by a power of 10) that can be very large or very small.

- LINT – A memory location for a 64-bit (two 32-bit words) number in the range of −9223372036854775808 to +9223372036854775807.

Additional predefined data types are also available such as CONTROL, COUNTER, TIMER, MESSAGE, PID, etc. They are used for storing and controlling data associated with a specific function such as timers, counters, and data manipulation instructions.

You should use DINT and REAL data types whenever possible as they use less memory and execute faster than other data types. They should be used for most numeric values and array indexes. Arrays will be covered later in this chapter. REAL data types should be used for manipulating floating-point data values.

SINT and INT should be used primarily in user-defined structures and when communicating with external devices that do not support the DINT or REAL data types.

It is best to group BOOL values into DINT arrays to save memory and to make the bits accessible to some specialized program instructions such as File Bit Comparison (FBC) and Diagnostic Detect (DDT).

Keep in mind that the minimum memory allocation for any tag is 32 bits (DINT). When assigning data types such as BOOL, INT, and SINT to a tag, the Logix controller still allocates a full 4 bytes (DINT or 32 bits) but only uses part of it, as shown in Figure 5–11.

Minimum Memory Allocation of One 32-Bit Word (DINT)		Data Type
31 30 29 28 27 26 25 24 23 22 21 20 19 18 17 16 15 14 13 12 11 10 9 8 7 6 5 4 3 2 1 0		
Unused Memory		BOOL
Unused Memory		SINT
Unused Memory		INT
		DINT
		REAL

Figure 5–11 DINT Memory Allocation

Arrays

An **array** allocates a contiguous block of memory in the controller of the same data type. Each data type in the array is a single tag and each tag is considered to be one element in the array. The elements in the array occupy memory in order, meaning the array starts at 0 and extends to the numbers of elements in the array.

For example, the array in Figure 5–12 is a one-dimensional, ten-element array of the data type DINT.

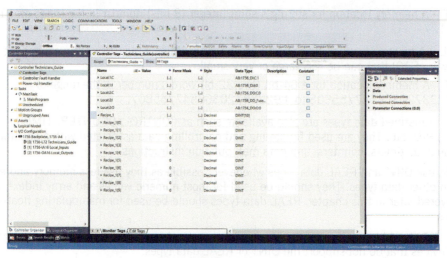

Figure 5–12 10-Element Array

In this example, the array is named "Recipe_1" and has ten elements or DINT tags assigned to the array. Since the elements in the array occupy memory in consecutive order, the array starts with tag Recipe_1[0] and ends with tag Recipe_1[9].

Since the Logix controller does not automatically group data of the same type in memory, you can use arrays to help group data of the same type. For example, in Figure 5–13 there is a 15-element timer array that was created called "Packing_Station_Timers." Now all of the timers associated with the Packing Station logic are grouped together. Single-dimensional arrays like this can help to organize data and conserve memory.

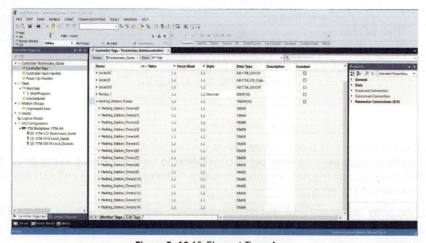

Figure 5–13 15-Element Timer Array

Another example of a single-dimensional array is shown in Figure 5–14. In this example, a BOOL array has been created called "Internal_Relays." The array size is 32 bits or one 32-bit word. If you recall a BOOL data type is one bit (1 = ON and 0 = OFF) and if you were to have created 32 individual BOOL tags you would have used 128 bytes of memory. By creating an array of 32 BOOL tags (one DINT) only 4 bytes of memory was used.

Note: BOOL arrays can only be used with bit instructions; if you need to use an array type instruction, then create a DINT array.

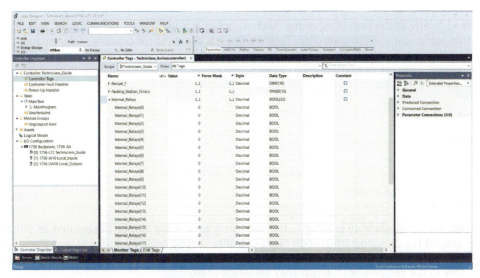

Figure 5–14 32-Element BOOL Array

Arrays can be one-, two-, or three-dimensional as illustrated in Figure 5–15. A three-dimensional array might store information such as part number, color, and size. As you can see, arrays open up all sorts of possibilities for storing specific data as a table of values.

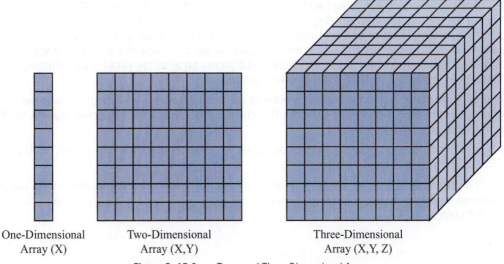

One-Dimensional
Array (X)

Two-Dimensional
Array (X,Y)

Three-Dimensional
Array (X,Y, Z)

Figure 5–15 One-, Two-, and Three-Dimensional Arrays

Note: Because of the many variations of creating and addressing arrays, the reader should consult the manufacturer's literature for information on configuring, addressing, and using arrays in the Logix family of PLC controllers.

Tags

As discussed briefly at the beginning of this section, a **tag** is a text-based name for an area of memory that stores data in the controller. Tags are the basic means for creating, referencing, and monitoring data. The controller stores tags in memory as they are created. When a tag is created, there are certain parameters that must be defined:

- Scope

- Name

- Tag Type

- Data Type

- Style

- Description

Scope – Defines the availability of a tag to the user programs. A tag can be designated as either a controller scoped tag or a program scoped tag. Controller scoped tags, such as I/O tags discussed earlier, are available to every task and program within the project, whereas program scoped tags are available only to the program with which they are associated.

Name – The tag name itself. Tag names can be up to 40 characters long, must start with an alphabetic character or an underscore (_) and cannot end with an underscore. The tag name can contain any combination of alphabetic and numeric characters. Spaces are replaced with underscores when you enter in a tag name.

Tag Type – The type of tag: Base, Alias, Produced, or Consumed. Base tags are tags that store a value for use by the logic within the project and are the actual named area of memory. Alias tags use a different or second name for an existing tag's data area of memory. Alias tags are commonly used to simplify long or complex naming structures, such as I/O tags, or to allow programming of the logic before the actual I/O drawings and wiring are complete. When the value of a base tag changes, so do all alias tags that reference the base tag. Figure 5–16 shows an alias tag (Start_Push_Button) for the first input point of the 1756-IA16 module discussed earlier.

A produced (broadcast) tag is any tag that is shared with other controllers or modules over the backplane, ControlNet™ network, or Ethernet/IP network, etc. A consumed (receive) tag is a tag that holds the value of a produced tag.

Data Type – The data type of the tag, which can be either a predefined data type (DINT, REAL, TIMER, etc.) or a user-defined data type.

Style – The display radix for the data type. This is an optional feature that allows the user to change to a different display radix such as decimal, binary, octal, etc. This feature is only available with certain data types.

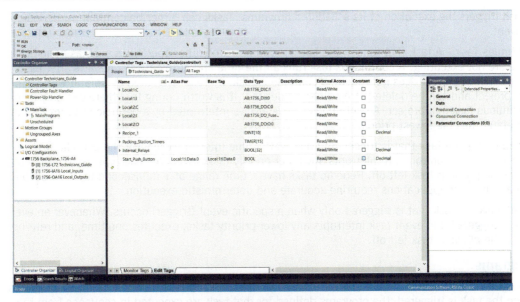

Figure 5–16 Alias Tag

Tasks

The Logix controller is a preemptive multitasking PLC system, meaning that only one task (a set of one or more programs) can be active at a time but has the ability to interrupt the active task, switch to a different task, and then return. A **task** contains programs, each with its own routines. The routines contain the executable code. Figure 5–17 shows the project organization and how tasks, programs, and routines fit together.

Figure 5–17 Project Organization Window

A task triggers the execution of its scheduled programs. Tasks can either be continuous, periodic, or event.

Continues Task – A task that runs in the background any time other operations such as periodic tasks are not executing. A continuous task runs all the time and automatically restarts after each completion. A project does not require a continuous task, but one is created by default with each new project and can be deleted or modified as desired. It is worth noting that there can only be one continuous task in a project. A continuous task is always interrupted by a periodic or event task and, by default, has the lowest priority.

Periodic Task – A task that is triggered at a repeated time interval. Whenever the periodic task is triggered by the controller, it interrupts any lower-priority tasks, executes one time, and returns to where the previous task left off. Periodic tasks have a time range of 1 millisecond to 2000 seconds and are used for applications requiring accurate and deterministic execution.

Event Task – A task that is triggered only when a specific event (trigger) occurs. Whenever an event task is triggered, the event task interrupts any lower-priority tasks, executes one time, and returns to where the previous task left off.

Programs

A program is the second level below the task. As you recall, each task can have up to 32 programs. When the task is triggered, the programs defined for that task are executed in sequence from the first scheduled to the last scheduled. The ladder logic with each program can modify controller scoped and local program scoped data. Developing multiple programs can be useful in helping to organize major equipment pieces or work areas. They can also be useful to isolate machine operations or during development by multiple programmers. When a Logix project is developed, a default program is created and scheduled in the default Main Task (Figure 5–18).

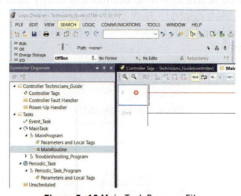

Figure 5–18 Main Task Program File

Programs are scheduled in a specific task or left unscheduled. For example, a technician may create a troubleshooting or test program and only schedule the program when needed. Another example might be that an original equipment manufacturer develops one project and then schedules or un-schedules programs based on the needs of that particular machine. In Figure 5–19 the author has created a troubleshooting program that is configured as unscheduled. The configuration screen on the right clearly shows how programs can be scheduled and unscheduled. If there are multiple scheduled programs, they can be arranged in the sequence desired, from the first scheduled to the last scheduled.

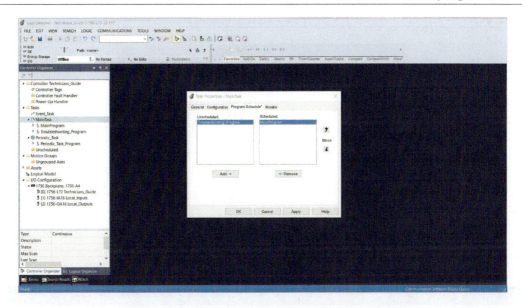

Figure 5–19 Configuring Unscheduled and Scheduled Programs

Routines

A **routine** contains the actual ladder logic within a program. The routine contains a set of logic instructions that provides the executable code to the controller. The routine is programmed in a single programming language, such as ladder logic. The Logix PLC is capable of four programming languages: relay ladder logic, function block diagram, sequential function chart, and structured text. Each of these programming languages will be covered in later chapters.

There is no limit to the number of routines that can be created under a program. Use routines to isolate machine functions or modularize code into subroutines. For example, you could develop a ladder logic routine that controlled the wash cycle and another for the rinse cycle. In this way, the logic is organized by function and can be executed only when needed.

A routine can be assigned as the main routine that executes automatically when the controller triggers the associated task and program, or a fault routine that executes if the controller finds a fault within any routines in the associated program. Routines can be created that are only executed when instructed by logic in the main routine. These types of routines are called subroutines.

In Figure 5–20 the author has created two routines: one is our rinse cycle routine and the other is our wash cycle routine. Since both routines are triggered from the main routine, as seen in Figure 5–20, they are called subroutines and are only executed when the inputs are *ON*.

Keep in mind that only one task can be assigned as the continuous task. Programs assigned to the continuous task will execute according to their assigned order, and only one routine in each program can be assigned as the main or continuous routine.

As the names and structure vary between PLC families, the only way to really understand the memory structure is to obtain the literature for the specific PLC that you are working with. Salespeople and technical representatives are all invaluable resources when trying to gather information or clarification about a particular PLC.

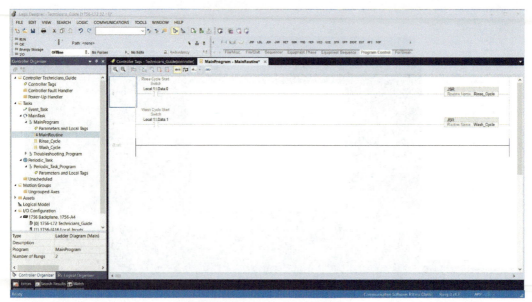

Figure 5–20 Routines and Subroutines

Chapter Summary

All data, logic, and numerics are stored with binary digits that are represented as either a 1 or 0. By storing binary data, the processor can rapidly scan and execute the user program and update the I/O section. In many cases, I/O addresses not only identify the word and bit that is associated with the I/O, but also indicates hardware location (chassis, module group, and terminal). The names given to memory sections or subsections are unique to each PLC manufacturer, but the memories all work in basically the same manner. The processor memory stores the I/O status, user program, and numeric data used by the processor. In tag-based memory, the user determines the memory structures and can adjust these to match the application.

Key Terms

memory	bits
byte	word
storage memory	user memory
producer/consumer	array
tag	task
routine	

Review Questions

1. The following types of information are normally found and/or stored in one of the PLC's two memory categories. Place the memory category after the information type to indicate in which category it is normally found and/or stored.
   a. status of discrete input devices
   b. preset values of timers and counters
   c. numeric values of arithmetic
   d. holding registers

2. Define the term *byte*.

3. What is a *tag*?

4. Describe the following data types:
   a. REAL
   b. BOOL
   c. SINT
   d. DINT

5. What is an *array*?

6. The acronym *bits* is short for what?

7. Explain how each of the following program tasks function in a Logix controller.
   a. Continuous
   b. Periodic
   c. Event

8. What are the two types of tag scopes and how are they different?

9. Describe the difference between a task, program, and routine.

10. Based on the I/O tag name "Local:3:I.Data.15", provide the following module information.
   a. Slot Number
   b. Type
   c. Bit
   d. Member Name
   e. Location

Chapter 6

Understanding and Using Ladder Diagrams

Learning Objectives

After completing this chapter, you should have the knowledge to:

- Identify a wiring diagram.
- Identify the parts of a wiring diagram.
- Convert a wiring diagram to a ladder diagram.
- List the rules that govern a ladder diagram.
- Identify digital logic gates.

There are basically two types of electrical diagrams: wiring diagrams and ladder diagrams. Having a basic understanding of ladder diagrams is essential in learning how to program PLCs in ladder logic code. Ladder diagrams originated as a graphical representation of electrical control circuits using relay-based logic and provided for the simplest form of showing and understanding electrical control circuits. Electricians and technicians have been using ladder diagrams to understand and troubleshoot electrical control circuits since the early days of hardwired relay controls. It is for these reasons that early PLC manufactures adopted ladder logic as the programming language of choice for PLCs. To this day, ladder logic remains the most widely used PLC programming language.

Wiring Diagrams

The wiring diagram shows the circuit wiring and its associated devices (relays, timers, motor starters, switches, and the like) in their relative physical locations (Figure 6–1). While this type of diagram assists in locating components and shows how a circuit is actually wired, it does not show the circuit in its simplest form. To simplify understanding of how a circuit works, and to show the electrical relationship of the components (not the physical relationship), a ladder diagram is used.

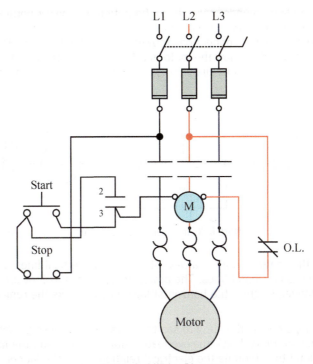

Figure 6–1 Wiring Diagram

Ladder Diagrams

The **ladder diagram**, also referred to as a schematic or elementary diagram, is used by the electrician or technician to speed their understanding of how a circuit works. Figure 6–2 shows the same circuit as Figure 6–1, but in ladder diagram form.

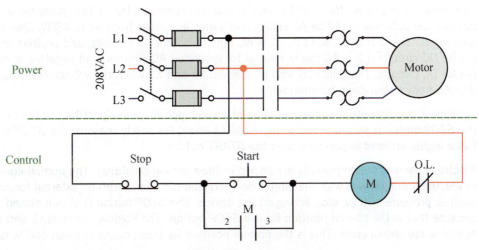

Figure 6–2 Ladder Diagram

To simplify the circuit and help to understand its configuration, the power portion of the circuit is shown separate from the control portion. No attempt is made to show the actual physical location of the components. Since the motor connections (power portion) are the same for any three-phase motor, it is common practice not to show the motor starter or the motor. By not showing the power portion of the circuit, a simplified ladder diagram is created, showing only the control circuit portion of the diagram (Figure 6–3).

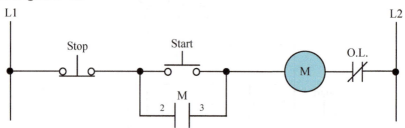

Figure 6–3 Simplified Ladder Diagram

The power required for the control circuit is always shown as two vertical lines, while the actual line(s) of logic are drawn as horizontal lines. The power lines, or rails as they are often called, are like vertical sides of a ladder, whereas the horizontal logic lines are like the rungs of a ladder and are referred to as rungs.

When referring back to Figure 6–1, it is easy to see the physical relationships between the *STOP/ START* station, the motor starter coil (M), the overload contacts (O.L.), and the holding contacts (2 and 3), but it is difficult to determine the electrical relationships. The ladder diagram in Figure 6–3, however, clearly shows the electrical relationships between all of the control circuit components.

Ladder Diagram Rules

Some basic rules for ladder diagrams are as follows.

1. A ladder diagram is read like a book: from left to right and from top to bottom.

2. The vertical power lines (rails) of the ladder diagram represent the *voltage potential* of the circuit. The potential could be AC or DC, and varies in voltage from 6V to 480V. Standard labeling for the rails is L1 and L2. L1 is AC high or hot for AC circuits, and positive or plus (+) for DC circuits. L2 is AC low or neutral for grounded AC circuits, and negative or minus (–) for DC circuits. The rails may also be marked X1 and X2 when the voltage potential is derived from a step-down transformer.

3. Devices or components are shown in order of importance whenever possible. In Figure 6–3 the *STOP* button is shown ahead of the *START* button. For safety reasons, the *STOP* button has a higher order of importance than the *START* button.

4. Electrical devices or components are shown in their normal condition. The normal condition of electrical diagrams is with the circuit de-energized (*OFF*) and with no external forces such as pressure or flow, etc., acting on the device. The *STOP* button is shown closed because that is the normal position for the *STOP* button. The holding contacts (2 and 3) of coil M are shown open. This is the normal position for these contacts when coil M is

de-energized. The normally open (N.O.) M holding contacts 2 and 3 do not close until there is a complete path for current flow to coil M. When coil M energizes, M contacts 2 and 3 close, providing a parallel path for current flow with the *START* button.

5. Contacts associated with relays, timers, motor starters, and the like always have the same number or letter designation as the device that controls them. This labeling method holds true no matter where the contacts(s) appear in the circuit. For example, in Figure 6–3 the normally open holding contacts 2 and 3 are controlled (activated) by motor starter coil M. Therefore, the contacts are identified with the letter M.

6. All contacts associated with a device change position when the device is energized. Figure 6–4 shows a control relay (CR) controlled by a switch (S1) on Rung 1 of the ladder diagram. Rung 2 shows a normally closed (N.C.) control relay contact in series with a green indicator lamp. Rung 3 shows a normally open control relay contact in series with a red indicator light.

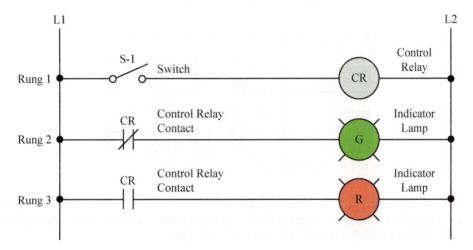

Figure 6–4 Three-Rung Ladder Diagram

When power is applied to the rails of the ladder diagram, the only device in the circuit that operates is the green indicator lamp. The green indicator lamp lights due to a complete path for current flow through the normally closed control relay contacts. These contacts are normally closed and only change position and open when the control relay in Rung 1 is energized. When switch S1 is closed, completing the path for current flow and energizing relay CR in Rung 1, the normally closed CR contacts in Rung 2 open, while the normally open CR contacts in Rung 3 close. The action of the contacts will turn OFF the green lamp in Rung 2 and turn ON the red lamp in Rung 3. As long as the control relay remains energized through S1, the normally closed contact in Rung 2 remains open, and the normally open contact in Rung 3 remains closed. When S1 is opened and relay CR de-energizes, the contacts controlled by relay CR will return to their normal state (normally closed in Rung 2 and normally open in Rung 3).

7. In a ladder diagram, devices that perform a *STOP* function are normally wired in series. Figure 6–5 shows two switches wired normally closed that control a green indicator lamp.

With the two switches wired in series, both A and B switches must remain closed for the lamp to remain lit. If either switch is opened, the green lamp will go out. When switches and/or contacts are wired in series, they are said to have an AND relationship. The AND relationship requires that both A *and* B switches must be closed for the lamp to light. A truth table for this concept is shown in Figure 6–6.

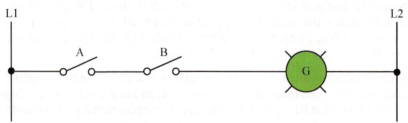

Figure 6–5 Two Switches Wired in Series

Switch A	Switch B	Indicator Lamp (G)
Off	Off	Off
Off	On	Off
On	Off	Off
On	On	On

Figure 6–6 Truth Table for Series Devices

8. Devices that perform a *START* function are normally wired in parallel. Figure 6–7 shows two switches (A and B) wired in parallel to control a red indicator lamp. In this configuration, if either switch A *or* B is closed, the red lamp will light.

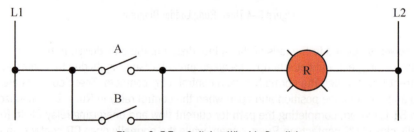

Figure 6–7 Two Switches Wired in Parallel

When switches or contacts are wired in parallel, they are said to have an OR relationship. The OR relationship requires that either A *or* B switch be closed for the red indicator lamp to light. A truth table for this concept is shown in Figure 6–8.

Switch A	Switch B	Indicator Lamp (R)
Off	Off	Off
Off	On	On
On	Off	On
On	On	On

Figure 6–8 Truth Table for Parallel Devices

With this understanding of what a ladder diagram is, and the rules that apply to it, a discussion of a basic motor *STOP/START* circuit (shown in Figure 6–2) can begin.

Basic *STOP/START* Circuit

As stated earlier in this chapter, the wiring diagram in Figure 6–1 is great for showing actual physical locations of the circuit wiring and the components. It does not, however, show the electrical relationships of the devices as simply as the ladder diagram. The wiring diagram is used for original installation and some troubleshooting, whereas the ladder diagram is used to show the electrical relationships of the components, and to speed understanding of how the circuit works as well as troubleshooting.

From viewing the ladder diagram in Figure 6–9(a), it can be seen that when power is applied to the circuit, the motor starter coil M cannot energize because there is an incomplete path for current flow due to the open *START* button and the normally open M contacts (2 and 3). The *START* button and the normally open M contacts are wired in parallel and have an OR relationship. When the *START* button is pushed, a path for current exists from the L1 potential through the normally closed *STOP* button, through the now closed *START* button, through the coil of the motor starter (M), and on through the normally closed overload contacts to the L2 potential as shown in Figure 6–9(b).

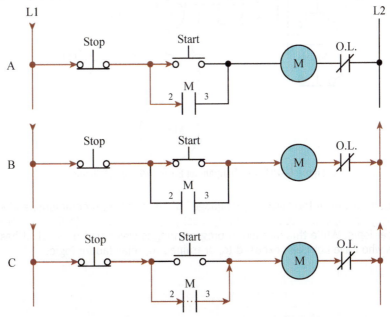

Figure 6–9 Ladder Diagram for Basic *STOP/START* Circuit

When the motor starter coil M energizes, the M contacts (2 and 3) close, providing an alternate path for current flow. At this point, the *START* button could be released, and the circuit would remain energized, or held in, by the holding contacts (2 and 3) of the motor starter as shown in Figure 6–9(c). When contacts from a motor starter or other device are wired in this fashion, they are often referred to as holding, maintaining, or sealing contacts as the circuit is held, maintained, or sealed-in after the *START* button is released.

When the holding contacts (2 and 3) are closed, the main motor contacts of the motor starter are also closed and the motor is started. The operation of the motor is normally taken for granted and is not shown on the ladder diagram. By keeping the ladder diagram as simple and uncluttered as possible, the explanation of the relationships between components and how the control portion of the circuit works is greatly enhanced.

Figure 6–10 again shows the wiring diagram of a motor *STOP/START* circuit. While this diagram looks entirely different from the ladder diagram, both are electrically the same. This comparison shows the electrician or technician why the ladder diagram is preferred.

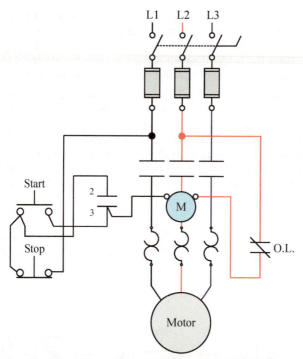

Figure 6–10 Wiring Diagram for Basic *STOP/START* Circuit

The ladder diagram has been the "working language" of electricians, technicians, and electrical engineers for many years, and helps explain why most programmable controllers are programmed using relay ladder logic. While this method of programming is welcomed by most, it has frustrated other PLC users who have not been exposed to, or trained in, relay ladder logic.

Sequenced Motor Starting

Relay ladder diagrams can become large and complex. It is not the purpose of this text to cover them in great detail, but instead to discuss the basic rules and present some concepts to enhance understanding of circuits that are discussed in later chapters.

Figure 6–11 shows a ladder diagram for a circuit that starts three motors.

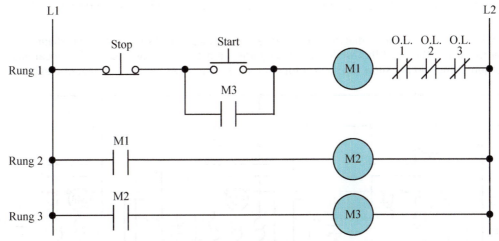

Figure 6–11 Three-Motor Start Circuit

Rung 1 contains the *STOP/START* buttons and the motor starter coil M1 for motor 1. Notice that the holding contacts wired in parallel with the *START* button are not M1 contacts, but instead are M3 contacts. With this arrangement, Rung 1 cannot be sealed in, or maintained, unless motor starter 3 energizes and closes its contacts. Additionally, the M1 contacts in Rung 2 must close to energize motor starter 2 (M2), and M2 contacts in turn must close in Rung 3 to energize motor starter 3 (M3). When the *START* button of this circuit is pushed, it operates as follows.

1. M1 energizes, closing the normally open M1 contacts in Rung 2, and M2 energizes.

2. M2 normally open contacts close in Rung 3 and energize M3.

3. M3 normally open contacts in Rung 1 close and act as holding contacts to keep the circuit energized after the *START* button is released.

| **Note:** This sequence happens almost instantaneously.

4. Pushing the *STOP* button de-energizes M1, which de-energizes M2 in Rung 2 when the normally open M1 contacts go open. M2's de-energizing opens the M2 contacts in Rung 3 and de-energizes M3. With M3 de-energized, the normally open M3 contacts in Rung 1 open.

5. By wiring all three overload contacts in series with M1 in Rung 1, it is ensured that an overload on any motor would shut down all motors. An open overload contact would have the same effect as pushing the *STOP* button.

It could be said that this circuit consists of basically three elements: inputs, outputs, and logic.

The inputs consist of the *STOP* button, the *START* button, and the overload contacts. The outputs are motor starters M1, M2, and M3. The logic that caused the sequential starting were normally open contacts M1, M2, and M3.

These three elements—inputs, outputs, and logic—also work well with PLCs. The inputs are wired to input modules, the outputs are wired to output modules, and the processor performs the logic functions.

Figure 6–12 shows the wiring diagram for the three-motor circuit just discussed. This diagram further illustrates the point that while wiring diagrams are great for giving the physical locations of components, they do not show the control function of the circuit as clearly as a ladder diagram does.

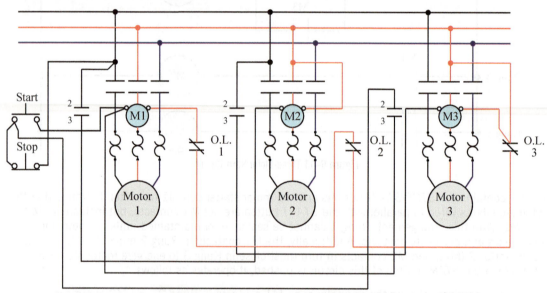

Figure 6–12 Wiring Diagram for Three-Motor Circuit

Digital Logic Gates

While the typical PLC is programmed using ladder logic symbols, some very early PLCs were programmed using digital logic notations such as AND, OR, NOT, etc. Those that work in the digital electronics field are very familiar with digital logic notations. To better understand these digital logic notations and to see how they compare to relay ladder logic, we will cover six basic digital logic gates.

Figure 6–13 shows a two-input AND gate.

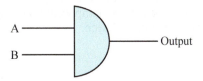

A	B	Output
0	0	0
0	1	0
1	0	0
1	1	1

Figure 6–13 Two-Input AND Gate with Truth Table

From the truth table, we see that both inputs A *and* B must be TRUE, or set to 1, before the output is turned *ON*, or set to 1. The AND gate functions like the two switches that were wired in series to a lamp in Figure 6–5. Both switch A and switch B had to be closed for the lamp to light. Figure 6–14 shows **AND logic** for two programmed input devices wired in series to an output device.

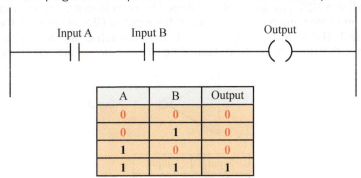

A	B	Output
0	0	0
0	1	0
1	0	0
1	1	1

Figure 6–14 Two Input Devices Wired in Series with Truth Table

Figure 6–15 shows a two-input OR gate.

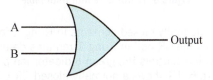

A	B	Output
0	0	0
0	1	1
1	0	1
1	1	1

Figure 6–15 Two-Input OR Gate with Truth Table

From the truth table, we see that if either input A *or* B is TRUE, or set to 1, the output will be turned *ON*, or set to 1. The OR gate functions like the two switches that were wired in parallel to a lamp in Figure 6–7. If either switch A *or* B was closed, the lamp would light. Figure 6–16 shows **OR logic** for two input devices wired in parallel to an output device.

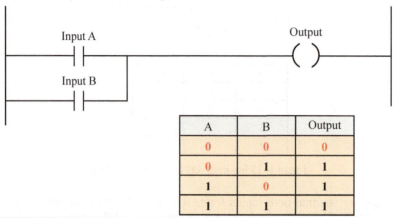

A	B	Output
0	0	0
0	1	1
1	0	1
1	1	1

Figure 6–16 Two Input Devices Wired in Parallel with Truth Table

The next gate is called a NOT gate and is often referred to as an inverter. The inverter, or NOT gate, will have only one input lead and one output lead. If the input is *OFF*, or set to 0, then the output will be *ON*, or set to 1. If the input is *ON*, or set to 1, then the output will be *OFF*, or set to 0. Figure 6–17 shows a **NOT logic** with a truth table. The circle in the output line is used to indicate an inverted function.

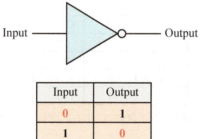

Input	Output
0	1
1	0

Figure 6–17 NOT Gate with Truth Table

The NOT gate functions like the normally closed contacts in Rung 2 for the three-rung ladder diagram in Figure 6–4. As long as the CR coil in Rung 1 remains de-energized, or *OFF*, the normally closed CR contact in Rung 2, which controls the green indicator lamp, will be TRUE and the indicator lamp will be *ON*. Figure 6–18 shows a normally closed CR contact controlling a lamp and the truth table for the circuit.

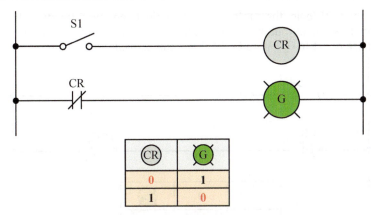

Figure 6–18 N.C. CR Contact Controlling a Lamp with NOT logic

As long as the single-pole switch (S1) that is wired in series with the CR coil is open, relay CR is *OFF*, or set to 0. With relay CR *OFF*, the logic for the normally closed contacts will be TRUE, and the lamp will be *ON*. When S1 is closed, relay CR will energize, the normally closed CR contacts will open, and the light will be turned *OFF*. The truth table reflects the action of the CR coil that controls the action of the CR contacts. It may help to understand the truth table if we think that the normally closed CR contacts are controlled by the CR coil. If the CR coil is *OFF*, or set to 0, we can think of the CR normally closed contacts also being closed or set to 1. As the input will be inverted when the CR coil is energized, or set to 1, then the output device controlled by the normally closed contacts will be FALSE or set to 0.

The programmable controller will use NOT logic in the same way as described above. If the output device is set to 0, or *OFF*, any normally closed contacts associated with the output device with the same address will also be set to 1. The NOT logic will be used in the next chapter for the Examine Off instruction. When a normally closed contact is addressed with the same address as an output coil, the normally closed contact will be true as long as the output coil is *OFF*, or FALSE.

By combining the NOT gate with the AND gate, we get what is called a NAND gate. Figure 6–19 shows a two-input NAND gate with a truth table.

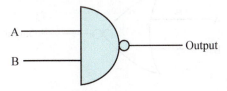

A	B	Output
0	0	1
0	1	1
1	0	1
1	1	0

Figure 6–19 NAND Gate with Truth Table

As discussed with the NOT logic, the circle is used to indicate an invert function. By placing the invert, or NOT, symbol at the output of the AND gate, the output can only be TRUE when one or both of the inputs are false or set to 0. Figure 6–20 shows the equivalent relay circuit.

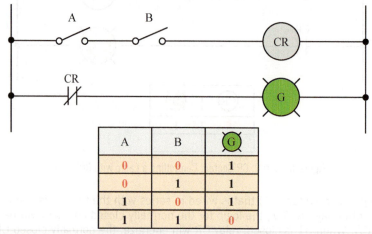

A	B	G
0	0	1
0	1	1
1	0	1
1	1	0

Figure 6–20 Ladder Diagram with NAND Logic

To help understand **NAND logic**, consider the ladder diagram in Figure 6–20.

With both inputs A and B *OFF*, or set to 0, the inverted output will be set to 1, or be turned *ON*. If only input A is set to 1, the inverted output will remain set to 1 because input B is still open, or set to 0. If input A is opened and input B is closed or is set to 1, the inverted output will again remain set to 1. However, if both A and B are closed, or set to 1, then the inverted output will be set to 0, or *OFF*. From the ladder diagram, we can see that with both input devices A and B open, relay CR would not be energized and the output would be set to 1, or *ON*. For the output device to go false, or to 0, both input switches A *and* B must be closed, or set to 1, as illustrated in the truth table.

When we combine the NOT gate with an OR gate, we get what is called a NOR gate. Figure 6–21 shows the NOR gate with a truth table.

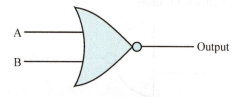

A	B	Output
0	0	1
0	1	0
1	0	0
1	1	0

Figure 6–21 NOR Gate with Truth Table

To help understand the **NOR logic** gate, consider the ladder diagram in Figure 6–22.

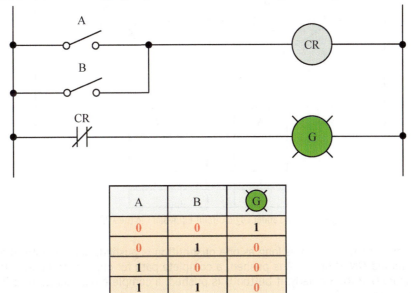

A	B	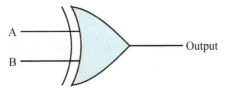 G
0	0	1
0	1	0
1	0	0
1	1	0

Figure 6–22 Ladder Diagram with NOR Logic

From this figure we can see that if either input A *or* B is closed the CR coil will energize and the normally closed contacts in Rung 2 will open and turn *OFF* the output. With this configuration, the only time the output lamp will light is if both switches A and B are open, or set to 0.

The final logic gate that will be covered is the exclusive OR gate (XOR). The XOR gate with truth table is shown in Figure 6–23.

A	B	Output
0	0	0
0	1	1
1	0	1
1	1	0

Figure 6–23 XOR Logic Gate with Truth Table

The **XOR logic** gate will only turn the output *ON* when *either* input A or B is *ON*, but *not both ON*. This logic gate can be compared to the two double-circuit pushbuttons shown in Figure 6–24a.

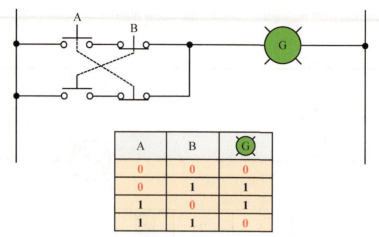

A	B	G
0	0	0
0	1	1
1	0	1
1	1	0

Figure 6–24a Ladder Logic Circuit Equivalent to the XOR Gate

This ladder diagram shows us that as long as one pushbutton is pushed, but *not* both, the output device will be turned *ON.* If button A is pushed, a complete path for current flow now exists, as shown in Figure 6–24b. Similarly, if button B is pushed, a complete path for current flow now exists, as shown in Figure 6–24c.

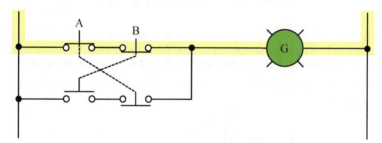

Figure 6–24b Input A Closed and Output is *ON*

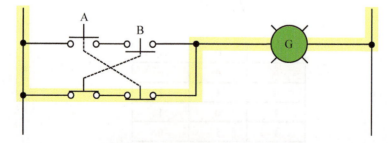

Figure 6–24c Input B Closed and Output is *ON*

Logic gates can be combined to create very complicated control logic. Figure 6–25 shows an OR gate combined with an AND gate to duplicate the logic of a ladder diagram that contains a start button, holding contacts, float switch, and pump starter. Compare the logic of the ladder diagram with the logic gate equivalent circuit and the truth table.

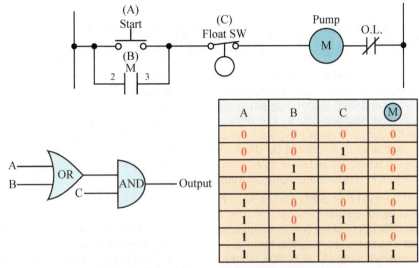

A	B	C	M
0	0	0	0
0	0	1	0
0	1	0	0
0	1	1	1
1	0	0	0
1	0	1	1
1	1	0	0
1	1	1	1

Figure 6–25 Combining an OR Gate and an AND Gate

Figure 6–26 shows a ladder diagram with four contacts that control an output device. We can see from the diagram that to turn on the output, any of the following combinations of contacts is required:

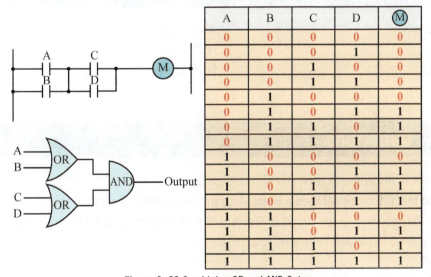

A	B	C	D	M
0	0	0	0	0
0	0	0	1	0
0	0	1	0	0
0	0	1	1	0
0	1	0	0	0
0	1	0	1	1
0	1	1	0	1
0	1	1	1	1
1	0	0	0	0
1	0	0	1	1
1	0	1	0	1
1	0	1	1	1
1	1	0	0	0
1	1	0	1	1
1	1	1	0	1
1	1	1	1	1

Figure 6–26 Combining OR and AND Gates

A and C
A and D
B and D
B and C

By combining two OR gates and an AND gate, we can duplicate the logic of the four contacts, as shown in Figure 6–26 and verified by the truth table.

Chapter Summary

There are basically two types of electrical diagrams: wiring diagrams and ladder diagrams. Wiring diagrams show actual physical locations and wiring, whereas ladder diagrams show electrical relationships. The simplified ladder diagram speeds understanding of circuit operation and is used for circuit design and troubleshooting. The vertical sides of the ladder diagram are referred to as *rails*, while the horizontal lines or logic are called *rungs*. On electrical diagrams, devices are always shown in their normal or de-energized condition. When two or more devices are wired in series, they perform an AND function, while two or more devices wired in parallel perform an OR function. The elements of the ladder diagram are inputs, outputs, and logic.

Logic gates can be used that duplicate the logic of the pushbuttons, contacts, and control devices typically used in motor control circuits. Common logic gates are AND, OR, NOT, NOR, NAND, and XOR (exclusive OR). Logic gates can be combined to create complex control circuits. The elements of the ladder diagram are inputs, outputs, and logic. Ladder diagrams are favored over wiring diagrams when a basic understanding of the control circuit is needed. The ladder diagram shows electrical relationships without regard to actual location and is the basis for PLC ladder logic.

Key Terms

ladder diagram

OR logic

NAND logic

XOR logic

AND logic

NOT logic

NOR logic

Review Questions

1. Define the terms *normally open* and *normally closed*.
2. Describe the difference between a wiring diagram and a ladder diagram.
3. Explain the operation of the circuit in Figure 6–9 if M contacts 2 and 3 do not close.
4. Contacts wired in parallel have what relationship?
   a.   AND
   b.   OR
5. Contacts wired in series have what relationship?
   a.   AND
   b.   OR

6. The two main vertical lines of a ladder diagram are often referred to as:

    a.   rungs

    b.   power ports

    c.   rails

    d.   tracks

    e.   none of the above

7. The horizontal lines of a ladder diagram are referred to as:

    a.   rungs

    b.   power ports

    c.   rails

    d.   tracks

    e.   none of the above

8. Devices that are intended to perform a *STOP* function are normally wired in_____ with each other.

9. Devices that are intended to perform a *START* function are normally wired in_____ with each other.

10. How are contacts that are associated with relays, motor starters, timers, and the like identified?

11. Convert the wiring diagram below into a ladder diagram.

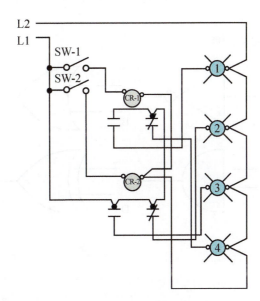

12. Convert the wiring diagram below into a ladder diagram.

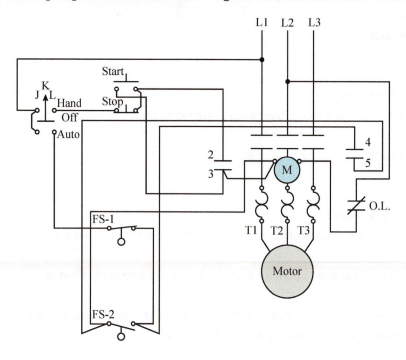

13. Identify the following logic gates.

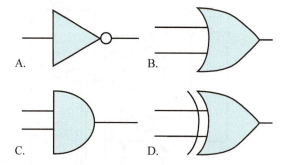

A.

B.

C.

D.

Chapter 7

Relay Type Instructions

Learning Objectives

After completing this chapter, you should have the knowledge to:

- Describe the Examine On instruction.
- Describe the Examine Off instruction.
- Write the logic for a standard *STOP/START* motor circuit.

The next step in understanding how the PLC works is to learn how ladder logic is changed into processor logic. The actual programming is accomplished using a PC and programming software. Figure 7–1 shows an Allen-Bradley MicroLogix PLC and PC used to program and monitor the PLC.

Figure 7–1 Allen-Bradley MicroLogix PLC and Computer Programmer

When programming in PLC ladder logic, some common electrical relay symbols are used. These symbols include N.O. contacts, N.C. contacts, and coils or outputs. There are no symbols for input devices such as push buttons, limit switches, pressure switches, etc. The contacts from all input devices are programmed using either the N.O. or N.C. relay contact symbols. The actual PLC programming will be covered in Chapter 8, but first the relay logic used for PLCs must be discussed and *understood*.

117

Programming Contacts

A PLC is traditionally programmed using a ladder logic-type language. Ladder logic is a good choice for a programming language because it closely resembles the way circuits are hardwired. Electricians and technicians feel comfortable with, and understand, ladder logic, so programming with a ladder logic-type language makes good sense. While there are many similarities between standard relay ladder logic and the ladder logic used for programming a PLC, there are some distinct differences.

Hardwired contacts in a motor control circuit control the path for current flow to the output (coil, light, solenoid, etc.). The contact symbols that are used when programming a PLC are actually logic instructions that the processor uses to make decisions.

The PLC N.O. contact symbol is actually an instruction that tells the processor to look for an *ON* condition at the memory address assigned to the symbol. If an *ON* condition exists, the instruction is said to be logically *true*, and logic continuity exists. This is much like saying that if a contact is closed, current can flow.

Note: As stated earlier in the text, the terms current flow and power flow are often used by the various PLC manufacturers to indicate that a circuit is complete, or logically true.

Figure 7–2 shows a simple circuit containing a single-pole switch and a lamp for an output. Figure 7–3 shows the equivalent circuit when programmed in a PLC. Addresses shown are in the Allen-Bradley ControlLogix format.

In Figure 7–2, if the switch is closed, current flows through the switch contacts and the lamp lights. In Figure 7–3, the single-pole switch is shown as an N.O. contact symbol with the memory address "Local:1:I.Data.0". The lamp, or output, is shown as a partial circle, and is given the memory address "Local:2:O.Data.1". The output is actually bit-1 in output memory word Local:2:O.Data.

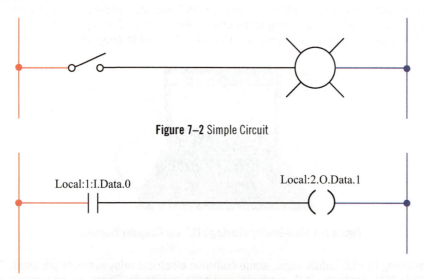

Figure 7–2 Simple Circuit

Local:1:I.Data.0   Local:2.O.Data.1

Figure 7–3 Equivalent Circuit Programmed in a PLC

Because of the way the program logic has been entered, the N.O. contact symbol tells the processor to look at input memory address "Local:1:I.Data.0" (input for single-pole switch); if a closed (*ON*) condition is found, then the logic of the circuit is true. When the logic of the circuit is true, the processor is instructed to turn *ON* output "Local:2:O.Data.1". In reality, there is no actual electrical connection between the switch (Local:1:I.Data.0) and the lamp (Local:2:O.Data.1). Instead, it is the processor that turns the lamp *ON* or *OFF* depending on the logic of the program that is written and the status of the input device.

Figure 7–4 shows the switch and lamp as they are wired to their respective I/O modules. It is the N.O. contact symbol in the program that tells the processor to examine the single-pole switch for an *ON* condition. If the switch is open (*OFF*), the program logic is not true and the processor will not turn on the lamp. On the next processor scan, however, if the switch has been closed, it will be *ON*, the logic of the circuit will be true, and the processor will turn *ON* the lamp.

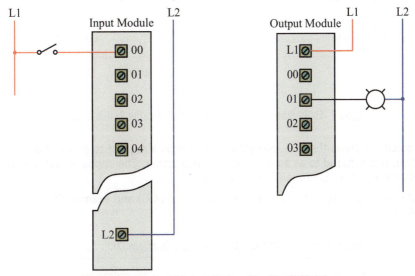

Figure 7–4 Input and Output Devices Wired to I/O Modules

Figure 7–5a shows the bit status of input word "Local:1:I.Data" when the switch is open. With the switch open (*OFF*), the logic of the circuit cannot be true so the lamp (Local:2:O.Data.1) will not be *ON*. The *OFF* condition of the switch and the lamp is indicated by a zero (0) in bit location 0 for the input and bit-1 for the output. When the switch is closed, the bit that represents the switch will change to a 1. This makes the circuit logically true, and the processor will turn the lamp to *ON* by placing a 1 in the output memory location for the lamp as shown in Figure 7–5b.

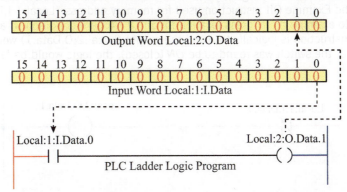

Figure 7–5a Bit Status for I/O Word Local:1:I.Data with Switch Open

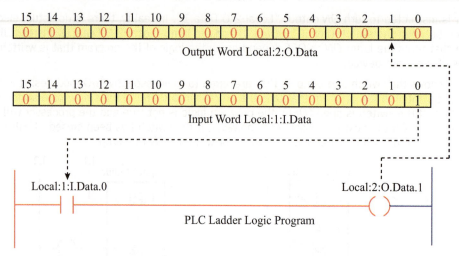

Figure 7–5b Bit Status for I/O Word Local:1:I.Data after Switch Is Closed

Because the processor views the N.O. contact instruction as a request to examine a given address for an *ON* condition, it is referred to as an **Examine On** instruction. The opposite instruction, the N.C. contact, is referred to as an **Examine Off** instruction.

These instructions are also referred to as examine if closed (XIC) and examine if open (XIO), as shown in Table 7–1.

Table 7–1 Examine if Closed (XIC) and Examine if Open (XIO).

Examine On	Examine Off
⊣ ⊢	⊣/⊢
Normally Open	Normally Closed
Examine if Closed	Examine if Open
XIC	XIO

The Examine Off instruction is only logically true when the device referenced is *OFF*, or open. Figure 7–6 shows the Examine Off instruction used for memory address "Local:1:I.Data.0". The processor is asked to examine the address location for an *OFF* (open) condition. If an *OFF* condition is found, then the instruction is logically true and the output (Local:2:O.Data.1) would be turned *ON*. If, on the other hand, the switch was found to be *ON* (closed), the logic would be false, and the lamp would be turned OFF.

Figure 7–6 Examine OFF Instruction

At first these two instructions may seem to be contrary to the logic of hardwired contacts, so it is important to remember that these are instructions to the processor and are not hardwired contacts. A review of both instructions seems appropriate.

Examine On

Whenever the processor sees an N.O. contact in the user program, it views the contact symbol as a request to examine the memory address of the contact for an *ON* condition. If the N.O. contact has an input memory address, and if the real-world input device is closed, or *ON*, the processor sets the appropriate bit in the input memory address to 1, or *ON*. The Examine On instruction is looking for an *ON* condition, a bit set to 1, as a true condition, and if found, a true logic path exists through the contact. If the real-world input device had been open, or *OFF*, the processor would have cleared the appropriate bit to 0, or *OFF*, and the contact would be false and there would not be a true logic path through the contact.

Examine Off

When the normally closed symbol (N.C.) is programmed in ladder logic, the processor views it as a request to examine the address of the contact for an *OFF* condition. Any address that is actually *OFF* becomes logically true and there is a true logic path through the instruction (contact). If an N.C. contact has an input memory address and the real-world input device is open, or *OFF*, the processor sets the bit to 0, or *OFF*. As the Examine Off instruction is looking for an *OFF* condition, a bit set to 0, or *OFF*, is a true condition and there would be logic continuity through the contact. If the input device had been closed, or *ON*, the bit would be set to 1, or *ON*. The Examine Off instruction can only be logically true when an *OFF* condition exists. Any bit set to 1 is viewed as an *ON* condition, which makes an Examine Off (N.C.) contact false, and no logic continuity exists.

To further reinforce the Examine On and Examine Off instruction concepts, a look at a standard *STOP/START* circuit may be helpful. Figure 7–7a shows a standard *STOP/START* circuit with overload contacts using a standard ladder diagram. Figure 7–7b shows the equivalent circuit programmed in a PLC.

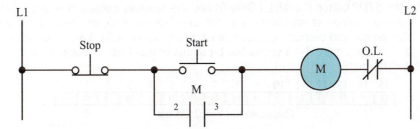

Figure 7–7a Standard *STOP/START* Ladder Diagram

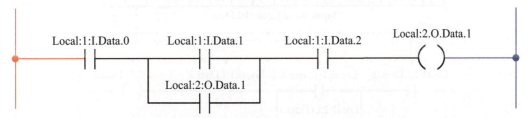

Figure 7–7b Equivalent *STOP/START* Circuit Programmed in a PLC

Note: Memory addresses shown are Allen-Bradley ControlLogix format. An "I" preceding ".Data" indicates an input data type, whereas an "O" indicates an output data type. Since the output is normally the last item programmed on a rung, the overload contacts (Local:1:I.Data.2) are programmed ahead of the motor output (Local:2:O.Data.1).

Once the input and output devices are wired to the I/O modules, as shown in Figure 7–7c, and the PLC controller is "powered up," or turned *ON*, the processor scans the inputs and sets the corresponding bits to 1 or 0 depending on the status of the real-world input devices. If an input is open, the corresponding bit is set to 0, or *OFF*, whereas any bit that represents a closed device will be set to 1, or *ON*.

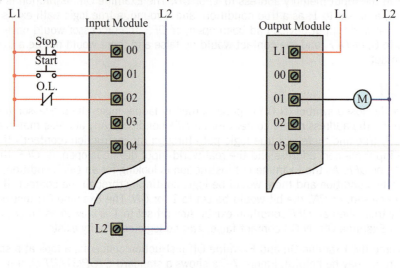

Figure 7–7c Input Devices Connected to an Input Module

In Figure 7–7c, the *STOP* button (Local:1:I.Data.0) and the overload contact (Local:1:I.Data.2) are wired as N.C. but programmed as normally open contacts (Examine On), as shown in Figure 7–7b. Because the *STOP* button and overload contacts are actually closed, bits 0 and 2 of word "Local:1:I. Data" are set to 1, or *ON*. Figure 7–8a shows the bit status of input word "Local:1:I.Data" with the processor in the *RUN* mode.

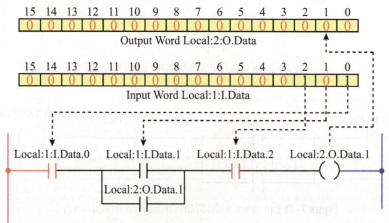

Figure 7–8a Bit Status for Input Word Local:1:I.Data and Output Word Local:2:O.Data with the Processor in *RUN* Mode

With the *STOP* button and the overload contacts (bits 0 and 2) set to 1, or *ON*, we need only press the *START* button to complete the circuit. When the *START* button (Local.1:I.Data.1) is pressed, bit-1 is set to 1 (*ON*) as shown in Figure 7–8b.

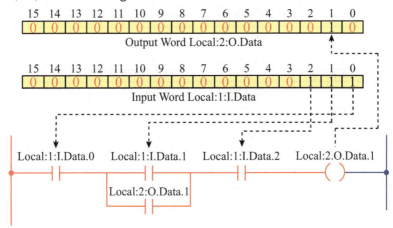

Figure 7–8b Bit Status for Input Word Local:1:I.Data and Output Word
Local:2:O.Data While *START* Button Is Being Pressed

With the circuit logic now complete (true), the processor sets bit-1 of output word "Local:2:O.Data" to 1, or *ON* (Figure 7–8c), and motor output "Local:2:O.Data.1" is *energized* and held energized by holding contacts "Local:2:O.Data.1". The holding contacts have the same address as the motor output (Local:2:O.Data.1) because they both reference the same bit (1) of word "Local:2:O. Data". The holding contacts do not actually exist as hardwired contacts but are used to maintain the logic path for the circuit. The Examine On instruction with the address "Local:2:O.Data.1" is the equivalent of holding contacts, and when the motor is energized (turned *ON*), bit-1 of word "Local:2:O.Data" is set to 1, or *ON*, and an alternate logic path around the *START* button exists.

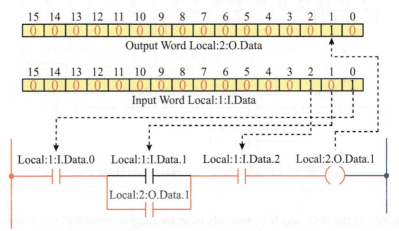

Figure 7–8c Bit Status for Input Word Local:1:I.Data and Output Word Local:2:O.Data after
Output Is Energized and the *START* Button Is Released

When the STOP button (Local:1:I.Data.0) is pressed, the processor clears bit-0 to zero (0), and the circuit logic is broken, or goes false. Bit-1 of output word "Local:2:O.Data" is cleared to zero (0), and the real-world output device (motor) connected to terminal 1 of the output module turns OFF.

Words "Local:1:I.Data" and "Local:2:O.Data" in the I/O memory now appear as shown in Figure 7–8d, with only bit-2, the overload contact, set to 1.

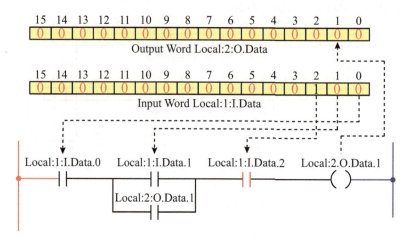

Figure 7–8d Bit Status for Input Word Local:1:I.Data and Output Word Local:2:O.Data with the *STOP* Button Pressed

When the *STOP* button is released or closed again, bit-0 is again set to 1, or *ON* (Figure 7–8e). The output (Local:2:O.Data.1) is not energized, however, as the *START* button (bit-1) is 0 (*OFF*), and the holding contact (Local:2:O.Data.1) is also 0 (*OFF*).

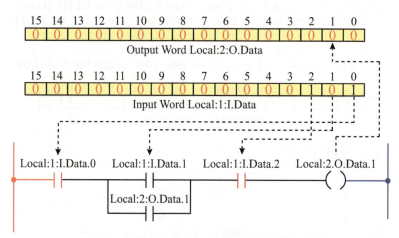

Figure 7–8e Bit Status for Input Word Local:1:I.Data and Output Word Local:2:O.Data with the *STOP* Button Released

Another way to look at the N.O. and N.C. symbols used for programming a PLC is called the relay analogy.

For the sake of discussion, imagine that each input device is connected to an invisible control relay inside the input module, and that each control relay has one N.O. and one N.C. contact, as shown in Figure 7–9.

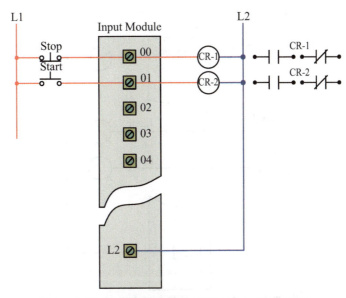

Figure 7–9 Imaginary Control Relays Wired to an Input Module

Connect one lamp to the N.O. open contact, and another lamp to the N.C. contact of CR-1, as shown in Figure 7–10.

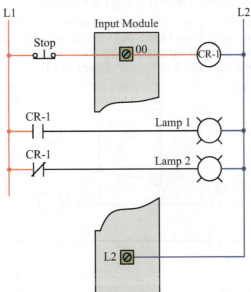

Figure 7–10 Lamps Wired to N.O. and N.C. CR-1 Contacts

When power is applied, relay CR-1 energizes through the N.C. contact of the *STOP* button. With relay CR-1 energized, the N.O. contact of CR-1 closes, and Lamp 1 lights, as indicated in Figure 7–11. The N.C. contacts of CR-1 are now open, so Lamp 2 *cannot* light.

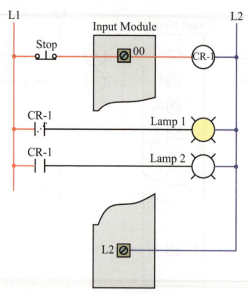

Figure 7–11 Lamp 1 Lights with Power Applied

Wired in this manner, N.O. CR-1 contacts controlled by an N.C. pushbutton will close or *conduct* when power is applied to the circuit. Likewise, N.O. contacts *programmed* to represent an N.C. pushbutton will be logically true when power is applied to the PLC.

A N.O. *START* button connected to an input terminal and an invisible or imaginary control relay (Figure 7–12) would not light Lamp 1 until the *START* button is pressed and would only stay lit if the button is held down. Lamp 2 would light as soon as power is applied but would go out when the *START* button was pressed and relay CR-2 energized (Figure 7–13).

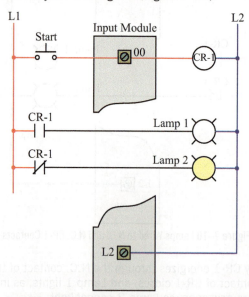

Figure 7–12 Power Applied—*START* Button Not Pressed

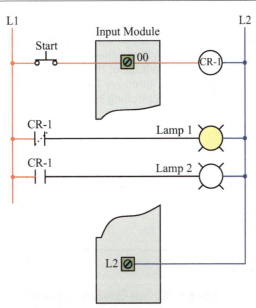

Figure 7–13 Power Applied—*START* Button Pressed

The rules for contacts that represent real-world input devices are shown in Figures 7–14a and 7–14b.

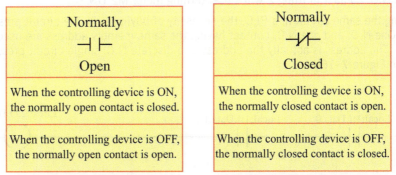

Normally ⊣ ⊢ Open	Normally ⊣／⊢ Closed
When the controlling device is ON, the normally open contact is closed.	When the controlling device is ON, the normally closed contact is open.
When the controlling device is OFF, the normally open contact is open.	When the controlling device is OFF, the normally closed contact is closed.

Figure 7–14a Normally Open Contact of an External Input   **Figure 7–14b** Normally Closed Contact of an External Input

There are no invisible control relays in the input modules, and there are also no symbols on the programming device for *STOP* buttons, *START* buttons, limit switches, and the like. If relay contact symbols must be used in place of standard electrical symbols, the relay analogy is an easy way to explain why N.O. contacts are programmed to represent N.C. input devices.

It does not matter which approach you use to understand the logic behind the way that PLCs are programmed, as long as you do understand it. The author believes that the Examine On and Examine Off approach is the easiest and clearest way to look at programming contacts. The conversion of a ladder diagram to a PLC program will be quite simple once you clearly understand the concept of Examine On and Examine Off.

Clarifying Examine On and Examine Off

From the previous discussion and examples, it appears that all input devices, whether they are N.O. or N.C., are programmed as N.O. contacts to achieve the desired results in the ladder logic. For many circuit applications this is true, and a big advantage of this programming technique is the ability to turn single-pole input devices into double-, three-, or four-pole devices in the PLC. Figure 7–15 shows a double-pole pressure switch (PS-1) controlling two outputs: motor M1 and motor M2.

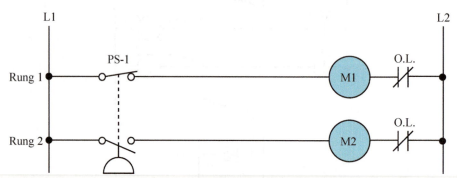

Figure 7–15 Double-Circuit Pressure Switch

When power is applied to the circuit, motor M1 will start through the normally closed contact of PS-1 in Rung 1. Motor M2 cannot start, however, due to the N.O. contact of PS-1 in Rung 2. When pressure switch (PS-1) is actuated, the N.C. contact in Rung 1 will open and motor M1 will turn *OFF*, while the N.O. contact in Rung 2 will close, turning motor M2 *ON*.

By programming the same circuit in a PLC, the necessity of buying a double-circuit pressure switch is eliminated. One N.O. and one N.C. contact having the same memory address are used (see Figure 7–16a). The address is actually the address of a discrete N.C. single-circuit pressure switch as illustrated in Figure 7–16b.

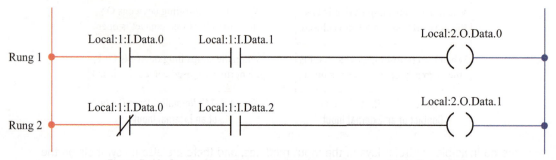

Figure 7–16a Double-Circuit Pressure Switch Circuit Programmed for a Typical PLC

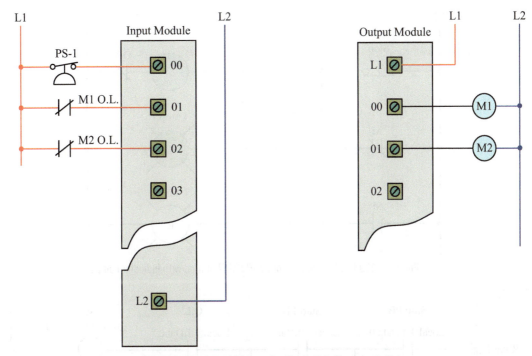

Figure 7–16b Actual Wiring of a Single-Pole Pressure Switch

When the PLC is placed in the RUN mode, it examines all N.O. contacts for an *ON* condition, or Examine On. As pressure switch (PS-1) and both overload contacts are closed (*ON*), bits 0, 1, and 2 of input word "Local:1:I.Data" are set to 1 or *ON*, making those portions of the ladder logic true. When the processor examines the N.C. contact in Rung 2 (Local:1:I.Data.0), it sees that bit-0 is set to 1 so this part of the ladder logic is false (Examine Off). Motor M1 (Local:2:O.Data.0) is *ON* and motor M2 (Local:2:O.Data.1) is *OFF*.

When the pressure switch is actuated, the processor continues scanning the user program and examines all programmed N.O. contacts for *ON*, and all N.C. contacts for *OFF*. Since the pressure switch (PS-1) is now actuated, the N.O. contact (Local:1:I.Data.0) in Rung 1 goes false, turning *OFF* motor M1 (Local:2:O.Data.0), whereas the N.C. closed contact (Local:1:I.Data.0) in Rung 2 goes true, turning motor M2 (Local:2:O.Data.1) *ON*.

Even though only double, three, and four poles were mentioned, there is no limit (except user memory size) to the number of times an input device can be addressed and used in a PLC program. This programming technique allows for six-pole, seven-pole, eight-pole, and so on, devices to be programmed using only a single-pole discrete device.

There are many more applications and circuits that normally require two- or three-pole devices that now only require single-pole devices when used in a PLC application. Examples are double-pole limit switches for forward and reversing circuits, and double-pole pressure switches for duplex controllers.

When programmed contacts (N.O. or N.C.) are controlled by outputs, the familiar standard relay logic is used. Figure 7–17a shows a standard *STOP/START* circuit with pilot lights. Lamp 1 (green) indicates power is available and the motor is *OFF*, and Lamp 2 (red) indicates the motor is *ON*. Figure 7–17b shows how the circuit is programmed in a PLC.

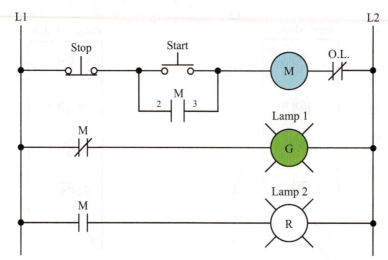

Figure 7–17a Ladder Diagram for *STOP/START* Station with Indicator Lamps

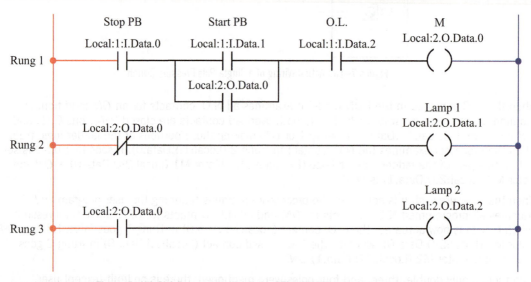

Figure 7–17b PLC Programmed Logic for *STOP/START* Station with Indicator Lamps

When the processor is placed in the RUN mode, Lamp 1 turns *ON* due to the N.C. contact "Local:2:O.Data.0". When the *START* button is pressed, output "Local:2:O.Data.0" energizes, N.C. contact "Local:2:O.Data.0" in Rung 2 goes false, and Lamp 1 (Local:2:O.Data.1) turns *OFF*. N.O. contact "Local:2:O.Data.0" in Rung 3 goes true and Lamp 2 (Local:2:O.Data.2) turns *ON*, and the holding contact (Local:2:O.Data.0) in Rung 1 goes true as well to complete the circuit logic.

Notice that the output, holding contact, N.C. and N.O. contacts for the lamps all have the same memory address (Local:2:O.Data.0). The address refers to bit-0 of word "Local:2:O.Data", and this same bit is referenced four times in the ladder logic program. The Examine Off, or N.C. contact, is logically true when bit-0 is cleared to 0 (*OFF*), whereas the Examine On, or N.O. contacts, are not logically true until bit-0 is set to 1 (*ON*).

Figure 7–18a shows the bit status for the circuit with power applied and the PLC in RUN mode; Figure 7–18b with the *START* button pressed; Figure 7–18c with the *START* button released; Figure 7–18d with the *STOP* button pressed; and Figure 7–18e after the *STOP* button is released.

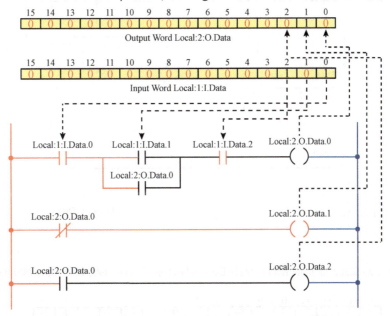

Figure 7–18a Bit Status for Input Word Local:1:I.Data and Output Word Local:2:O.Data with Power Applied

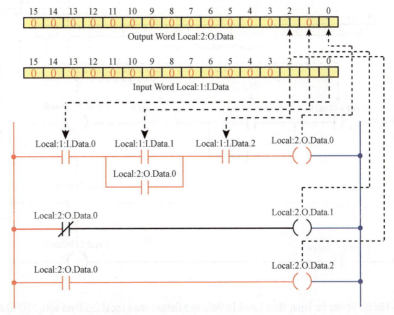

Figure 7–18b Bit Status for Input Word Local:1:I.Data and Output Word Local:2:O.Data with *START* Button Pressed

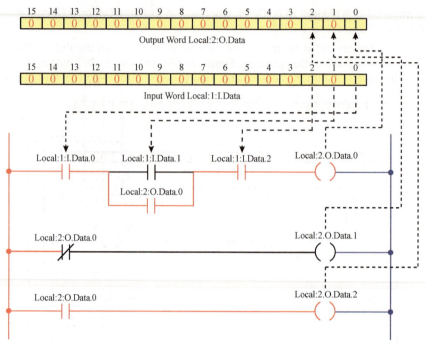

Figure 7–18c Bit Status for Input Word Local:1:I.Data and Output Word Local:2:O.Data with *START* Button Released

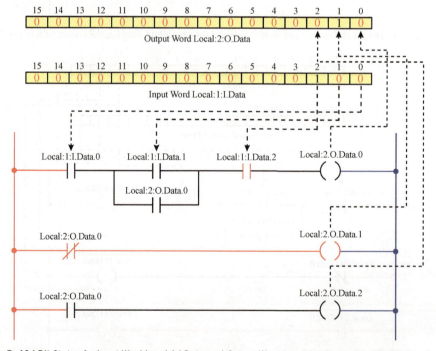

Figure 7–18d Bit Status for Input Word Local:1:I.Data and Output Word Local:2:O.Data with *STOP* Button Pressed

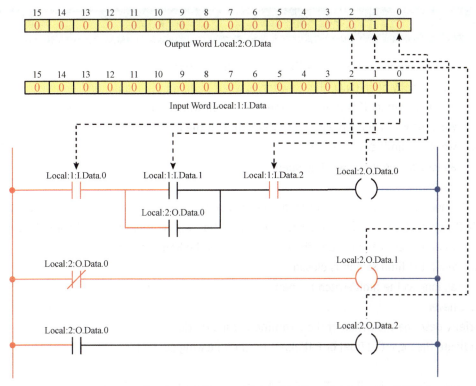

Figure 7–18e Bit Status for Input Word Local:1:I.Data and Output Word Local:2:O.Data with *STOP* Button Released

Chapter Summary

The relay logic used for programming the PLC at first seems to be in conflict with standard ladder diagram logic. But once the concepts of Examine On and Examine Off are understood, the logic process is easy to understand. An Examine On instruction looks for an *ON* condition and will be logically true when an *ON* condition (1) is found. The Examine OFF instruction will be logically true when an *OFF* condition (0) is found. Looking at the actual status of the bits within individual I/O memory locations (words) is another way to help understand how the processor evaluates the logic. Using the XIC or examine if closed (Examine On) and the XIO or examine if open (Examine Off) approach may be helpful. Others find that the relay analogy approach to understanding the N.O. (Examine On) and N.C. (Examine Off) symbols used for programming is better. Another approach is to accept the fact that a normally closed *STOP* button, or a similar closed input device, must be programmed using an N.O. contact symbol, and just have the philosophy that logic is relative to application.

Key Terms

Examine On   Examine Off

Review Questions

1. Briefly describe the action of the Examine On instruction.
2. When a normally open (N.O.) limit switch is wired to an input module, and programmed using a N.O. contact symbol (Examine On), the instruction will be true (check all correct answers):
   a. when power is applied and the key switch is in the RUN position
   b. when the limit switch is closed
   c. as long as the limit switch is open
   d. never
3. If the normally open (N.O.) limit switch in Question 2 is programmed using a normally closed (N.C.) contact symbol (Examine Off), the instruction will be true (check all correct answers):
   a. when power is applied and the key switch is in the RUN position
   b. when the limit switch is closed
   c. as long as the limit switch is open
   d. never
4. Briefly describe the action of the Examine Off instruction.
5. Indicate the logic (T [True] or F [False]) for the following contacts:

CONDITION OF INPUT DEVICE	PROGRAM INSTRUCTION	LOGIC TRUE-FALSE
		T F
		T F
		T F
		T F

6. The XIO instruction is the same as:
   a. Examine On
   b. Examine Off
7. The XIC instruction is the same as:
   a. Examine On
   b. Examine Off

Chapter 8

Programming a PLC

Learning Objectives

After completing this chapter, you should have the knowledge to:

- Explain the term *on-line programming*.
- Describe basic programming techniques.
- Explain the term *download*.
- Describe the *force on* and *force off* features, and the hazards that could be associated with forces.

PLC programming software can create, modify, monitor, and load programs into the PLC memory, and can also make changes to the program while the processor and driven equipment are running. This feature is often referred to as **on-line programming**. Changing the PLC program while the processor is running must *only* be done by persons with a thorough understanding of the circuit operation and the process or driven equipment. To prevent unauthorized *on-line programming* changes, a key switch is provided on the front of most PLCs to restrict the access to a monitor-only mode. In addition, on many PLCs, a **password** can be configured through the programming software and may be used to limit the access to the PLC and programs. Passwords act like a key switch. If the wrong password is entered when requested, the user is denied further access to the PLC or program(s).

Off-line programming means that the program is being developed without being connected to the PLC and is the most common method of initially programming a PLC and, of course, the safest. Since few PLC programs are ever created without mistakes, it is always best to create the program *off-line*. Once the program is completed, it should be tested while still in the *off-line* mode if this is possible using emulation software. After testing and verifying the PLC program(s) in the *off-line* mode, the project can be downloaded to the PLC for final verification, testing, and operation.

Most PLCs are designed so that any digital input or output can be forced on or off. The **force on** feature (and troubleshooting aid) allows the technician to *force on* or make an input or output point go *ON* regardless of actual status or circuit logic of the input or output device. Similarly, inputs and outputs can be forced off (**force off**), or turned *OFF,* regardless of the actual device status or circuit logic.

Caution: The *force on* and *force off* features should *never* be used except by personnel who completely understand the circuit *and* the process machinery or driven equipment. An understanding of the potential effects that forcing a given input or output has on machine operation is essential if hazardous and/or destructive operation is to be avoided. Forcing inputs or outputs *ON* or *OFF* should only be done temporarily during system checkout or troubleshooting, and by qualified personnel.

Note: The programming examples in this chapter and elsewhere in this text are based solely on one manufacturer's software and hardware (Rockwell Automation). With so many different PLC manufacturers and programming software on the market, it is not practical or possible to show them all. The basic concepts will, however, be very similar. This is also not a definitive guide to the programming software. At the time of this publication, the latest software was used for the examples in this chapter and elsewhere. Because software is continually being updated, the version of software you are using may appear different than the examples shown in this text. Only after studying and using a particular manufacturer's software will one gain the knowledge necessary to adequately and safely program and communicate with a PLC.

Programming an Allen-Bradley Micro810 PLC

Most PLC manufacturers have developed their own software for programming their PLCs. For the purposes of illustrating how PLC programming software works, and the relative ease of programming, the author has selected the *Connected Components Workbench* and *Studio 5000* software created by Rockwell Automation, Inc. for programming the Allen-Bradley family of PLCs in this chapter. The *Connected Components Workbench* software can be used to program the Micro800 family of processors and the *Studio 5000* software can be used to program the Logix family of processors. Both are Windows based and very user friendly. For the first programming example, we will use *Connected Components Workbench* software to program a Micro810 controller. The Micro810 controller is the smallest of the Micro800 family of controllers. The Micro810 has 12 embedded digital I/O points: 8 inputs and 4 outputs. The Micro810 is a small standalone PLC ideally suited for small machine control applications. It is also referred to as a smart relay, but do not let that fool you. This little controller packs a lot of the same features that the larger PLCs have. Figure 8–1 shows the Micro810 controller.

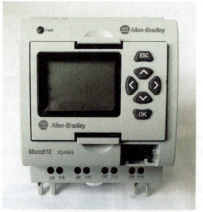

Figure 8–1 Micro810 PLC Controller

The example circuit to be programmed will be a simple *STOP/START* circuit that uses a *START* button, *STOP* button, holding contact, and output to control an exhaust fan motor. The ladder diagram of the circuit is shown in Figure 8–2.

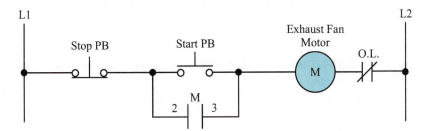

Figure 8–2 Basic *STOP/START* Circuit

To start the *Connected Components Workbench* software, double-click on the *Connected Components Workbench* icon. The opening window is shown in Figure 8–3. The opening window displays the "Start Page" which allows you to start a new project, open an existing project, open a recently opened project, and a "Getting Started" section with some very helpful tools for the person new to the software. The "Discover..." selection under "Project" will be covered later in this chapter. For this first demonstration, we will program our simple *STOP/START* circuit in the *off-line* programming mode.

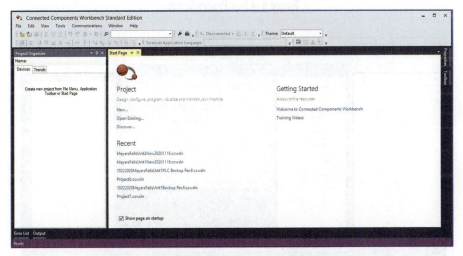

Figure 8–3 Connected Components Workbench Opening Screen

Using the computer mouse, left-click on "*New...*" on the Start Page. You should now see a pop-up window titled "New Project" as shown in Figure 8–4. On this pop-up window, enter the name of your PLC project and select the location on your computer to store the project file(s). After entering the project name and file location, left-click on the "*Create*" button. You should now see a pop-up window titled "Add Device" as shown in Figure 8–5. At this point, select the PLC controller model you will be using. In our example, we will be programming a Micro810 controller so we have to expand the folders as shown in Figure 8–5 to display the available Micro810 controllers and select the *2080-LC10-12AWA* controller. After selecting the controller, the selection window to the right will show a detailed description of the controller and allows you to select the version of controller you have. After you have selected the correct version, left-click on the "*Select*" button and then left-click on the "*Add To Project*" button in the lower-right corner of the pop-up window.

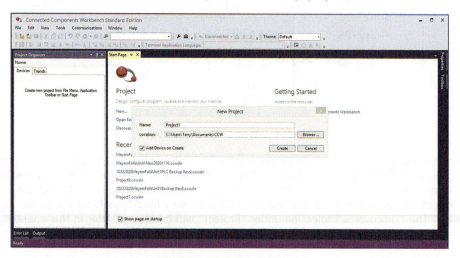

Figure 8–4 New Project Screen

Figure 8–5 Add PLC Controller

After selecting the correct PLC controller and adding the project, the main screen will be displayed as shown in Figure 8–6. The main screen is divided into three windows. The far-left window is called the "Project Organizer", sometimes referred to as the project tree. The Project Organizer window allows you to select the controller area you wish to view and edit/configure. The center window is referred to as the workspace window and will change based on your selection in the Project Organizer window. As you can see in Figure 8–6, the "Micro810" is selected by default in the project window and the center window displays information and configuration settings specific to the controller. The far-right window titled "Properties" will display the properties of the selected item in the workspace window (center window). All three windows can be increased in size, minimized, or closed. It is recommended that you leave the windows in their default locations until you become familiar with the software.

Figure 8–6 Main Screen After Adding PLC Controller

Under the Project Organizer window and Devices tab, the main folder is our Micro810 controller with four subfolders titled: Programs, Global Variables, User-Defined Function Blocks, and Data Types. Not all of these folders will be discussed since the intent of this section is merely to illustrate the relative ease of programming using the *Connected Components Workbench* software, not to be a definitive programming guide.

Since the Micro810 PLC comes with a power supply and I/O section already integrated into one small self-contained unit, there is no I/O modules to add or configure, making it nice for our first programming project. As mentioned earlier, we are using a Micro810 PLC with 12 embedded I/O points: 8 digital inputs and 4 digital outputs. All I/O points are mapped into the "Global Variables" area of memory in the PLC. Global Variables are memory locations that belong to the controller and are available to any program in the project. By double-clicking on the "*Global Variables*" in the Project Organizer for our Micro810, you should see all Global Variables in the workspace window as shown in Figure 8–7. By default, I/O variables and system variables are created automatically by the controller. Keep in mind that variables are memory locations that are used to store and process information that we can access and use in our PLC program(s). The term *tag* is most often used to describe a variable. The variable name or tag is what identifies that memory location including its memory type. As we can see in Figure 8–7, the first column lists the variable name (tag) followed by columns labeled "Alias", "Data Type", "Dimension", "Initial Value", and "Comment", and "Retained" and "String Size" which are not shown. At this point, we are only going to focus on those variable attributes shown in Table 8–1.

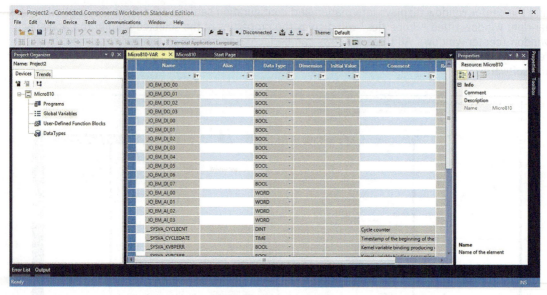

Figure 8–7 Global Variables

Table 8–1 Variable Attributes

Column	Description
Name	The name of the variable or tag. Tag names cannot exceed 40 characters. The first character must be a letter (user created tags), and subsequent characters can be letters, numbers, or the underscore character.
Alias (optional)	An Alias tag lets you create one tag that represents another. Both tags share the same value. Alias tags are useful when assigning a descriptive name to an I/O device or providing a simple name for a complex tag.
Data Type	Data Type defines the type of data the tag is. For example: Boolean, word, integer, double integer, real (floating point), etc.
Initial Value	The Initial Value of the tag when the controller first starts execution of the program. If left blank, the initial value defaults to zero or false.
Comment	A description of the tag or variable.

The first 12 variables or tags listed are the embedded digital I/O memory locations. These tags store the value of each of the 12 digital I/O points on the PLC. For example, the first tag name is "_IO_EM_DO_00" and indicates that this tag represents the first digital output point. You will notice that the data type is Bool which stands for Boolean (bit) and will only have a value of 1 and 0 or true and false. By looking at the tag name you can identify this as an embedded digital output by the letters used in the name. IO stands for Input/Output, EM stands for embedded, and DO stands for digital output. Can you identify the tag name for the first digital input point? If you identified tag name "_*IO_EM_DI_00*" you would be correct. Did you notice the four tag names after the last digital input point (_*IO_EM_DI_07*)? The last four I/O tags represent the analog inputs that are available on some Micro810 PLCs. ***Hint:*** look at the data type for these four analog input tags. Notice that they are Word data types and not Bool. Analog readings must be represented by a digital value that corresponds to the actual voltage or current being measured by the controller, not just *ON* or *OFF*.

For our basic STOP/START circuit of Figure 8–2, we will plan to wire our start button to digital input *00*, stop button to digital input *01*, and the exhaust fan motor starter to digital output *00* as shown in Figure 8–8.

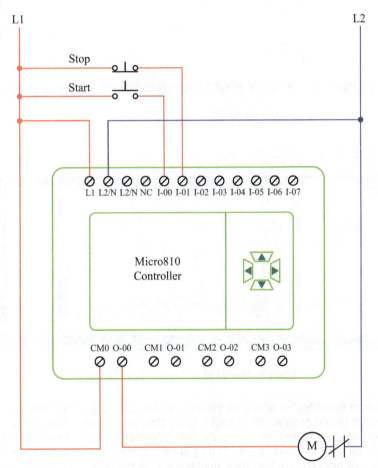

Figure 8–8 Micro810 I/O Wiring

Now that we have identified the digital I/O points on the Micro810 that we plan to use for our field devices and the corresponding variables or tags, we can begin programming our ladder logic circuit. The first step is to create a ladder diagram program. In the Project Organizer window under our Micro810 PLC, right-click on "*Programs*" and then under "Add", left-click on "*New LD: Ladder Diagram*" as shown in Figure 8–9a. You should now see a folder under Programs titled "Prog1" with a subfolder titled "Local Variables" as shown in Figure 8–9b.

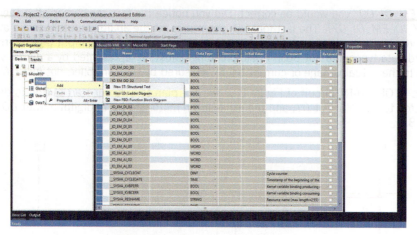

Figure 8–9a New Ladder Diagram Selection

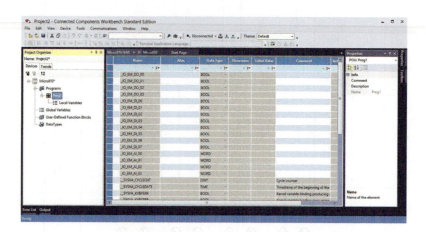

Figure 8–9b New Ladder Diagram "Prog1" Added

The subfolder "Local Variables" is where we can create user variables or tags that can only be used in that specific ladder program file (Prog1). Every time you create a new program file, a local variables subfolder will be created automatically. The difference between local and global is local variables can only be used in the program file they are associated with and global variables can be accessed and used in any program file. Now, double-click on the default ladder diagram program file we created, *"Prog1"*, and you should see the ladder diagram editor in the center workspace window for our program file as shown in Figure 8–10. This is our ladder logic programming window that allows us to create the PLC logic needed to control our field devices. Notice the dark horizontal line with the number "1" to the left of it? This is our first rung of logic and we are now ready to begin adding the logic necessary to create the circuit of Figure 8–2. Before we begin, look at the top of the workspace window and notice that there is a row of graphical symbols. These symbols are the logic symbols we will use to create the PLC ladder logic necessary to control our exhaust fan motor. Directly below the logic symbols is a second row with words like "Favorites", "Rung/Branch", "Bit", etc. These are the categories of logic symbols and instructions we can use in developing our ladder logic program or circuit. By left-clicking on any of these key words you will notice that the logic symbols above will change, allowing you to select a logic symbol from that category.

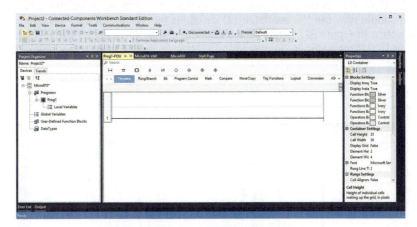

Figure 8–10 Ladder Diagram Editor Window

The first step is to program the stop button. This is accomplished by clicking on the *Normally Open* contact symbol (—] [—) at the top of the workspace window while dragging your mouse pointer down to the left end of the horizontal black line on Rung 1. You should see a black circle with cross-hairs appear on the rung, your normally open contact symbol will be placed to the right of this anchor symbol. A pop-up window titled "Variable Selector" will appear after inserting the normally open contact. Figure 8–11 shows the screen after inserting the normally open contact symbol. You can move the Variable Selector window as needed to see your contact if desired.

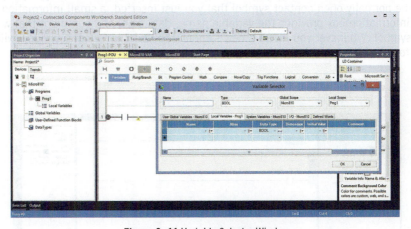

Figure 8–11 Variable Selector Window

The Variable Selector window allows you to select an existing variable (tag) or create a new one. The variable you select is the memory location that the PLC will use when evaluating the logical state of the normally open contact symbol during execution of the ladder logic program. The Variable Selector window has five tabs labeled "User Global Variables - Micro810", "Local Variables - Prog1",

"System Variables - Micro810", "I/O - Micro810", and "Defined Words". Since our normally open contact represents the stop button that is wired as a digital input to the PLC, the variable we need to select is _IO_EM_DI_01, as seen in Figure 8–8. To find the variable, select the tab labeled "I/O - Micro810". You should now see the available I/O variables for our Micro810 PLC as shown in Figure 8–12. You will notice in Figure 8-12 that I have added a comment to each of the I/O variables we have wired to our PLC. This helps in quickly identifying what device is wired to which digital I/O point.

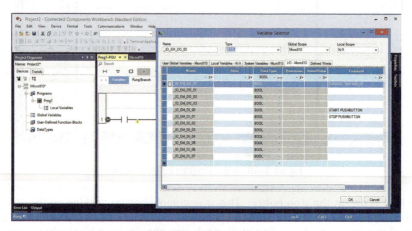

Figure 8–12 Available I/O Variables

Click on the stop push button variable (_IO_EM_DI_01) to highlight it and then click the "OK" button at the bottom right of the Variable Selector pop-up window. You should now see the variable name we selected above the normally open contact as shown in Figure 8–13. If you put your mouse pointer over the normally open contact symbol that you just inserted, a small information box will appear. This information box provides the variable name, data type, and comment that was assigned to that logic symbol. Whenever you insert a new symbol, the Variable Selector window will pop-up allowing you to select the variable for that logic symbol.

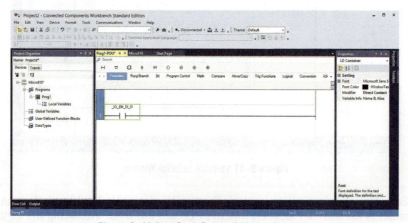

Figure 8–13 Stop Push Button Added to Program

The next step will be to add the start push button. Since the start push button is in parallel with a holding contact, we will first insert a branch symbol to the right of the normally open contact we just inserted. We can do this by first clicking on the normally open contact symbol so that it is highlighted (green box) as shown in Figure 8–13. Next, left-click on the *Branch* symbol in the tool bar at the top and the branch symbol will appear on the rung to the right of the normally open contact. The branch symbol ⊤⊐ is the second symbol from the left above the Favorites tab. After the branch symbol has been selected, your screen should look like the one in Figure 8–14.

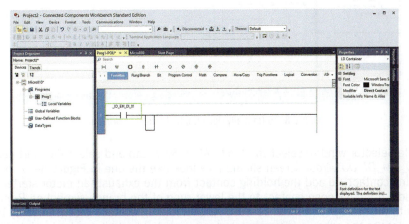

Figure 8–14 Branch Symbol Inserted into Rung

Next, we will add our start push button by clicking and holding the left mouse button down on the *Normally Open* contact symbol (—] [—) at the top of the workspace window while dragging your mouse pointer down to the branch symbol on Rung 1, you should see four black circles with cross-hairs appearing on the branch symbol. Make sure your mouse pointer is pointing to the top-center black circle before releasing your left mouse button as shown in Figure 8–15. If done correctly, you should see the normally open contact inserted into the top rung of the branch symbol as shown in Figure 8–16 as well as the Variable Selector pop-up window.

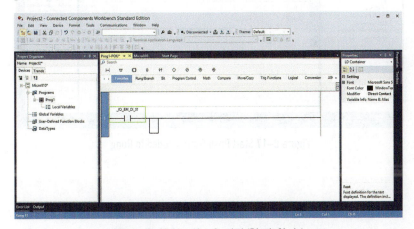

Figure 8–15 Insertion Symbol (Black Circle)

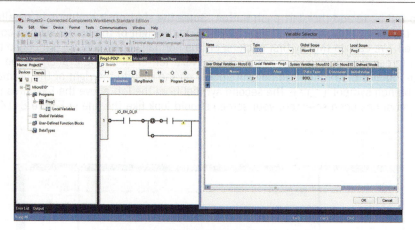

Figure 8–16 Normally Open Contact Added to Branch

In the Variable Selector window select the "*I/O - Micro810*" tab and select the start push button variable "*_IO_EM_DI_00*". Your screen should now look like the one in Figure 8–17. Next, we will add the holding contact. To add the holding contact from the exhaust fan motor starter output, select and add a normally open contact to the lower rung of the branch symbol on Rung 1, just like we did with the start push button contact. As the holding contact is controlled by the output, which in this case is the exhaust fan motor starter connected to digital output point *00* and will have the same variable name "*_IO_EM_DO_00*". Your program should now look like the one in Figure 8–18.

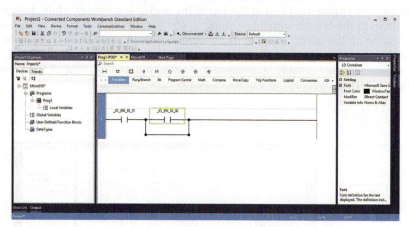

Figure 8–17 Start Push Button Added to Rung

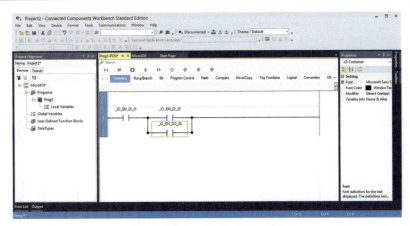

Figure 8–18 Holding Contact Added

The last step necessary to complete our simple STOP/START circuit is to program the output device, which in this case is our exhaust fan motor starter output, variable *"_IO_EM_DO_00"*. To add the output, click and drag the *Output Energize* symbol (—( )—) to Rung 1 and point to the black circle to the right of our branch we inserted previously. See Figure 8–19. The output energize symbol will appear on the far-right side of Rung 1. In the Variable Selector window, find and select the exhaust fan motor output variable *"_IO_EM_DO_00"*. Your PLC program logic should now look like the one in Figure 8–20.

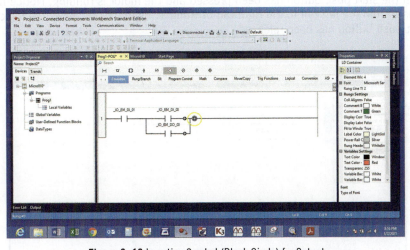

Figure 8–19 Insertion Symbol (Black Circle) for Output

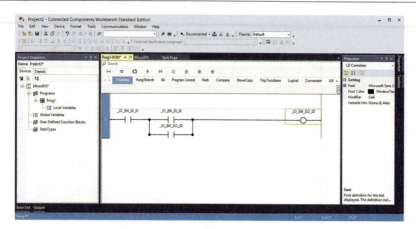

Figure 8–20 Motor Coil Output Added to Rung

Congratulations! You have just programmed your first PLC ladder logic program to control a motor (exhaust fan) using a simple START/STOP circuit. After saving the program, the only remaining action is to download the program to the Micro810 PLC (covered later in this chapter) and place the PLC into run mode. Figure 8–21 shows what the ladder logic programming screen would look like with the controller in run before the start push button is pressed. Notice how the stop push button contact and connecting lines have turned red, indicating that there is a true logic path or logic continuity through those devices, and blue indicates no logic path. Figure 8–22 shows what it would look like when the start push button has been pressed and the motor is running. Notice how we have a true logic path from the left rail to the right rail of Rung 1. Figure 8–23 shows the logic circuit after the start push button has been released.

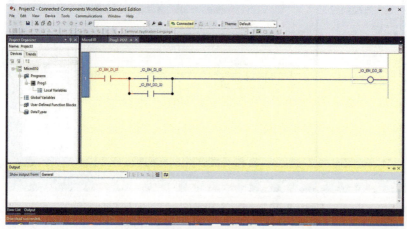

Figure 8–21 Controller in Run Mode Before Start PB Pushed

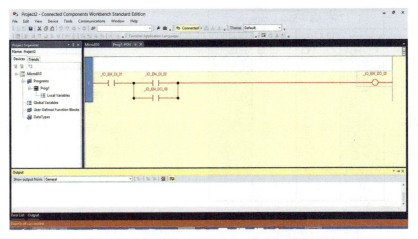

Figure 8–22 Controller in Run Mode with Start PB Pushed

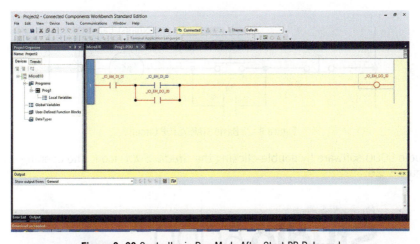

Figure 8–23 Controller in Run Mode After Start PB Released

Programming a ControlLogix PLC with Studio 5000

For this programming example, we will use the Rockwell Automation Studio 5000 and Logix Designer software to program a ControlLogix 5570 PLC mounted in a 4-slot chassis, as shown in Figure 8–24.

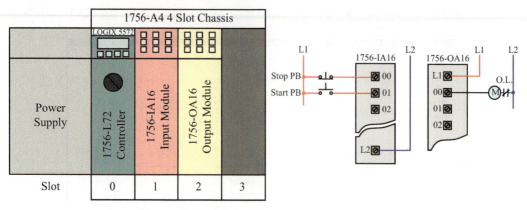

Figure 8–24 ControlLogix 5570 Controller & 4 Slot Chassis

The circuit to be programmed will be the same simple STOP/START circuit that was used in the first example. The ladder diagram of the circuit is shown again in Figure 8–25.

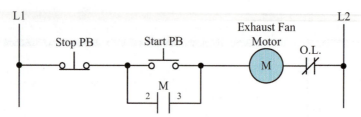

Figure 8–25 Basic START/STOP Circuit

Start the Studio 5000 software by double-clicking the *Studio 5000* icon. The opening screen is shown in Figure 8–26.

Figure 8–26 Studio 5000 Opening Screen

Since we will be creating a new project, left-click on *New Project*. The window that appears, titled New Project, lists the available PLC controllers that can be programmed with the Logix Designer software. Figure 8–27 shows the controller selection window.

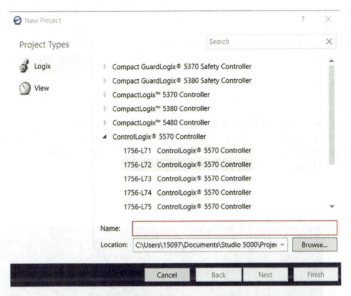

Figure 8–27 Controller Selection Window

The first step is to select the PLC controller type. Under "Project Types" in the left pane, make sure that "Logix" is selected and in the right pane select your controller type. In our example, we will be programming a ControlLogix 5570 PLC part number 1756-L72. If you are not using the 1756-L72 controller, then select the controller type you are using. After selecting the controller type, enter your project name in the "Name" box and select the location to store the project files. The project name is typically the name of the process or machine that the PLC will be controlling. In our example, we have named our project "Technicians_Guide". After selecting the controller type, project name, and location, click the *Next* button at the bottom of the screen. You should now see the controller configuration window as shown in Figure 8–28.

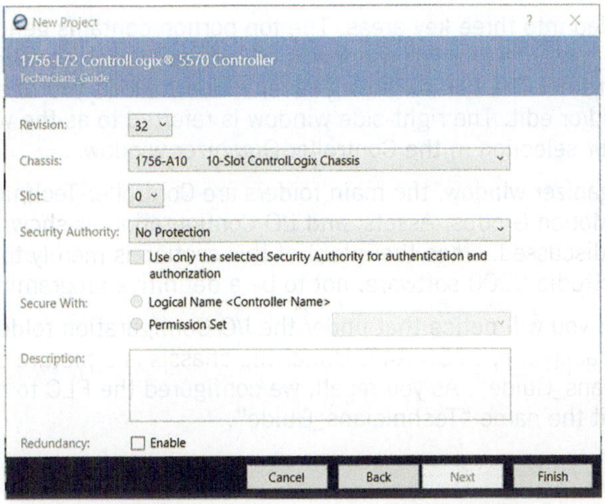

Figure 8–28 Controller Configuration Window

The first configuration to select is the controller revision number. Using the drop-down box, select the revision number for the controller you are using. From the "Chassis" pull-down menu, choose the appropriate chassis size in which the controller will reside. In our example, we will select the 1756-A4, 4-slot chassis. In the "Slot" field, enter the chassis slot number that the controller will reside in.

Note: In ControlLogix, controllers can be placed in any slot. It is also possible to place multiple controllers in the same chassis. Most often, the controller is placed in slot 0 of the local chassis.

After entering the processor slot number, you can enter a short description in the "Description" box if desired as this is optional. All remaining configuration fields can be left in their default setting. Click the *Finish* button at the bottom of the window and you should now see the main Logix Designer programming screen as shown in Figure 8–29.

Figure 8–29 Logix Designer Programming Main Screen

The main screen is divided into three key areas. The top portion contains general information, program navigation, and tools. The far-left window is called the "Controller Organizer", sometimes referred to as the project tree. The Controller Organizer window allows you to select the controller area you wish to view and/or edit. The right-side window is referred to as the workspace window and will change based on your selection in the Controller Organizer window.

Under the Controller Organizer window, the main folders are Controller Technicians_Guide (name of your project), Tasks, Motion Groups, Assets, and I/O Configuration as shown in Figure 8–30. Not all these folders will be discussed, since the intent of this section is merely to illustrate the ease of programming using the Studio 5000 software, not to be a definitive programming guide.

In studying Figure 8–30, you will notice that under the I/O Configuration folder there is a picture of a chassis titled "1756 Backplane, 1756-A4". Under the chassis is a picture of a controller titled "[0] 1756-L72 Technicians_Guide". As you recall, we configured the PLC to reside in slot [0] of the 4-slot chassis and gave it the name "Technicians_Guide".

Figure 8–30 Controller Organizer Window

The next step in the configuration process is to configure the I/O modules that will be installed in the local chassis with the controller. In reviewing Figure 8–24, you will notice that we showed a 1756-IA16 input module located in slot 1 and a 1756-OA16 output module in slot 2. To enter the correct I/O modules that are installed in slots 1 and 2 of the 4-slot chassis, right-click on the picture of a chassis titled "*1756 Backplane, 1756-A4*" and left-click on "*New Module*". Figure 8–31 shows the window after clicking on New Module.

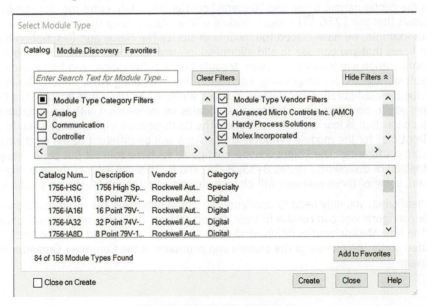

Figure 8–31 Adding New I/O Module

The bottom window in Figure 8–31 lists the available I/O modules based on the filter selections in the upper two windows. By default, all chassis-based modules are displayed. You can use the filters to help narrow the list to just the type of modules you are looking for. Using the scroll button on the right side of the lower window, scroll down until you find the digital input module you are using in slot 1 of your chassis (1756-IA16) and then highlight the module by left-clicking on it. Once it is highlighted, left-click on the *Create* button at the bottom of the screen. You should now see the New Module screen for our 1756-IA16 input module as shown in Figure 8–32.

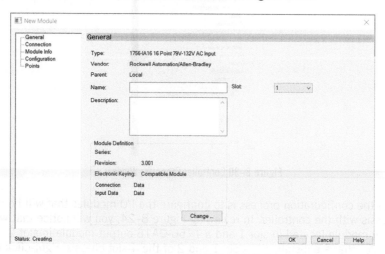

Figure 8–32 New 1756-IA16 Input Module

To complete the configuration, under the "General" configuration tab, enter the slot number of the local chassis that the 1756-IA16 input module will reside in, using the drop-down box titled "*Slot*". In our example, we have placed the module in slot 1. The Name and Description boxes are optional text boxes that you can use to add additional information about the new module if desired. The other configuration windows (Connection, Module Info, Configuration, and Points) can be left in their default settings for our programing example. It is worth taking a few minutes however, to briefly describe the configuration parameters in these windows. As the name implies, the Connection window allows you to configure when the module updates its information with the controller and what happens if a connection is lost with the controller. The Configuration window allows you to select the input filter times for the module's inputs. This window will be different based on the type of module you are configuring. The Points window configures each module's input point change of state condition (enabled or disabled). Figures 8–33, 8–34, and 8–35 show each of these configuration screens. Again, some of these windows will change depending on the module type you have selected.

As already mentioned, you only need to configure the slot number that the input module will reside in. The other configurations can remain in their default setting. Once you are finished, click the *OK* button and the New Module configuration window should close. You should now see the 1756-IA16 module located under the picture of the chassis and controller in the Controller Organizer window of the main screen.

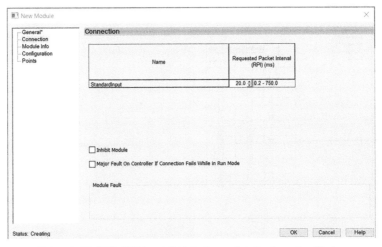

Figure 8–33 1756-IA16 Input Module Connection Configuration Window

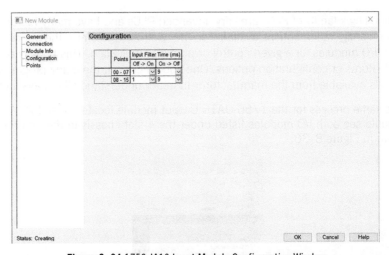

Figure 8–34 1756-IA16 Input Module Configuration Window

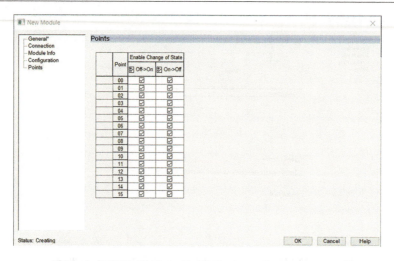

Figure 8–35 1756-IA16 Input Module Points Configuration Window

Note: The ControlLogix family of PLCs are very advanced PLCs and have many features that others do not. There are configuration options available so the user can customize the operation of the controller and/or I/O modules for a given control or process application. This book is not intended to show all the features or configuration options. One should consult the many programming and reference manuals available from the manufacturer to better understand all the available features.

Now, repeat the same process for the 1756-OA16 Output module located in slot 2. When you are finished you should see both I/O modules listed under the 4-slot chassis in the Controller Organizer window as shown in Figure 8–36.

Figure 8–36 Controller Organizer Window After I/O Modules Added

Open the controller tags by double-clicking on the *Controller Tags* folder in the Controller Organizer window. After the controller tags window opens, left-click on the arrow sign to the left of where it says "Local:1:I". The "Local" stands for local chassis, the "1" stands for slot 1, and the "I" stands for input. If you recall, we configured slot 1 for the digital input module 1756-IA16, which has 16 digital inputs. Now click on the arrow sign to the left of "Local:1:I.Data" and you should see the screen in Figure 8–37.

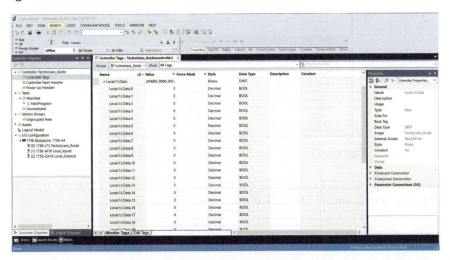

Figure 8–37 I/O Tag "Local:1:I.Data" Boolean Tags

From this screen you can see all the input data points (tags) for the 1756-IA16 input module located in slot 1. After studying Figure 8–37, you probably noticed that there are more than 16 input data tags. How can there be more than 16 input data tags when the input module only has 16 input points? If you recall from Chapter 5, the ControlLogix controller is a 32-bit word processor and allocates memory in full word increments. Since the input module only has 16 inputs, the first 16 input data tags (bits) are used (Data.0...15) and the remainder are unused.

When the processor is in the *on-line* mode, the actual status of the input devices, 1 or 0, can be determined by looking at the bit status in the column titled "Value". Also from this window, each input data tag point can be assigned a description in the "Description" column to the far right of the tag name.

In our example, the stop push button has been wired to digital input point 0 (Local:1:I.Data.0) and the start push button to input point 1 (Local:1:I.Data.1). In Figure 8–38, we have added the descriptions for each button in the description field for each input tag.

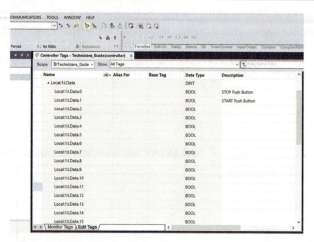

Figure 8–38 Start & Stop Button Descriptions Added to I/O Tags

Minimize the input tags for slot 1 by clicking on the arrow symbol to the left of the tag name Local:1:I.Data. Your screen should now look like the one in Figure 8–39.

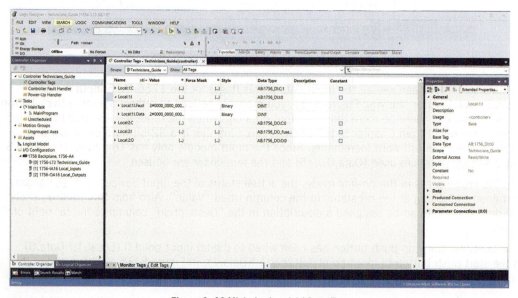

Figure 8–39 Minimize Local:1:I.Data Tag

In Figure 8–40, we have expanded the output data tags associated with the 1756-OA16 output module located in slot 2 of the chassis. Since this is an output module, we have expanded "Local:2:O" to reveal the output data tags for each output point on the module.

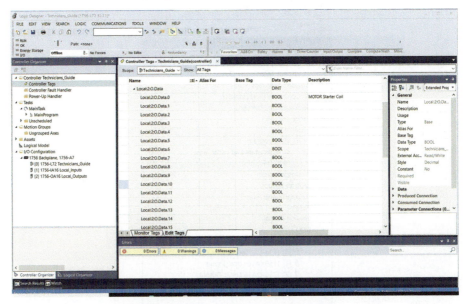

Figure 8–40 Tag "Local:2:O.Data" Expanded

The exhaust fan motor is wired to output point 0 on the 1756-OA16 module (Local:2:O.Data.0) and has been given the description "MOTOR STARTER COIL" as shown in Figure 8–40. It is good practice to take the time to add all your descriptions for the I/O points that are used before starting to program the ladder logic. In this way, your description will already be there when you enter the tag address while programming and there is less chance for mistakes.

Now that we have configured our I/O modules and entered a description for each of the I/O points we will be using, it is time to program our START/STOP circuit. If you recall from Chapter 5, the ControlLogix controller divides logic into Tasks, Programs, and Routines. By default, when you create a new project, a main task, main program, and main routine are automatically created. The routine is where the ladder logic is entered. Figure 8–41 shows the Tasks, MainTask, and MainProgram folders expanded to reveal the MainRoutine folder. When you double-click on the "*MainRoutine*" folder, your screen should look like the one in Figure 8–42.

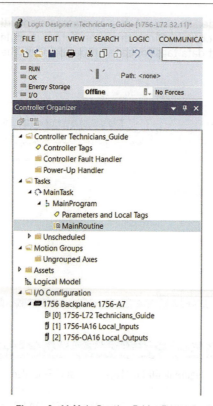

Figure 8–41 Main Routine Folder Exposed

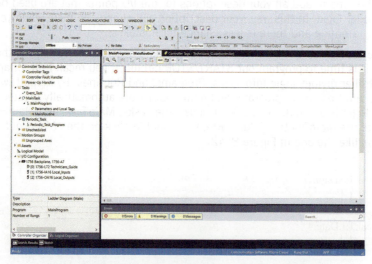

Figure 8–42 Main Routine Ladder Logic Programming Window

Note: Refer to Chapter 5 and the manufacturer's literature for a detailed description of creating and using Tasks, Programs, and Routines.

The "MainRoutine" ladder logic programming window (Figure 8–42) is where the ladder logic program will be entered for our STOP/START circuit. The box in the upper left-hand corner of the programming screen has a zero (0) inside the box indicating that Rung 0 is selected. Notice the red circle with a white X to the right of the "0"? This indicates that the rung is incomplete and contains errors (blank rung).

Just like in the first example, the first step will be to program the stop push button. This is accomplished by single clicking on the *Normally Open* contract symbol (—] [—) in the instruction toolbar located above the programming window. Ladder programming elements can also be dragged from the instruction toolbar to a valid placement location (green target circle). The green target circles indicate where a ladder logic element will be inserted when the mouse button is released. Figure 8–43 shows the screen with the normally open (—] [—) contact inserted.

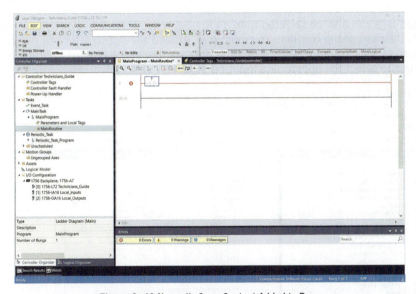

Figure 8–43 Normally Open Contact Added to Rung

Double-click on the question mark (?) above the contact, enter the tag name for the stop push button, and then press the *Enter* key. If you recall, the tag for the stop push button was Local:1:I. Data.0. Figure 8–44 shows the tag entered above the normally open contact for our stop push button. Notice that the description "STOP Push Button" also appeared above the tag. This is a good indicator that we have entered the correct tag address.

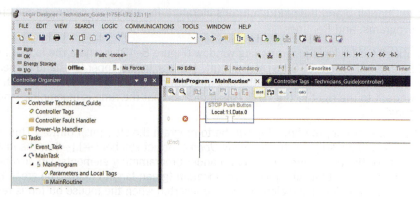

Figure 8–44 Stop Push Button Tag Name Added to Normally Open Contact

Already created tags can also be selected from a drop-down list in the text box by clicking on the black arrow to the right of the text box. You must then expand the tag structure as necessary to find the tag you wish to select. Figure 8–45 shows the tag structure and a bit grid for the input module that you can use to select the appropriate input bit or point.

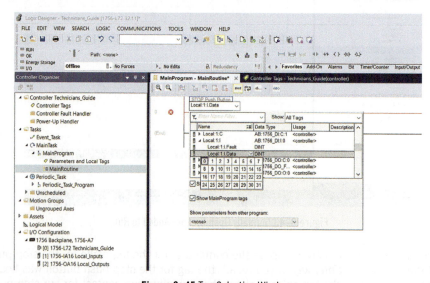

Figure 8–45 Tag Selection Window

If a tag is not already created, it can be created from the tag text box by simply entering in a name for the tag and pressing the *Enter* key. The tag will be "Undefined" until you right-click on the tag box, at which time a menu will appear allowing you to create the new tag. This is commonly done with internal memory tags rather than I/O tags. See the manufacturer's programming reference manuals for details on creating tags.

The next step is to add the start push button. Since the start push button is in parallel with the holding contact, single click on the *Branch Start* symbol on the toolbar. After the branch start symbol is clicked, your screen should look like the one in Figure 8–46.

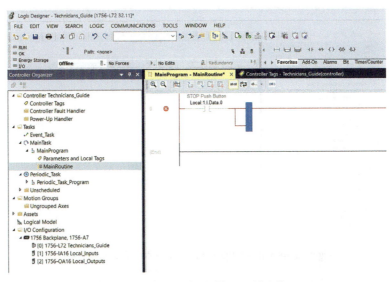

Figure 8–46 Branch Symbol Inserted into Rung

Notice that a dark vertical line appears on the right side of the branch symbol. It is necessary to move this line to the upper left-hand corner of the branch symbol before inserting the start push button contact. To move the line, point the arrow of your mouse at the upper left-hand corner of the branch symbol and click. The heavy line should now move to the upper left corner of the branch symbol, as shown in Figure 8–47.

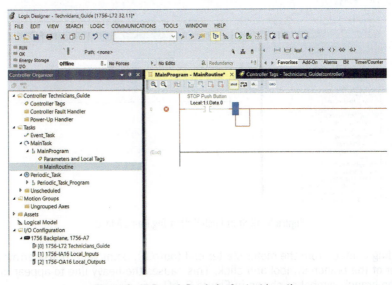

Figure 8–47 Branch Ready for Contact Insertion

Next, single click on the *Normally Open* contact symbol (—] [—) to add the next contact. Figure 8–48 shows the screen after the normally open contact has been added.

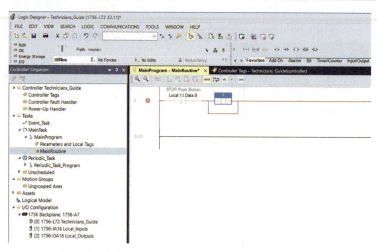

Figure 8–48 Normally Open Start Push Button Contact Added

Enter the tag name for the start push button by double-clicking on the question mark (?) above the contact and entering the tag address. When finished, press the *Enter* key. If you recall, the tag for the start push button was Local:1:I.Data.1. You can also use the drop-down list in the text box. Figure 8–49 shows the screen with the tag entered above the normally open contact for our start push button. Notice that the description "START Push Button" also appeared above the tag.

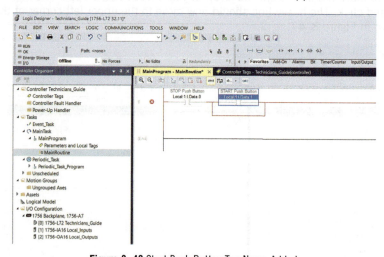

Figure 8–49 Start Push Button Tag Name Added

To add the holding contact from the motor starter coil (output), point your mouse arrow at the lower left-hand corner of the branch symbol and click. This causes the heavy line to appear in the lower left corner of the branch symbol as shown in Figure 8–50.

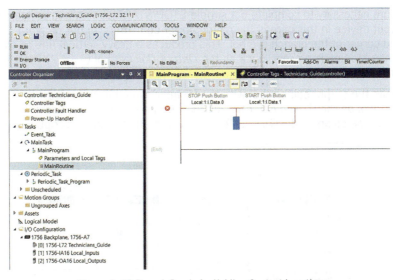

Figure 8–50 Branch Ready for Holding Contact Insertion

Select and click on the *Normally Open* symbol to add the holding contact that is in parallel with the start push button. As the holding contact is controlled by the output, the holding contact will have the same address as the motor starter coil output, Local:2:O.Data.0. Figure 8–51 shows the screen after the holding contact has been added. Notice that the description you entered earlier appears above the tag address Local:2:O.Data.0.

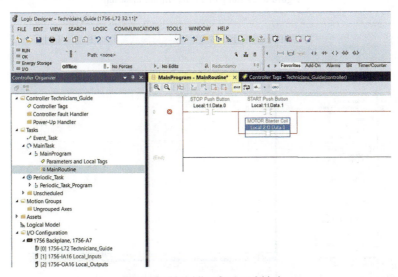

Figure 8–51 Holding Contact Added

The last procedure necessary to complete our simple STOP/START circuit is to program the output device. The output device in this case is the motor starter coil. To finish our program, click the upper right-hand corner of the branch symbol. This produces a heavy vertical line as shown in Figure 8–52. Once the line is present, click on the *Output Coil* symbol (—( )—) in the toolbar above the programming window.

Figure 8–53 shows the output coil symbol with the (?) indicating that the symbol needs a tag address. Double-click on the (?) and enter the tag for the motor starter coil, Local:2:O.Data.0. Figure 8–54 shows the completed STOP/START circuit with all tags and descriptions added.

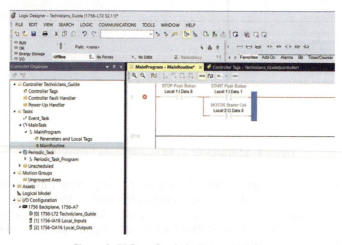

Figure 8–52 Rung Ready for Output Insertion

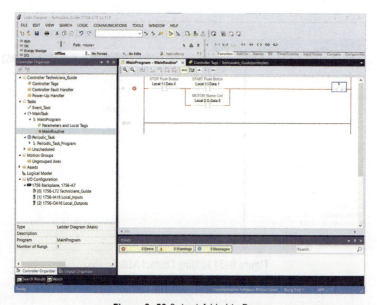

Figure 8–53 Output Added to Rung

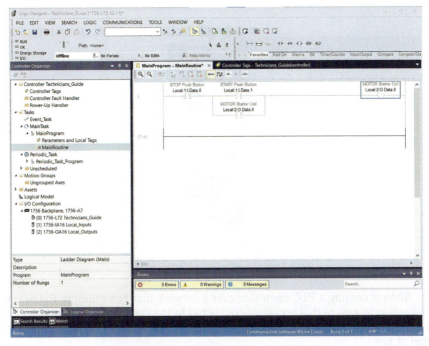

Figure 8–54 Completed Start/Stop Logic Circuit

If there are no errors, the red circle will disappear. The ControlLogix software can check for errors and provide a description by right-clicking on the rung number to the left of the finished rung and selecting "*Verify Rung*". If there are any errors, an error window will appear below the programming window describing the errors. Congratulations, you have just programmed your first ControlLogix PLC program using Studio 5000.

Controller Communications

All PLCs have what is referred to as a programming port. The programming port is typically located on the front of the PLC as shown in Figure 8–55. Depending on the model of PLC controller, the programming port will either be serial, USB, or Ethernet. On some controllers there may be more than one communications port. In fact, many PLCs today have more than one communications port. It is not uncommon to find PLCs with a USB or serial port for local programming and one or two Ethernet ports for communications with other devices and PLCs on a network. If you have an existing PLC connected to a network along with your computer workstation, your workstation can typically access all PLCs that are connected to that network without having to connect directly to the PLC's programming port.

Courtesy of Rockwell Automation, Inc.

Figure 8–55 PLC Controller Programming Port

⚡ **Caution:** When accessing a PLC controller over a network that has multiple PLCs, you must use extreme caution to ensure that you have connected to the correct or intended PLC. If you connect to the wrong PLC, unintended consequences could occur putting equipment and/or personnel at risk.

When programming a new PLC for the first time, or to ensure you are working with the correct controller, use the dedicated programming port on the front of the PLC to connect your workstation (computer) too. In this section, we will focus strictly on using the USB port on the PLC to download, upload, and go on-line with the PLC. The USB port is intended only for temporary local programming and monitoring of the PLC, and not intended for permanent connection. PLCs with a USB programming port typically have a Type B receptacle and are USB 2.0 compatible. Use a USB cable to connect one end (Type B) to the PLC and the other end to your PC's USB port. Do not use a USB cable longer than 10 feet.

Once connected to the PLC, you can *download*, *upload*, or go *on-line* with the PLC to monitor the project (program). Before you can *download*, *upload*, or edit the project *on-line*, you must place the key switch on the PLC in the Program or Remote (REM) position if so equipped. Remote position is the preferred mode. When you **download** a project to a PLC, the programming software copies the project files from your workstation application into the PLC controller memory. Conversely, when you **upload** a project from a PLC, it copies the project files from the PLC to your workstation application. *On-line* is used to monitor or edit an existing project that has already been *downloaded* to a PLC. As a technician, *on-line* monitoring is where you will spend most of your time when maintaining existing PLC controllers. The ability to monitor the real-time status of inputs, outputs, and logic makes for an extremely powerful troubleshooting tool for the technician.

⚡ **Caution:** You should always use caution when downloading a project to an existing PLC controller. If the PLC is in *run* mode and actively controlling equipment, the *download* process will cause the PLC to stop executing the current project and overwrite the existing project with the new one. It is always advisable to *upload* an existing project and save a copy first, before *downloading* a new project. In this way, if the new project has issues or you run into problems, you can always recover the last project and download it to the PLC.

Communicating with Allen-Bradley PLCs

To communicate with Allen-Bradley PLCs using Rockwell Automation software requires using software called RSLinx. RSLinx is a Windows-based OPC data server that normally comes standard with Studio 5000 and Connected Components Workbench software. RSLinx can be configured with various communications drivers based on your application and equipment.

As stated at the beginning of this chapter, this is not a definitive guide to the programming software. At the time of this publication, the latest software was used for the examples in this chapter and elsewhere in this textbook. Because software is continually being updated and new software comes out faster than one can keep up with, the version of software you are using may appear different than the examples shown here. With that said, the basic concepts will most likely be the same.

Communicating with a PLC can be challenging at times, especially when you must manually configure a specific communication's driver, or you are working with various versions and revisions of software and PLCs. This is true when you are attempting to use an older version of programming software with a newer PLC controller. In fact, you may have to update your software or, in the case of an older PLC, upgrade the controller's firmware. This text and the examples that follow will assume that you have the latest software and PLC controllers. Should you have problems communicating with a PLC, the help files are an invaluable troubleshooting tool for the technician. In fact, because there are so many different automation devices, and software, found in modern industrial plants, the technician cannot realistically know all the intimate details of the various devices and software required to program and/or maintain them. This is especially true when these devices operate trouble free for years. The Internet and help files are invaluable tools for the technician when additional help is needed.

In the two examples that follow, we will be connecting our workstation (computer) directly to the PLC controllers with a USB cable. This is the easiest and safest method of communicating with a PLC controller. We will *download* the two PLC projects (programs) we previously developed in this chapter as well as showing how to go *on-line* with an existing project.

 Caution: The examples that follow should only be undertaken on PLCs that are configured and used in a teaching environment (lab), not on actual operating equipment.

Allen-Bradley Micro800 and Connected Components Workbench Software

Step 1. Start the Connected Components Workbench software, if not already started, and open the project we previously developed.

Step 2. With the Connected Components Workbench software up and running, and our project opened, as shown in Figure 8–56, connect your USB cable between your computer and the Micro800 PLC. Depending on the model of Micro800 PLC you are using, it may require the installation of a USB adapter (2080-USBADAPTER) if not already equipped. Turn on power to the Micro800 PLC.

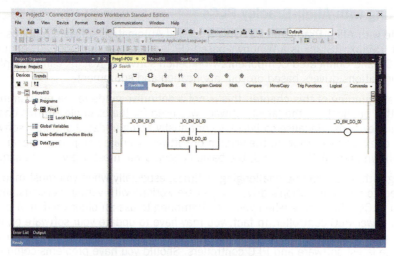

Figure 8–56 Micro800 Program Ready for Download

Step 3. With your project open as shown in Figure 8–56, double-click on the *Micro Controller* (Micro810) in the Project Organizer window so that your main window displays the controller configuration and monitoring window as shown in Figure 8–57. Check to see if there is a connection path displayed for the controller in the upper right-hand corner of the window as shown in Figure 8–58a. If so, click the eraser icon to the right of the connection path to delete it. Your screen should now look like the one in Figure 8–58b. It is best to clear any connection path the first time you *download* a project. If you do not clear the connection path, when you click *download* it will use that path to attempt to connect with the controller, which may or may not be the right path. After the first time you have connected to the Micro800 PLC, the connection path will be updated in your project files and there will be no need to clear the connection path in the future unless the path is no longer valid.

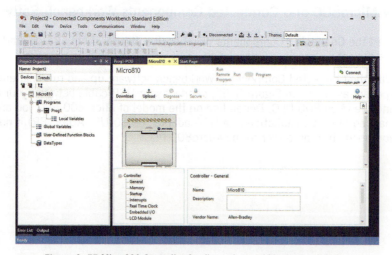

Figure 8–57 Micro800 Controller Configuration and Monitoring Window

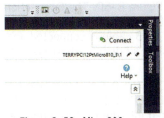

Figure 8–58a Micro800
Connection Path Shown

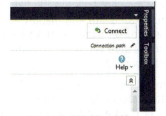

Figure 8–58b Micro800
Connection Path Cleared

Step 4. Once the path is deleted, or if no path is displayed to begin with as shown in Figure 8–58b, click the *Download* button and the Connection Browser window will open allowing you to select the PLC controller to *download* your project files to, as shown in Figure 8–59.

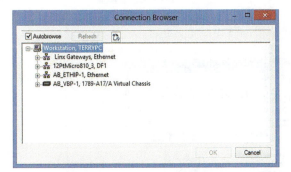

Figure 8–59 Connection Browser Window

With the Connection Browser selected for Autobrowse (default), it will attempt to search and display all PLCs connected to your computer. When using the USB direct connection method, you do not need to configure a communication's driver. By default, the Connection Browser will check your computer's USB ports and Ethernet port for any Allen-Bradley PLCs that might be connected. Find and click on your Micro800 PLC in the Connection Browser window. You will need to expand the connection drivers to see the PLCs connected to that driver as shown in Figure 8–60.

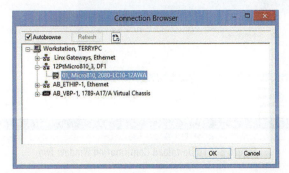

Figure 8–60 Connection Browser Communications Driver Expanded

Your screen may or may not look exactly like the one in Figure 8–60. The Connection Browser may take several minutes to locate and display all connected PLCs. If you do not see your Micro800 PLC in the Connection Browser window, make sure that your USB cable is connected and power is *ON* to your PLC. Also, make sure you have expanded and checked all displayed drivers in the Connection Browser window for your Micro800 PLC.

Once you have found and clicked on your Micro800 PLC in the Connection Browser window, then click the *OK* button. Depending on your model of Micro800 PLC, the mode the PLC is in (run/program/remote), or if an existing project is already loaded into the PLC, you may see one or more download confirmation windows like the ones in Figure 8–61a and 8–61b. Proceed to select and click on the *YES* buttons to confirm the download process. If you see a download confirmation window that says "Download overwrites the project in the controller with current project contents" and gives you two selections, "Download" and "Download with Project Values", select the "*Download with Project Values*" and proceed.

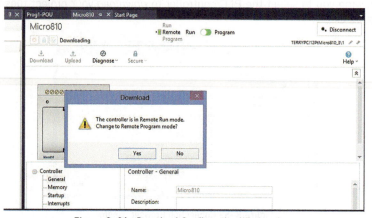

Figure 8–61a Download Confirmation Window One

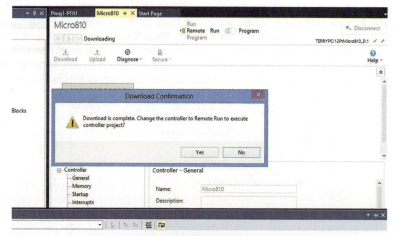

Figure 8–61b Download Confirmation Window Two

Step 5. If you were successful in downloading your project to the Micro800 PLC, you should see either the screen in Figure 8–62 or Figure 8–63. You are now *on-line* with the Micro800 PLC and you can confirm that by looking at the top of the screen where it says "Connected" and is highlighted. Congratulations, you have successfully downloaded your first PLC project to a Micro800 PLC controller.

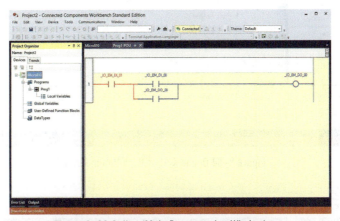

Figure 8–62 Online (Main Programming Window)

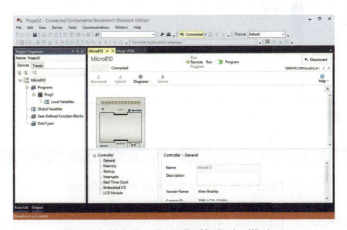

Figure 8–63 Online (Controller Monitoring Window)

From the screen in Figure 8–62 you can monitor your ladder logic program and the status of the inputs and outputs by looking at the displayed colors. Red indicates a true logic path as evident by looking at our programmed stop push button contact and address _IO_EM_DI_01. Figure 8–64 shows the same circuit logic with the start push button pressed and the motor output coil *ON*. Figure 8–65 shows the circuit after the start button has been released.

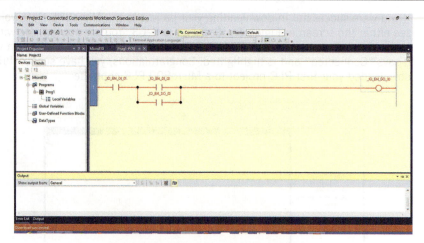

Figure 8–64 Online & Start Push Button Pushed

Figure 8–65 Online & Start Push Button Released

Step 6.   Double-click on the *Micro Controller* (Micro810) in the Project Organizer window so that your screen displays the controller configuration and monitoring window as was shown in Figure 8–63. By looking at this screen, you can see the mode that the controller is in, the communications path to the controller, and a selection to disconnect from the controller. Click on the "*Disconnect*" button above the communications path. Your screen should now look like the one in Figure 8–66. Notice at the top of the screen that you are no longer connected to the PLC (disconnected) and you are now in the *off-line* mode. The project you downloaded to the PLC is still in the PLC and running. If you want to go back on-line with the PLC, you simply go to the controller configuration and monitoring window (Figure 8–63) and click on "*Connect*" in the upper-right corner. There is no need to use the Connection Browser to locate your Micro800 PLC since you have already established a communications path for the project to that PLC controller.

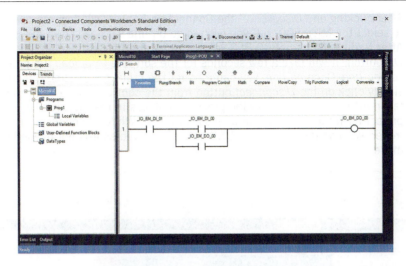

Figure 8–66 Off-line with Micro800 Controller

Depending on the version of Connected Components Workbench software you have installed, you may or may not be able to make edits to the project (program) while *on-line*. If you make changes to your PLC program while in the *off-line* mode and the connection path is still valid, simply click on the *download* button on the controller configuration and monitoring window.

If there is already an existing Micro800 PLC controller running and you wish to go on-line to monitor the project (program), or you would like to save an existing project that is already loaded into a PLC, follow these two steps.

Step 1.   Open the Connected Components Workbench software and on the main opening screen click on "*Discover...*" under Project as shown in Figure 8–67.

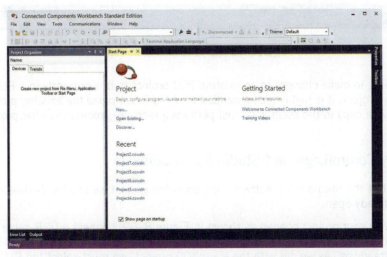

Figure 8–67 Connected Components Workbench Main Opening Screen

Step 2. After clicking on "*Discover...*" the Connection Browser window will open allowing you to select the Micro800 PLC you wish to go on-line with as shown in Figure 8–68. After finding the Micro800 PLC you wish to go *on-line* with, select it and click on the *OK* button. It will take a few seconds for the software to *upload* the project and take you *on-line*. Once *on-line*, you can monitor the project as desired. When finished, simply click on the *Disconnect* button from the controller screen and you will be taken off-line with a copy of the program that was in the PLC. Now you can save the project for future reference or as a backup copy. While disconnected from the controller, you can disconnect the USB cable if desired as it is not needed in the *off-line* mode.

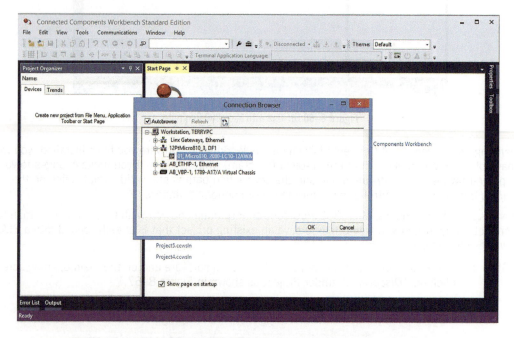

Figure 8–68 Connection Browser Window

If you are planning to make changes to an existing PLC project that is currently in a PLC or to *download* a new project, it is highly recommended that you first *upload* the existing project that is in the PLC and save a copy in the event you must perform a recovery procedure of the project at a later date.

Allen-Bradley ControlLogix and Studio 5000 Software

Step 1. Start the Studio 5000 software and open the project we previously developed if not already open.

Step 2. With your project opened as shown in Figure 8–69, connect a USB cable between your computer and the ControlLogix PLC controller. Turn power on to the PLC if it is not already on. As we did with the Micro800 project, we must select the PLC connection path. In Studio 5000, the connection browser is called Who Active (RSWho). You can open the Who Active window by either clicking on the Who Active symbol or by clicking on "Communications" and selecting Who Active (RSWho), both are located at the top of the screen as shown in Figure 8–70.

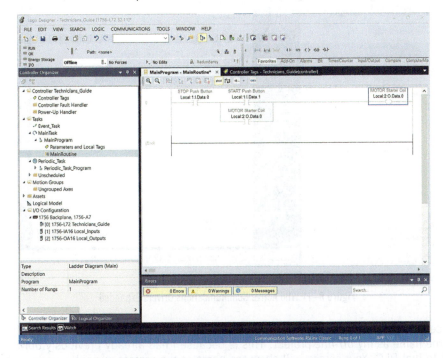

Figure 8–69 Logix Designer Software Open & Start/Stop Project Open

Figure 8–70 Selecting Who Active Window

Step 3. With the Who Active window open and selected for Autobrowse (default), the software will attempt to search and display all PLCs connected to your computer. When using the USB connection method, you do not need to configure a communications driver. By default, the Who Active software will check your computer's USB ports and Ethernet port for any Allen-Bradley PLCs that might be connected. This process may take several seconds or more. Find and click on the ControlLogix PLC in the Who Active window for the PLC you plan to *download* the project files to. You will need to expand the communications paths to see the PLCs connected to that path as shown in Figure 8–71.

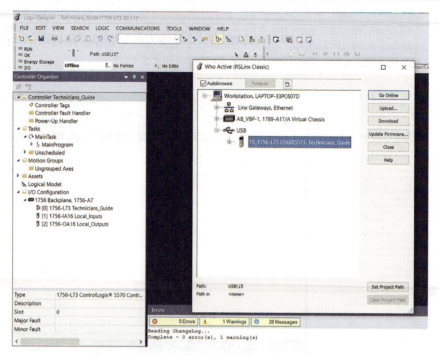

Figure 8–71 Who Active Communications Drivers Expanded

If you notice in Figure 8–71 that the PLC controller connected to the USB communications path has the same name as my *off-line* project. This is because I had already previously *downloaded* my project to that PLC. Your screen will show the name of the project currently loaded into that PLC if there is a project loaded. It is worth noting that the PLC will also be shown under the "Virtual Chassis" path. They are, in fact, the same PLC and you can use either communication path. It is recommended, however, that you select the communication path appropriate for your physical connection, in this case, the USB communications path as was shown in Figure 8–71. After selecting the PLC controller to download to, click the *Download* button. You can also set the project path for your off-line project by selecting "Set Project Path" at the bottom of the Who Active window, but this is not required to download a project. In fact, the project path will automatically be updated once on-line. If you upload and save the project, the project path will also be saved. After clicking on the *Download* button, you should see a download warning message like the one in Figure 8–72. This warning message is prompting you to confirm that you have selected the correct PLC (controller information) and that there may be potential hazards with downloading to a PLC controller. If you have the correct PLC and are prepared to continue, click on the "*Download*" button at the bottom of the window to continue.

If you receive an "Unable to Connect" message, follow the instructions in the message(s) to proceed.

Figure 8–72 Download Warning Message

If the PLC was in the run mode when you started the download, you should see a window like the one in Figure 8–73 confirming that you wish to place the controller back into run mode or not. If the controller was not in run mode when you started the download, you may not see this message.

Figure 8–73 Place Controller in Run Mode Warning Message

Once the download is complete, you will be taken on-line with the project as shown in Figure 8–74. If you were successful, congratulations you have just downloaded your first PLC project to a ControlLogix PLC controller.

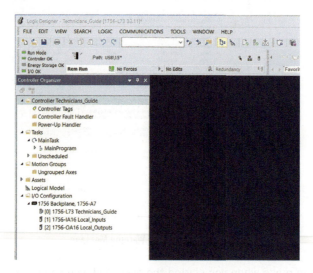

Figure 8–74 Online with Logix Controller

In viewing Figure 8–74, you will notice that the current status of the controller is shown in the upper-left corner of the screen. Information like the current mode, processor status, connection path, etc. will be displayed. If the controller's key switch is in the remote (REM) position, you can change the controller's mode (run/program) or go off-line or on-line as shown in Figure 8–75.

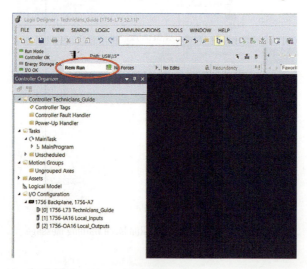

Figure 8–75 Controller Status and Mode Selection

Step 4. With the project downloaded, controller in run mode, and our sample project wired to the I/O modules, double-click on the *Main Routine* (project organizer) and you should see your ladder logic program as shown in Figure 8–76. Notice how a true logic path is highlighted in green. If we push the start push button, the motor output should come on and remain on until we press the stop push button as shown in Figure 8–77.

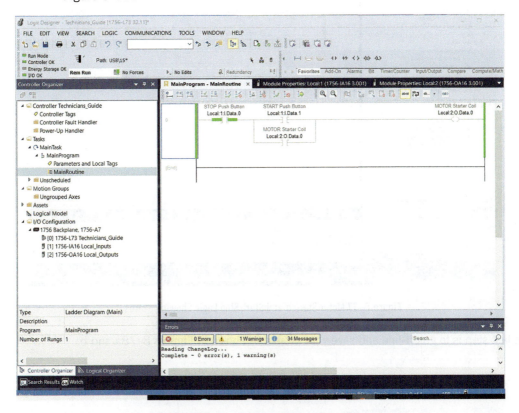

Figure 8–76 Main Routine Ladder Logic Program

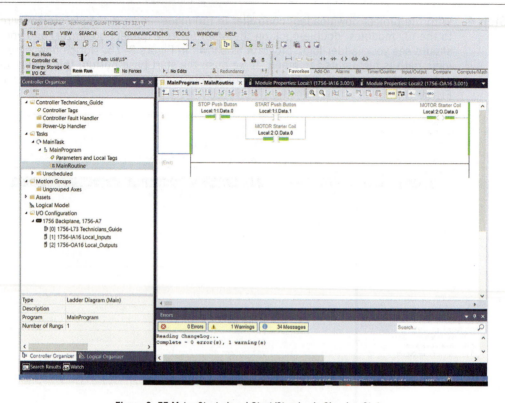

Figure 8–77 Motor Started and Start/Stop Logic Showing Status

The last step is to go off-line and save your project as shown in Figures 8–78a and b.

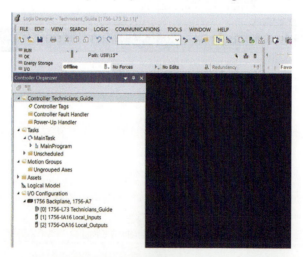

Figure 8–78a Going Offline with ControlLogix Controller

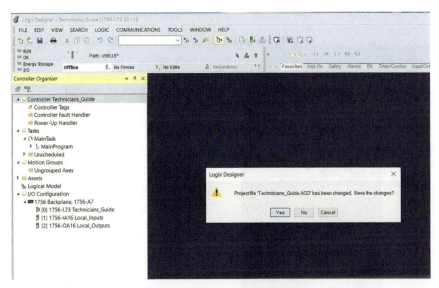

Figure 8–78b Saving Offline Project

If there is already an existing PLC running and you wish to go on-line to monitor the project (program), or you would like to save an existing project that is already loaded on a PLC controller, follow these two steps.

Step 1. Open the Studio 5000 software and click on "*From Upload*" under Open as shown in in Figure 8–79.

Figure 8–79 Studio 5000 Opening Screen & Selecting "From Upload"

Step 2. After clicking on "*From Upload*" the Logic Designer software will open and the Who Active (RSWho) window will appear allowing you to select the ControlLogix PLC controller you wish to go *on-line* with as shown in Figure 8–80. After finding the PLC you wish to go on-line with, select it and click on the "*Go Online*" button. It will take a few seconds for the software to upload the project and take you *on-line* as shown in Figure 8–81. Once on-line, you can monitor the project as desired.

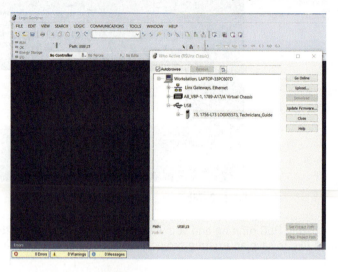

Figure 8–80 Who Active Window

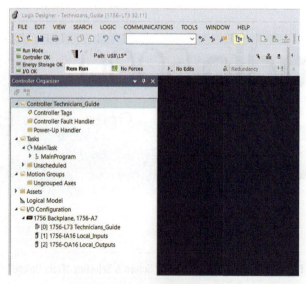

Figure 8–81 Online with PLC Controller

If you are planning to make changes to an existing PLC project that is currently in a PLC or to download a new project, it is highly recommended that you first *upload* the existing project that is in the PLC and save a copy in the event you must perform a recovery procedure at a later date.

Chapter Summary

The PLC is programmed by using a PC that uses custom programming software that has been developed by the PLC manufacturer for their specific brand of PLCs. Logic symbols and function blocks are used during the programming to tell the processor what to do, while addresses and tags are used to tell the processor where to do it. The programming device (programmer) is used to enter, modify, and monitor the user program(s). PLC programs are initially developed in the *off-line* mode. Once the PLC program(s) is complete, it can be tested in the *off-line* mode using simulation software, if available, to ensure correct operation before the program is downloaded to the controller for final testing and verification. Keys or passwords are used to prevent unauthorized access or modifications to the PLC controller and programs. While *on-line,* and from the programming device, contacts and coils can be forced *ON* or *OFF* while the PLC program is running or operational. The FORCE ON and FORCE OFF capability should be restricted to personnel who have a complete understanding of the circuit and the driven equipment. The programming software can be used as a troubleshooting aid when connected to a PLC controller by monitoring the status of contacts, coils, etc. Programming a PLC is not difficult, but time must be spent to become familiar with the programming software and the programming techniques used by the various PLC manufacturers.

Key Terms

on-line programming   password
off-line programming   force on
force off   download
upload

Review Questions

1. What does the term *on-line* programming mean?
2. What is the FORCE feature used for?
3. What does *off-line* programming mean?
4. What is the difference between a global and local variable (tag)?
5. Contacts and coils are Boolean symbols.

    T   F

6. What is *download* used for?

7. In both the Micro and ControlLogix processors, I/O variables or tags can be found where?

8. In a ControlLogix processor, ladder logic is entered in:

   Tasks, Routines, or Programs

Chapter 9

Programming Considerations

Learning Objectives

After completing this chapter, you should have the knowledge to:

- Define a *network*.
- Describe the term *dummy relay*.
- Describe the horizontal and vertical contact limits.
- Define the term *nesting*.
- Correctly wire and program *STOP* buttons.
- Describe the difference between logical and discrete holding contacts.
- Define the term *buffering*.
- Correctly program for motor overloads.

Network Limitations

A **network** is defined as a group of connected logic elements used to perform a specific function. Figure 9–1 shows a typical network consisting of seven series contacts and three parallel branches. A network also constitutes one rung of a ladder diagram that starts at the left rail and ends at the right rail.

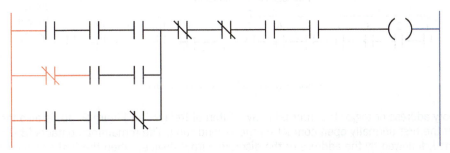

Figure 9–1 Network (Rung)

187

Many PLC manufacturers have virtually no network limitations, whereas other PLC systems are limited by the number of contacts or other logic elements that can be included on the horizontal line of a network, and the parallel branches (lines) that make up one network. A typical PLC network limitation of ten series contacts per line and seven parallel lines, or branches, is shown in Figure 9–2. Additionally, some PLCs are further limited because they only allot one output per rung or network, and the output must be on the first line.

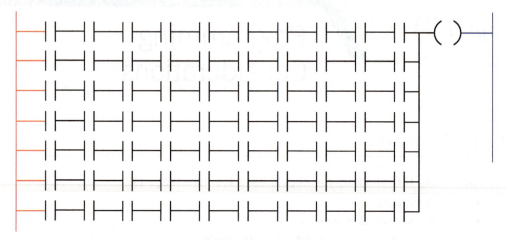

Figure 9–2 Network Limits

Note: Special functions and other logic symbols alter the network limitations and requirements; check the operation or programming manual of the specific PLC for additional information.

While the number of elements and lines within a network may be limited, only the size of user memory limits the number of networks or rungs. When a network is complex with many logic elements and/or parallel branches, consideration should be given to dividing the logic into multiple smaller networks or rungs. This will make it easier to maintain and view.

When a circuit requires more series contacts than the network allows or to limit the number of series contacts (Figure 9–3a), the contacts are split into two rungs (Figure 9–3b). The first rung contains part of the required contacts and is programmed to an internal, or "dummy," relay. Internal relays are actually a bit location in memory.

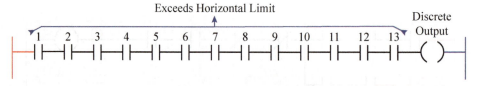

Figure 9–3a Contacts Exceed Horizontal Limit

The memory address or tag of the internal relay, "Internal Relay1" in Figure 9–3b, is also the address of the first normally open contact on the second rung. The remaining contacts (8–13) are programmed, followed by the address of the discrete output device. When the first seven contacts close, the internal output (Internal Relay1) is set to 1. This makes the normally open contact "Internal Relay1" in Rung 2 true. If the other six contacts (8–13) are closed, the rung is true, and the discrete output is turned *ON*.

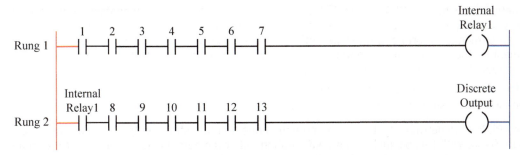

Figure 9–3b Contacts Split into Two Rungs

Note: It is not necessary to split the contacts in any ratio. If the network allows 10 horizontal contacts, 10 could be placed on the top rung, and 3 could follow the normally open contact of the internal relay (Internal Relay1) on Rung 2. This technique applies not only to normally open contacts, but to normally closed (or combinations of normally open and normally closed) contacts as well.

The internal relay just used does not exist as a real-world device that has to be hardwired but is merely a bit in memory that performs the logic of a relay. In actual programming, internal control relays that do not actually exist, except in memory as bits, are extensively used. The use of these internal relay equivalents is what makes the programmable logic controller unique, eliminating hours of hardwiring and shortening installation and maintenance time.

When a program requires more parallel branches than the network allows, the circuit can be split into two networks, or rungs. The first six parallel contacts are programmed to an internal relay, as shown in Figure 9–4. A contact with the same address as the internal, or dummy, relay is then programmed in parallel with the remaining contacts to control the output.

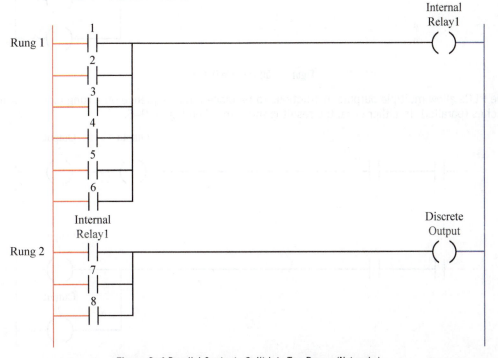

Figure 9–4 Parallel Contacts Split into Two Rungs (Networks)

Sometimes it is not the PLC that limits the number of horizontal or parallel contacts, but rather the ability to easily view a rung and all its elements. If all the elements cannot be easily seen on the computer monitor or printout, it can be difficult to troubleshoot. The ability to view each contact or element on the monitor and to know the status (*ON* or *OFF*) of each contact or element, as well as the status of output elements, is what makes the PLC such a powerful tool.

Note: Most programming software allows the electrician or technician to "shift" the screen to monitor all the instructions, even when the normal screen is filled with instructions.

Many PLCs have networks that allow for more than one output instruction (parallel outputs), see Figure 9–5a. With this parallel output configuration, all of the outputs are *ON* or *OFF* at the same time, based on the network logic. Other PLCs allow multiple outputs that can be *ON* or *OFF* at different times, depending on the network logic as shown in Figure 9–5b.

Figure 9–5a Parallel Output Format

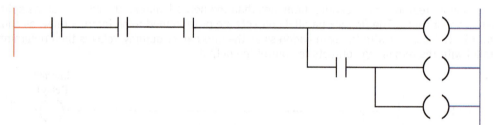

Figure 9–5b Multiple Outputs

Some PLCs allow multiple outputs instructions to be placed in a sequence on a rung (series) or in branches (parallel). In either case, the result is the same. See Figure 9–6.

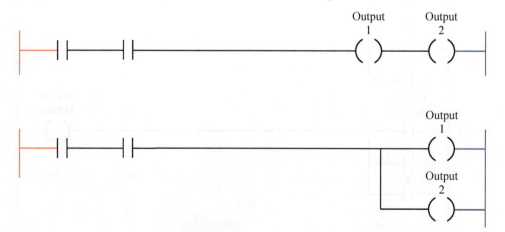

Figure 9–6 Series Sequence Outputs

Note: The author has found that placing output instructions in parallel branches rather than in sequence is more consistent with hardwired ladder logic circuits and therefore much easier for electricians and technicians to understand.

Programming Restrictions

In addition to the number of horizontal contacts on one line, and the number of lines in a network or rung, the PLC does not allow for programming vertical contacts (Figure 9–7). In the real world, one could wire the circuit as shown in the figure, but programming restrictions would not allow the PLC to be programmed in this manner.

Figure 9–7 Vertical Contacts

If one analyzes the logic of the circuit in Figure 9–7, the circuit logic shows that output F can be energized by any of the following contact combinations: A, B (Figure 9–8a); A, C, E (Figure 9–8b); D, C, B (Figure 9–8c); and D, E (Figure 9–8d).

Figure 9–8a Path A, B

Figure 9–8b Path A, C, E

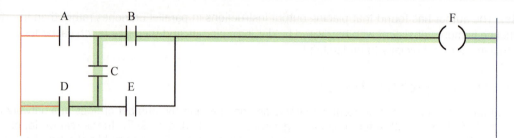

Figure 9–8c Path D, C, B

Figure 9–8d Path D, E

To duplicate the logic, the circuit could be programmed as shown in Figure 9–9.

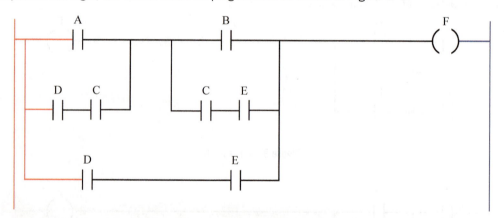

Figure 9–9 Equivalent Circuit without Vertical Contacts

This circuit maintains the circuit logic. Contact combinations A, B; A, C, E; D, C, B; and D, E all energize output F.

Another limitation to circuit programming is the way in which the processor considers power flow, or logic continuity, when it scans a rung of logic. Flow is from left to right *only,* and vertically *up* or *down.* The processor *never* allows logic continuity (power flow) from right to left.

Normally, relay logic for the circuit shown in Figure 9–10 would indicate the following possible contact combinations to energize output G: A, B, C; A, D, E; F, E; and F, D, B, C.

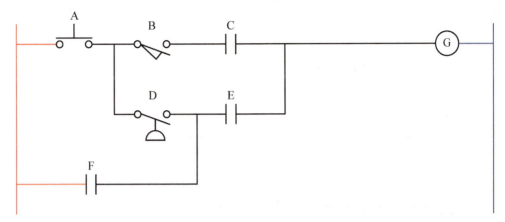

Figure 9–10 Hardwired Circuit

If the circuit shown in Figure 9–10 was programmed into user memory as shown in Figure 9–11a, the processor would ignore contact combination F, D, B, C because it would require power flow (logic continuity) from right to left. If combination F, D, B, C was required, the circuit would be reprogrammed as shown in Figure 9–11b.

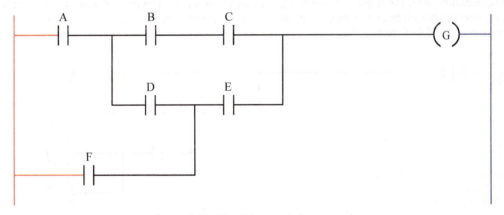

Figure 9–11a Circuit Improperly Programmed

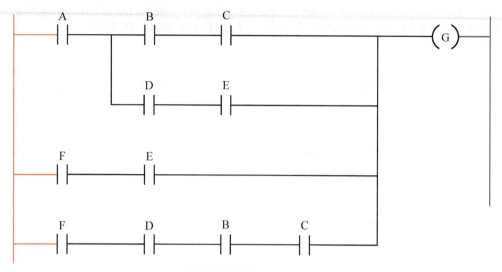

Figure 9–11b Circuit Properly Programmed

The last restriction placed on the programming of circuits into user memory by some—*but not all*—PLCs is the use and number of "a branch circuit within a branch circuit," or the **nesting** of branches. Figure 9–12 is an example of a circuit that has nested branches. For example, Allen-Bradley Logix 5000 controllers limit the number of nested branches to six levels. The easiest way to avoid nesting is to remember that all branches must start at a common point, and all of the branches must end at a common point. See Figure 9–13.

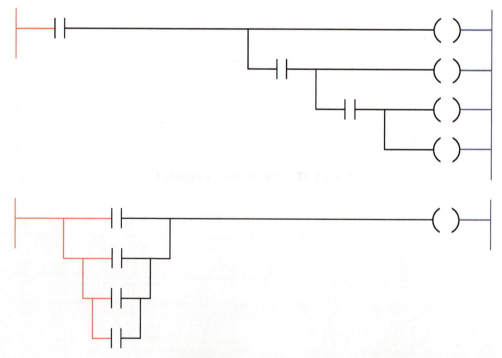

Figure 9–12 Branches within a Branch

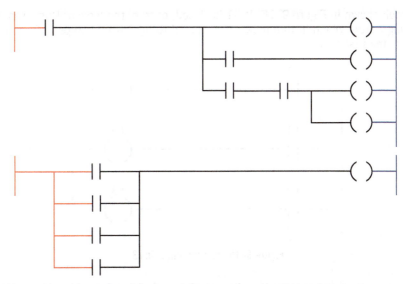

Figure 9–13 Programmed to Eliminate Branches within a Branch

Program Scanning

When the PLC updates I/O synchronous to the program scan, the processor first determines the status of the input devices, next it scans the user program, and then updates (turns *ON* or *OFF*) the outputs. The way the processor scans the program varies from PLC to PLC. One common method is to scan the program from left to right and top to bottom, similar to the way in which a book is read. In this method, the processor scans the first rung of the program from left to right, then the second rung from left to right, and continues in this fashion until all the rungs have been scanned. In the next scan, the processor returns to the first rung and starts all over again, scanning each rung in order from top to bottom. Figure 9–14 illustrates the order of scanning for a processor that scans from left to right and top to bottom. This is the scanning method used by Allen-Bradley for their family of PLCs as well as most others.

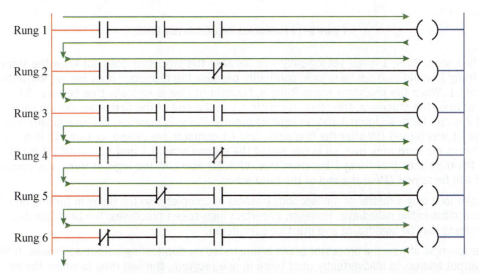

Figure 9–14 Processor Scan Left to Right, Top to Bottom

Look at the circuit shown in Figure 9–15. If S1 is closed, or true, the logic of Rung 1 is true, making the logic of Rung 2 true, which in turn makes the logic of Rung 3 true. Lamps 1, 2, and 3 are turned *ON* at the end of the first scan.

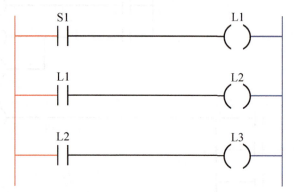

Figure 9–15 One Scan Turns On L3

If, however, the circuit was programmed as shown in Figure 9–16, L3 would *not* turn *ON* until the third scan of the program was completed.

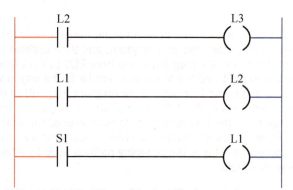

Figure 9–16 Three Scans Required to Turn On L3

In the first scan, Rung 1 is not yet logically true because the processor does not know the status of L2. Rung 2 would also not have logic continuity because the processor does not yet know the status of L1. When the processor scans Rung 3, however, it now is logically true because S1 is closed or true. The processor, therefore, turns *ON* L1 at the end of the first scan. On the second scan, L2 is still false, so Rung 1 has no logic continuity. Rung 2, however, is now logically true because L1 was turned *ON* after the first scan, and L1 contacts are closed, or true. S1 is still closed, which keeps the third rung true, so at the end of the second scan, L1 and L2 are *ON*. It is only during the third scan that Rung 1 becomes true. With L2 now *ON*, the logic of Rung 1 is complete and L3 will be turned *ON* at the end of the third scan.

Each scan took only a matter of milliseconds (msecs) to complete, so the delay in turning on L3 is not perceptible to the naked eye. However, in certain high-speed processes, the time lost due to poor programming may be significant and must be considered.

Another example of how the processor scans is illustrated by duplicating output addresses. If the same output address is inadvertently used twice in one program, the last rung in which the address is used will indicate the status of the output. For example, if the address is used in Rung 5 and the logic of Rung 5 is true (which would tell the processor to turn *ON* the output), and the address is used again in Rung 13 and the logic of Rung 13 is false, the output will not be turned *ON* because the processor scanned Rung 13 last, and the last state of the output (true or false) will be based on the last rung scanned.

Another aspect to program scanning is to understand and consider rung conditions. When the processor scans rungs from left to right, which most often is the case, the processor evaluates the ladder instructions based on the rung condition preceding the instruction. What this means is that the processor will evaluate the instruction if the rung condition preceding the instruction is true (the in condition). If it is false, the processor sets the rung out condition to false and moves on to the next instruction. By placing instructions that are most likely to be false on the left to least likely on the right (when placed in series), will improve processor performance and decrease scan times.

Buffering I/O

When the PLC controller updates its I/O asynchronous to the execution of the program (logic), I/O values may change during the execution. This can be problematic when I/O data values can change in the middle of a program scan, creating unpredictable results. For example, if you evaluate an input condition in one rung of code and then several rungs of code later you evaluate that same input expecting the same condition, the input condition may have changed leading potentially to unexpected results. In order to address this issue, a technique called buffering is used.

Buffering is a technique in which your program logic does not directly reference or manipulate the actual I/O data (tags) stored in memory. Rather, the program logic uses a copy of the I/O data. This technique requires that the actual I/O data be copied into a set of base tags or internal memory locations that will not change state during the program's execution. When buffering to base tags or internal memory locations, the actual input data is copied before executing the program logic, and the buffered output data is copied to the corresponding output tags at the conclusion of the program logic. Think of buffering as taking a snap shot of the inputs at the beginning of the program scan and then updating all the outputs at the end of the program scan. This buffering technique duplicates the synchronous scanning of early PLC processors as was discussed in Chapter 4.

There are several methods of buffering I/O: buffering to base tags or program parameter buffering. Program parameter buffering is a method that is available on Allen-Bradley Logix version 24 and later controllers. Program parameter buffering will not be covered in this text; refer to Allen-Bradley's Logix 5000 Controllers Program Parameters Manual for additional information on program parameter buffering.

Follow these steps when buffering I/O to base tags.

1. On the rung or subroutine before the program logic, copy the data in the input tags to be used into their corresponding buffer tags.

2. In your program logic routine, reference the buffer tags.

3. On the rung or subroutine after the program logic, copy the data from the output buffer tags used in your program to their corresponding output tags.

The following example uses three ladder logic routines to easily segment and execute the three steps just discussed. The Logix's family of Allen-Bradley controllers gives you the ability to organize your ladder logic into routines. Refer back to Chapter 5. This ability to have separate routines makes buffering I/O much easier and well organized. The first subroutine reads the input values and buffers them for use within the logic routine. The second subroutine contains the executable logic that controls the machine or process. The third subroutine writes the output buffered values to the output tags. Figure 9–17 shows an example of each subroutine. All three subroutines are executed in sequential order by programming Jump to Subroutine (JSR) instructions in the main routine (continues routine) as shown in Figure 9–18.

| **Note:** Jump to Subroutine (JSR) instructions will be covered in Chapter 10.

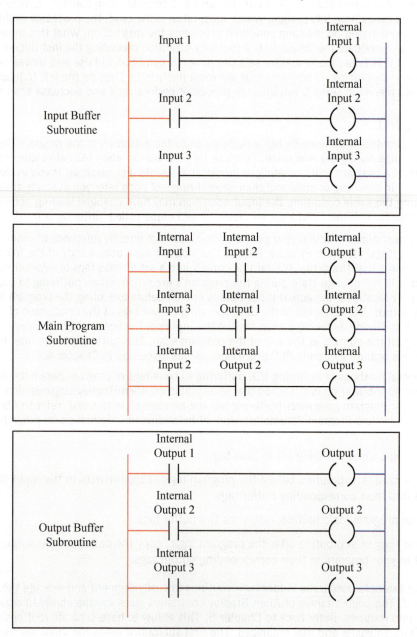

Figure 9–17 Buffering I/O

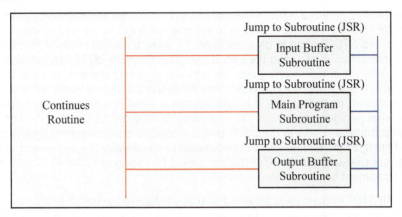

Figure 9–18 Main Routine and Jump-to-Subroutines

The logic shown in Figure 9–17 for buffering the I/O is really only practical if you have just a few inputs and outputs. A better approach is to use the Synchronous Copy File (CPS) instruction. The CPS instruction copies multiple data values from one memory location (source) to another memory location (destination) uninterrupted. The CPS instruction will be covered in Chapter 15.

Programming *STOP* Buttons

During the early days of PLCs, it was common for salespeople to use a demonstrator model that had the *STOP* buttons wired normally open. This technique was used so the salespeople did not have to explain why a *STOP* button was shown as a normally open contact in a PLC program.

Figure 9–19a shows a standard *STOP/START* ladder diagram, and Figure 9–19b shows the equivalent diagram used by some PLC salespeople.

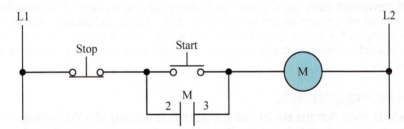

Figure 9–19a Standard *STOP/START* Ladder Diagram

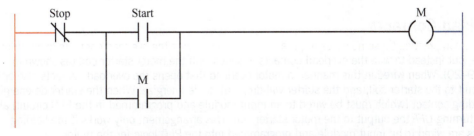

Figure 9–19b Incorrectly Programmed *STOP/START* Circuit

From an understanding of how a *STOP/START* circuit works, and an understanding of the EXAMINE ON and EXAMINE OFF instructions, it is easy to see that the only way the circuit could be logically true would be for the *STOP* button to be wired open. By using a normally open *STOP* button, the EXAMINE OFF instruction is true, and the circuit energizes when the *START* button is pressed. The problem is that once the circuit is energized, the only way it can be stopped, or turned *OFF*, is if the *STOP* button is pushed and the contacts close. While this circuit will work, there is a built-in *safety hazard* that must be considered. If a *STOP* button is wired in a normally open position, and the button becomes jammed and cannot close its contacts, it would be impossible to de-energize the circuit. Similarly, if a wire breaks in the *STOP* button circuit, it is possible to complete the logic of the circuit and energize the equipment, but *impossible* to de-energize the equipment. With the wire broken, changes in the status of the *STOP* button cannot be conveyed to the PLC processor, and the circuit and/or equipment cannot be de-energized.

Safety Note: All *STOP* buttons must be wired so that a failure of the switch or a broken wire will automatically break logic continuity and turn the circuit *OFF*. A good programmer will always wire the devices and program the circuit so that if the real-world device fails, it creates a safe condition, not a safety hazard.

This practice of wiring *STOP* buttons in the normally open position was common during the 1980s and may still be found today. If an electrician or technician finds equipment wired in this fashion, they should change the *STOP* buttons to normally closed and the PLC program from EXAMINE OFF to EXAMINE ON.

The *START* button should be wired normally open and programmed with an EXAMINE ON (XIC) instruction. As a general rule all input devices, except *STOP* buttons and other safety related devices, are wired normally open and given EXAMINE ON instructions in the program.

Logical Holding Contacts

In previous programming examples we have used the output address to address the holding contacts. This method of providing holding logic works well in many applications and eliminates the need to actually wire the holding contacts on the motor starter. When the output address is also used for the holding contacts (logic), the circuit is maintained logically because the output point has been turned *ON*. This is no guarantee, however, that the actual motor starter connected to the output module has been energized.

Discrete Holding Contacts

The only real way to know that the starter has energized is to actually wire the holding contacts of the motor starter to an input module and use that address when programming the holding contacts or when programming motor fault logic. This method has many advantages, and, short of installing a motor sensor, is the best way to verify that the motor starter has been energized or pulled in.

Overload Contacts

It is common practice (in some cases required by code) *not* to wire the overload contacts to an input module, but instead to wire the overload contacts in series with the motor starter coil (as shown in Figure 9–20). When wired in this manner, a motor overload that opens the overload contacts also opens the circuit to the starter coil, and the starter will drop out, or de-energize. When the starter de-energizes, the holding contact (which must be wired to an input module and programmed in the PLC circuit) also opens, turning *OFF* the output to the motor starter coil. This arrangement only works if the holding contacts are wired to an input module and programmed into the PLC logic for the motor.

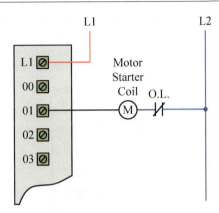

Figure 9–20 Overload Contacts Wired in Series with the Starter Coil

If the holding contacts are not wired to an input module, but instead the output address is used as the holding logic, an overload condition would not be detected and the motor output would remain *ON* even though the motor starter has been de-energized by the overload contacts. This wiring scheme can cause a safety hazard as well as other potential problems. With the motor output remaining *ON*, the motor will restart the instant that the overload is reset.

Some applications require the addition of a hardwired backup control circuit for PLC-controlled motors. These hardwired backup control circuits provide a means to control a motor or other output independent of the PLC, if for any reason the PLC is unable to control its outputs. Many of these hardwired backup control schemes can be found in critical processes such as water and wastewater systems, environmental systems, emergency power, cooling systems, etc. When a hardwired backup control circuit is used in conjunction with a PLC to control a motor, the overload contact on the motor starter must be wired in series with the motor coil to provide protection for the motor under both means of control, PLC *and* hardwired. Figure 9–21 shows a motor controlled from a PLC *and* hardwired control circuit.

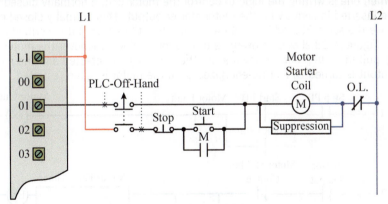

Figure 9–21 PLC and Hardwired Controlled Motor

By applying common sense and good PLC programming practices, many of the previously mentioned problems concerning overload contacts can be resolved. One method that the author and many control engineers use to solve these problems and provide additional benefits is as follows.

Step 1. Hardwire the overload contacts in series with the motor starter coil, as was shown in Figure 9–20. This will protect the motor regardless of the method being used to control the motor, PLC *or* hardwired, and will not depend upon software for motor protection.

Step 2.   Wire the holding contacts to an input module (as shown in Figure 9–22). The holding contact input can be used to monitor the motor starter for two abnormal conditions: failure of the motor starter to pull-in when the output to the motor starter coil is *ON*, and failure of the motor starter to drop-out when the output to the motor starter coil is *OFF or* de-energized.

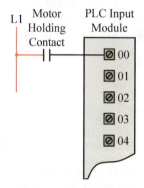

Figure 9–22 Motor Holding Contacts Wired to Input Module

Step 3.   Write a rung of PLC logic that will monitor the motor starter for either of the two abnormal conditions described above, failure to pull-in *or* drop-out (see Figure 9–23, Rung 2). If an abnormal condition is detected, the logic in Rung 2 will turn *ON* an internal memory bit labeled "Motor Fault" that will seal in and remain *ON* until an operator or maintenance person reattempts to start the motor. The first two branches in Rung 2 monitor the motor for the two abnormal conditions, and the third branch acts as a holding circuit if a motor fault is detected. The On-Delay timer in Rung 2 is used to allow for the physical movement of the motor starter when energizing and de-energizing the motor starter (PLC timers will be discussed in Chapter 11).

Step 4.   When one is writing the logic to control the motor coil, a normally closed contact would be inserted in series with the motor starter output. This normally closed contact would have the same PLC address as the "Motor Fault" memory bit described above. Rung 1 in Figure 9–23 shows an example of this logic. Now, whenever the motor starter fails to pull in or drop out for any reason, the condition is detected and the motor starter output is turned *OFF* or de-energized until the motor is again restarted.

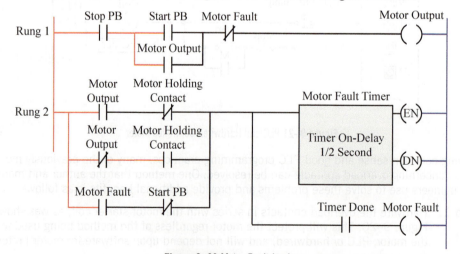

Figure 9–23 Motor Fault Logic

Note: The "Motor Fault" memory bit can also be used to flash a pilot light, such as the motor's run light, to alert the operator and maintenance personnel of the problem. See Chapter 22, "Programming Examples," for examples of this logic.

Chapter Summary

Each PLC has a maximum network size, or matrix, that limits the number of horizontal and vertical contacts for any one network or rung. The only limitation to the number of rungs (networks) is memory size. Since the processor reads program logic (power flow) from left to right *only*, and vertically either *up* or *down*, the logic of a relay circuit must be examined carefully to ensure that the logic is maintained when the circuit is programmed into the PLC. The programming device does not allow contacts to be programmed vertically, but the logic of a ladder diagram with vertical contacts may be duplicated by adding additional contacts. Depending on the PLC, contacts may or may not be programmed as a "branch-within-a-branch," or nested.

Consideration must be given to the way that the processor scans the program to eliminate unnecessary scans before a rung of logic goes true. Buffering I/O may be required when the I/O is updated asynchronous to the program scan to ensure logic continuity.

STOP buttons must always be wired as normally closed and use an EXAMINE ON (XIC) instruction to work correctly and safely. Holding contacts can be either logical or discrete (real world). The discrete method has the added advantage of verifying that the motor starter has indeed energized. Holding contacts wired to a PLC input provide the added advantage of monitoring the motor starter for abnormal conditions.

Key Terms

network   nesting

buffering

Review Questions

1. Define the term *network*.
2. Draw and label a diagram to show how a network that requires 14 series contacts to control a discrete output can be programmed in a PLC that limits series logic elements to 10 per rung.
3. Draw a circuit with nested contacts.
4. Draw a circuit that retains the logic of the accompanying figure, which has a vertical contact (D), so the circuit could be programmed into a PLC.

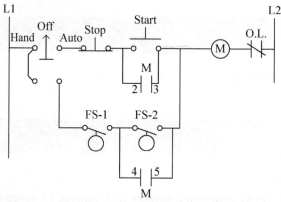

5. Power flow, or logic, in a PLC is considered to be (check all correct answers):
   a.   up to down only
   b.   up or down only
   c.   left to right
   d.   right to left
   e.   up or down and from left to right
   f.   up, down, and from right to left
   g.   up to down only and left to right
   h.   up to down only and right to left
   i.   up or down and left to right or right to left
6. Write a program for a PLC that does *not* allow nested contacts, for the Hand-Off-Auto circuit shown in the accompanying figure.

7. Explain how a *STOP* button must always be wired, and why.
8. List two ways that holding contacts can be programmed.
   a.   _____
   b.   _____
9. Explain two advantages of wiring the holding contacts to an input module and then programming the holding contact address into the PLC program.
10. Define the term *buffering I/O* and explain why it is used.
11. What is a dummy relay?
12. Explain motor overloads and how to program for an overload condition.

Chapter 10

Program Control & Miscellaneous Instructions

Learning Objectives

After completing this chapter, you should have the knowledge to:

- Write a program using a *latching* instruction.
- Describe the term *retentive*.
- Describe how a *master control relay instruction works*.
- Describe how a *jump to subroutine* instruction works.
- Describe how the *jump and label* instructions work.
- Describe a reason for using the *temporary end* instruction.
- Describe how the *ONS* instruction works and where it might be used.

Master Control Relay Instructions

In hardwired relay control systems, a master control relay is often used to control power to the entire control circuit or selected circuits (rungs). This allows selected circuits, or the whole circuit, to be de-energized by turning off the master control relay (MCR). Figure 10–1 shows a typical hardwired master control relay that controls power for the entire circuit.

MCRs are often used with circuits that have off-delay timers so the circuit can be shut down completely without waiting for the timers to time out.

With PLCs, an MCR function can be programmed to control an entire routine or just selected rungs in the routine. When the MCR instruction is programmed as shown in Figure 10–2, any rungs that follow the MCR can only be true if the MCR instruction is *ON* or true.

Note: MCRs are sometimes called Master Control Resets because they reset the outputs to zero or *OFF*.

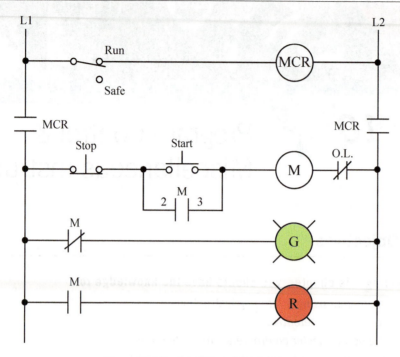

Figure 10–1 Hardwired Master Control Relay

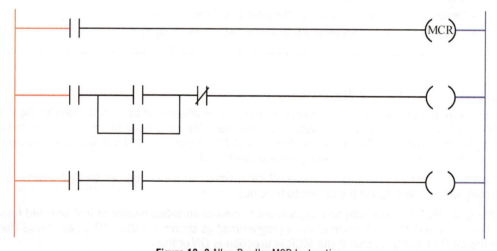

Figure 10–2 Allen-Bradley MCR Instruction

Allen-Bradley MCR Instruction

As shown in Figure 10–2, when the MCR instruction is true, the outputs in the rungs that follow are executed normally. If the MCR instruction is false, the outputs in the rungs below the MCR are also false and cannot be set true or *ON* even if the programmed logic for each rung is true. The exception is retentive outputs such as latching relay instructions. Latching relay instructions will remain *ON* even when the MCR instruction is false. Latching relay instructions are covered later in this chapter.

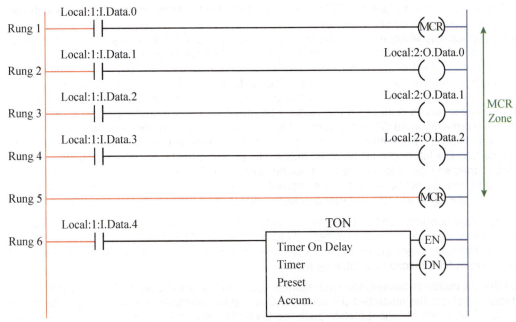

Figure 10–3 Two MCR Instructions Used to Create an MCR Zone

To create a zone within a program routine that will turn *OFF* all non-retentive outputs within the zone, two MCR instructions are used. Figure 10–3 shows how the two MCR instructions are programmed. Rung 1 has the first MCR instruction that is controlled by input device Local:1:I. Data.0. Note that the second MCR instruction has been programmed in Rung 5 with no logic elements preceding the output instruction. When an instruction is programmed in this manner, it is said to be programmed *unconditional.* This second MCR instruction is used to end the zone controlled by the first MCR instruction in Rung 1.

When input device Local:1:I.Data.0 is true, the MCR instruction will also be true and the rungs between the two MCRs instructions (rungs 2, 3, and 4) will be actively scanned by the processor. As long as the MCR in Rung 1 remains true, the rungs below the MCR instruction will control their outputs. When the MCR instruction in Rung 1 goes false, all output instructions (except retentive) between the two MCR instructions will be turned *OFF*, regardless of the logic of the individual rungs. Rung 6 is outside the zone controlled by the MCR instructions and is not affected.

Additional MCR zones can be created within the program routine by using pairs of MCR instructions. An MCR zone cannot be nested within another MCR zone.

Safety Note: A programmed MCR must never be used to replace a hardwired emergency stop or master control relay that provides emergency shutdown functions. You should still install a hardwired master control relay to provide output power shutdown during an emergency.

Note: While MCR instructions are still available, they are rarely used today. They are shown here to demonstrate how they function in the event a technician encounters MCR instructions in an existing program.

Latching Relay Instructions

Before discussing how a latching relay instruction works in a PLC, it may be helpful to review traditional hardwired latching relays.

Latching relays are used when it is necessary for contacts to stay open and/or closed although the coil is only energized for a short time by a momentary signal. Latching relays are often used for lighting applications where multiple circuits are needed for the lights in a large room or auditorium. Instead of having a switch for each circuit, a multiple contact latching relay is used. Each lighting circuit is wired to a set of contacts on the latching relay. One switch now controls the latching relay, which in turn controls several circuits of the lighting load. The latch and unlatch feature of the relay only requires three wires, so wiring is greatly reduced, and controlling the latching relay from multiple locations is quite simple. Another advantage of using a latching relay occurs when lighting is normally turned *ON* at the start of the work day and left *ON* until quitting time. Under these circumstances, if a normal relay is used, the coil needs to be energized all day long to keep the lights *ON*. The latching relay however, needs only to be momentarily energized to latch, or turn *ON* the connected load. The relay remains latched even though the coil to the latching relay is no longer energized. By not having the coil energized all day, there is a small energy savings and more importantly, the lights will turn on after a power outage if they were *ON* before the outage.

Latching relays normally use two coils: one to *latch* and one to *unlatch*. There is a mechanical linkage that holds the relay in the latched, or closed, position. When the unlatch coil is energized, the coil action disengages the mechanical latch and allows the relay to open. Figure 10–4 shows the wiring diagram for a mechanical latching relay.

When the *ON* button is pushed, the latch coil energizes and opens the normally closed latch (L) contacts and closes the unlatched (U) contacts. Opening the normally closed L contacts de-energizes the L coil. The length of time it took to push the *ON* button and to energize the latch coil was only a fraction of a second. During the short time the latch coil energized, it closed the normally open CR contacts, completing the circuit to the lamp. The CR contacts remain closed even though the latch coil de-energized because of the mechanical latch mechanism. To open the mechanically latched contacts to turn the light *OFF* requires pushing the *OFF* button. The U contacts in the unlatch coil circuit are now closed. Pushing the *OFF* button energizes the unlatch coil, which in turn closes the normally closed L contacts, and opens the U contacts, which de-energizes the unlatch coil. For the brief instant that the unlatch coil is energized, it releases the mechanically latched CR contacts so they can open and turn the light *OFF*.

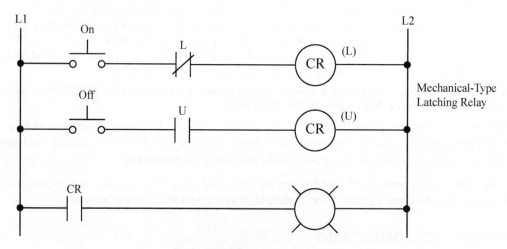

Figure 10–4 Mechanical Latching Relay

Figure 10–5 Programmed Latch and Unlatch Rungs

Mechanical latching relays can be replaced by programming PLC latching and unlatching instructions. Like the internal (dummy) relays discussed earlier, the programmed internal latching relays do not exist as real-world devices but can perform all the same functions as hardwired latching relays.

Programmed latch and unlatch instructions, like their physical real-world counterparts, are **retentive** during a power failure. Should the PLC lose power or be taken out of run mode, all discrete outputs are turned *OFF* but the state of the latch instructions are retained in memory. When power is restored or the PLC is switched back to the run mode, the outputs that were *ON* previously will return to their *ON* state.

Figure 10–5 shows latch and unlatch rungs as they would be programmed using the Allen-Bradley Output Latch (OTL) and Output Unlatch (OTU) instructions.

The OTL instruction is a retentive output instruction that can be programmed to turn *ON* an output. This instruction cannot be used to turn an output *OFF*. Once an output has been turned *ON* by a latch instruction, the output unlatch instruction (OTU) must be used to turn the output *OFF*. As these two instructions must be used in pairs, it follows that they will be programmed with the same memory address. Notice in Figure 10–5 that both the latch (L) and unlatch (U) outputs have the same address (Local:2:O.Data.0).

Output Local:2:O.Data.0 will be set to 1, or *ON,* when the latch rung is true (input Local:1:I.Data.5 true) and will be cleared to 0 or turned *OFF* when the unlatch rung is true (Local:1:I.Data.6 true). Like hardwired latching relays, only a momentary closure of input device Local:1:I.Data.5 is required to latch output Local:2:O.Data.0 *ON.* Output Local:2:O.Data.0 remains latched, or *ON*, until the unlatch output (U) rung is true by closing input Local:1:I.Data.6.

Normally, an internal memory bit (dummy relay) is used for the latch and unlatch memory address, rather than an actual discrete output address. If a discrete output address is used, the output, once latched, remains *ON,* even in an MCR zone.

Jump and Label Instructions

Used in combination, these two instructions allow for skipping over portions of the program logic to improve scan times when the logic is not being used. If there is a portion of the program logic that is not needed, it can be jumped over or bypassed until it is needed again. By jumping over portions of the program, overall program scan times can be reduced in applications where scan time may be critical for reliable operations. The jump instruction (JMP) tells the processor to jump over a portion of the program. Where to jump is controlled by the label instruction (LBL). Figure 10–6 shows a jump and label instruction. When the JMP instruction is true, the processor will jump over Rungs 3 and 4 and go directly to Rung 5, as shown.

The label (LBL) and jump (JMP) instructions must be assigned a unique label name within a routine. The label name can have as many as 40 characters. The label name can have no spaces, only letters, numbers, and underscores. In Figure 10-6, when the JMP instruction is true, the processor is instructed to jump all successive rungs until it reaches the rung that contains the label instruction (LBL) with the same label name as the jump instruction. In this illustration, only two rungs are jumped. In actual practice, any number of rungs can be jumped in the same routine. The LBL must be the first instruction on the rung.

The jump and label instructions can be used to jump forward or backward in the program, depending on need. Jumping backward increases the total scan time.

Note: Jumping backward an excessive number of times could increase the scan time to a point where the watchdog timer will time out (the processor has a watchdog timer that is reset on each scan). If the scan time exceeds the watchdog timer's preset time, the PLC will declare a major fault.

Safety Note: When a portion of the program is jumped over, the outputs located within that portion will remain in their last state until scanned again by the processor.

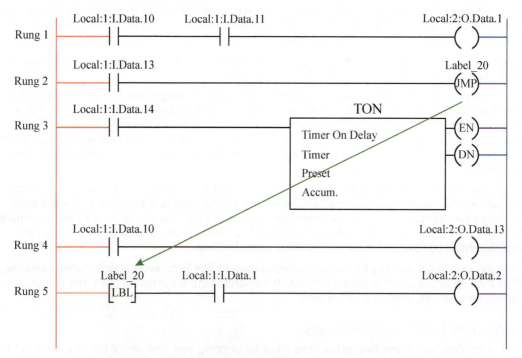

Figure 10–6 Programming the Jump (JMP) and Label (LBL) Instructions

Jump to Subroutine, Subroutine, and Return Instructions

The jump to subroutine, subroutine, and return instructions are used to direct the processor to a different routine (subroutine), scan it, return to the sending routine and continue scanning. The formats used by the various PLC manufacturers to program these instructions vary widely, and for this reason, only instruction blocks are shown in Figure 10–7. The blocks are identified using the Allen-Bradley mnemonics: jump to subroutine (JSR), subroutine (SBR), and return (RET). Subroutines are very valuable for program organization, and for using blocks of programming logic over and over by simply changing the variables used in the subroutine. To use this group of instructions, consult the programming instruction guide for the PLC system you are using.

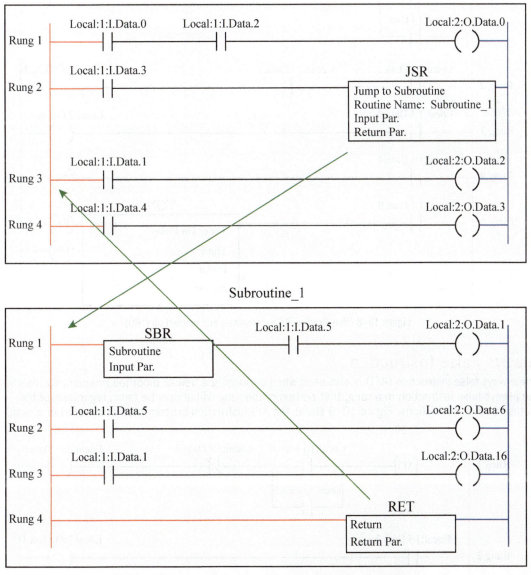

Figure 10–7 Jump to Subroutine, Subroutine, and Return Instructions

Temporary End Instruction

The temporary end instruction (TND) is used to place a temporary end to a routine. When the TND instruction is true, the processor stops scanning the routine and moves to the end of the current routine. This instruction is often used when a new program is being debugged for the first time. It allows for portions of the routine to be checked without running the entire routine. Figure 10–8 shows a TND using the Allen-Bradley format. When the TND instruction is true, the processor stops scanning at Rung 4 and does not scan Rung 5.

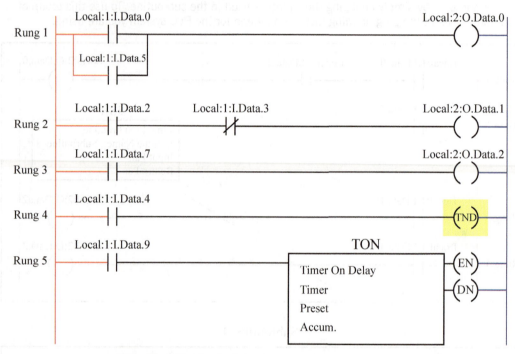

Figure 10–8 Allen-Bradley PLC—Temporary End Instruction (TND)

Always False Instruction

The always false instruction (AFI) is also used when debugging a new or modified program. By inserting the always false instruction in a rung, that portion of the rung will always be false, regardless of the status of other instructions. Figure 10–9 shows the AFI instruction programmed at the start of a rung.

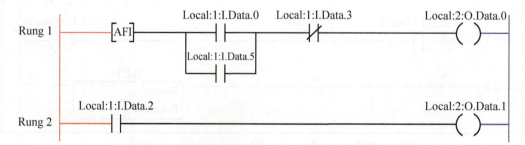

Figure 10–9 Always False Instruction

One-Shot Instruction

The one-shot instruction (ONS) is an input instruction that makes the rung true for just one program scan, based on a false-to-true transition of the instruction(s) that precedes the one-shot instruction. Figure 10–10 shows a rung of logic with the one-shot instruction programmed after input device Local:1:I.Data.4, which is controlling output Local:2:O.Data.12.

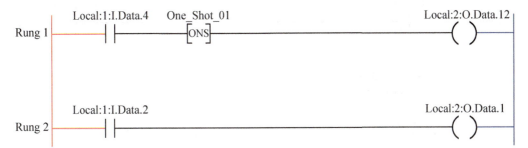

Figure 10–10 One-Shot Instruction

When the rung is programmed as shown, the output is turned *ON* for one scan, and one scan *only*, when input Local:1:I.Data.4 makes a false-to-true transition. The output cannot be turned *ON* again until the input device is first opened, then closed again, making a false-to-true transition. With the next false-to-true transition, the output device is again only turned on for one scan of the program. This is a beneficial instruction when an output operation is to be executed only once (one scan). The ONS instruction must be assigned a unique Boolean tag address. This tag address is used by the PLC to store the status of the ONS instruction and is referred to as the storage bit.

Math operations, data or word moves, and the like, are completed only once if a one-shot instruction is put in series with these output instructions. The one-shot instruction is often used with timers and counters for changing preset and accumulated values. This technique is discussed further in Chapter 13.

One-Shot Rising Instruction

The one-shot rising instruction (OSR) is an output instruction that sets its output bit *ON* for one program scan when the rung condition preceding the OSR instruction transitions from false-to-true. This instruction functions in the same way that the ONS instruction did, except it is an output instruction. It requires two parameters (bit tags) to be assigned: one is the "Output Bit" and the other is the "Storage Bit". The output bit is used within your program and the storage bit is used by the PLC to store the status of the instruction. The benefit of the OSR instruction over the ONS instruction is that the output bit of the OSR can be used anywhere within your program routine as needed. Figure 10–11 shows a rung of logic with the OSR instruction programed as an output instruction after input device Local:1:I.Data.0. The next rung controls output Data:2:O.Data.0 when the output bit OSR_01 of the OSR instruction is true for one scan only.

One-Shot Falling Instruction

The one-shot falling instruction (OSF) is an output instruction that sets its output bit *ON* for one program scan when the rung condition preceding the OSF instruction transitions from true-to-false. This instruction is identical to the OSR except the output bit is only *ON* after the rung condition preceding the OSF instruction transitions from true-to-false. Thus, the name one-shot *falling*. Figure 10–12 shows the OSF instruction.

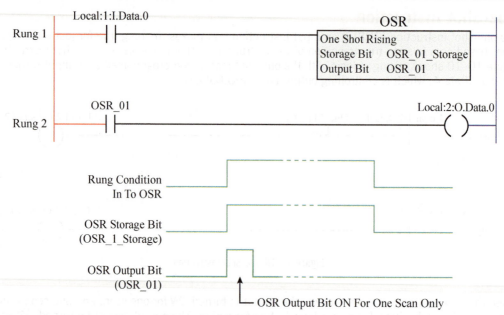

Figure 10–11 One-Shot Rising Instruction

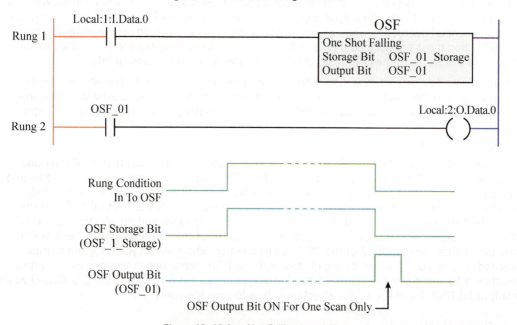

Figure 10–12 One-Shot Falling Instruction

Note: This chapter has covered some of the basic instructions that are available for programming a PLC. As the instruction sets vary with each manufacturer, it is necessary that the programming manuals be consulted to determine what instructions are available, what their mnemonics or designations are, and how to properly use them in a program.

Chapter Summary

Latching and master control instructions can be programmed to serve the same control functions as their real-world counterparts. Where personnel safety is a factor, a hardwired safety circuit should be used instead of depending on latching or MCR instructions alone. By using various PLC control type instructions, the user can program the processor to jump between specific rungs of logic in a routine or jump to subroutine programs and then return when finished. The temporary end and always false instructions can assist the user when testing and debugging the user's programmed logic. When used, the one-shot instructions limit the execution of logic to just once or one program scan. Although the mnemonics used by the various manufacturers may differ for each type of instruction, they however function generally in the same way. Once the technician has become familiar with instructions for one type of PLC controller, the transition to other PLC controllers is much easier.

Key Terms

unconditional
unlatch

latch
retentive

Review Questions

1. Will both programmed and real-world latching relays, if latched, remain latched if power is lost and then restored?

2. Latching relays are normally used when it is necessary for:
   a. contacts to open and/or close only while the coil is energized.
   b. contacts to open and/or close every 30 seconds.
   c. contacts to stay open and/or closed even though the coil is only energized a short time.
   d. none of the above.

3. Explain why an MCR is often used with off-delay timers.

4. An MCR can be used to control (check all correct answers):
   a. selected circuit rungs (networks).
   b. entire circuits.
   c. individual contacts within a rung (network).
   d. all of the above.

5. Define the term *unconditional*.

6. How does the jump to subroutine, subroutine, and return instructions work?

7. Define the term *retentive*.

8. What does the *jump and label instruction* do?

9. Give one reason why you might use a *temporary end instruction*.

10. What is the function of the *always false instruction*?

11. What is the function of a *one-shot instruction* and provide an example on how it could be used?

Chapter 11

Programming Timers

Learning Objectives

After completing this chapter, you should have the knowledge to:

- Describe how *pneumatic time delay relays* work.
- Write a program using *on delay* and *off delay* timers.
- Describe the difference between an *on delay timer* and a *retentive timer*.
- Describe how to extend the time range of timers by *cascading*.

Pneumatic Timers (General)

Although pneumatic time-delay relays have since been replaced by solid-state timers and PLCs, they were once used in vast numbers to provide timing functions in industrial controls. To better understand how PLC timers work, it is in the author's opinion that first understanding both the basic pneumatic time-delay relay and the standard symbols used can be very helpful.

Figure 11–1 shows a complete Allen-Bradley pneumatic timing relay, and Figure 11–2 shows a cutaway view of the contact and timing mechanism.

For the timer to time when power is applied (coil energized), the solenoid unit (coil), core piece, and armature are mounted so that the natural weight of the armature (10) pushes down on the operating plunger (11). This causes the bellows (1) and bellows spring (3) to collapse the bellows and dispel the air out through the release valve (9). When the coil is energized, the armature is attracted magnetically to the pole pieces, and lifts up and off the bellows assembly. Air now comes in through the air inlet filter, past the needle valve (2), and fills the bellows with air. The incoming air expands the bellows upward, pushing on the timing mechanism

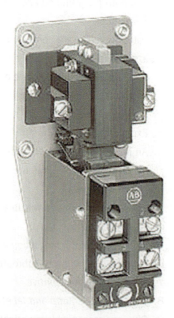

Figure 11–1 Pneumatic Timing Relay

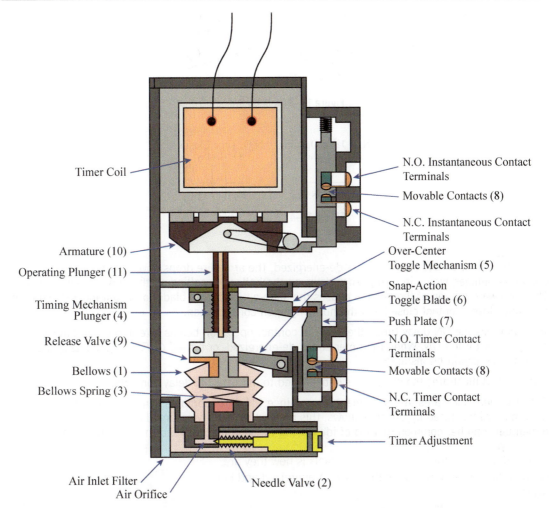

Timer Coil

N.O. Instantaneous Contact Terminals

Movable Contacts (8)

N.C. Instantaneous Contact Terminals

Armature (10)

Operating Plunger (11)

Over-Center Toggle Mechanism (5)

Snap-Action Toggle Blade (6)

Timing Mechanism Plunger (4)

Push Plate (7)

Release Valve (9)

N.O. Timer Contact Terminals

Bellows (1)

Movable Contacts (8)

Bellows Spring (3)

N.C. Timer Contact Terminals

Timer Adjustment

Air Inlet Filter
Air Orifice

Needle Valve (2)

Figure 11–2 Cutaway View of Pneumatic Timing Relay

plunger (4). As the plunger rises, it causes the over-center toggle mechanism (5) to move the snap-action toggle blade (6) upward. This picks up the push plate (7) that carries the movable contacts (8) to open the normally closed contact and close the normally open contact. The time it takes for the bellows to fill with air and activate the contact mechanism is controlled by adjusting the needle valve in the air orifice. The valve is adjusted with a screwdriver. A counterclockwise rotation moves the needle valve further into the air orifice, restricting airflow into the bellows, slowing the airflow, and increasing the time it takes for the bellows to expand and operate the contact mechanism. Conversely, clockwise adjustment of the needle valve decreases the time it takes the bellows to fill with air and activate the contacts after the armature has been lifted off the bellows mechanism.

When the contact action is delayed after the coil has been energized and the armature is lifted up and off the bellows mechanism, it is called ON delay.

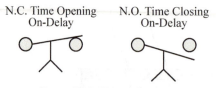

Figure 11–3 ON Delay Symbols

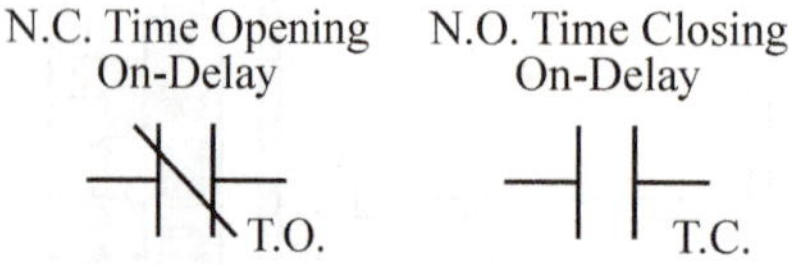

N.C. Time Opening
On-Delay

N.O. Time Closing
On-Delay

T.O.

T.C.

Figure 11–4 ON Delay Symbols

When the coil of an **ON delay timer** is de-energized, the armature drops down, pushing on the operating plunger, which in turn pushes down on the bellows expelling air through the release valve. The downward motion of the bellows causes the snap-action toggle blade to instantaneously snap the normally closed contact closed and the normally open contact open.

To summarize the ON delay timer, the delay in contact operation begins *after* the timer coil has been energized or turned *ON*. When the timer coil is de-energized, or turned *OFF*, the contacts go back to their normal condition instantly.

Figure 11–3 illustrates the electrical symbols used to indicate ON delay contacts.

The arrowhead indicates that movement is up. Since ON delay contacts can only time *after* the armature has lifted up off the bellows, this method of identifying timed contacts is easy to remember. Another common method of identifying timed contacts is shown in Figure 11–4.

Note: Remember that "normal" for contacts is how they are, open or closed, with the coil of the relay de-energized and time expired.

For a pneumatic timer to time when power is removed from the relay coil **(OFF delay timer),** the solenoid unit is mounted as shown in Figure 11–5. With a spring holding the armature up, no weight is applied to the bellows assembly, and the bellows are filled with air in a fully extended position.

When the coil is energized and the armature moves down, the armature pushes on the operating plunger. The plunger pushes on the bellows assembly, and all air is immediately forced out of the bellows through the release valve. This causes the snap-action contact assembly to instantly open the normally closed contact and close the normally open contact. The contacts stay in this configuration as long as the coil is energized, and the armature is holding the bellows mechanism down (compressed).

Figure 11–5 OFF Delay Pneumatic Timer

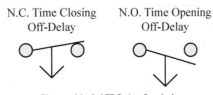

Figure 11–6 OFF Delay Symbols

When the relay coil is de-energized, or turned *OFF*, the spring on the armature lifts it up and off the operating plunger, which allows the bellows to start to fill with air. The normally closed contact remains open, and the normally open contact remains closed until the bellows are filled with enough air to activate the snap-action contact mechanism. When the contact mechanism has been activated, the normally closed contacts go closed and the normally open contacts go open.

Figure 11–6 shows the electrical symbols for OFF delay contacts.

To avoid confusion when reading electrical drawings with OFF delay contacts, it must be remembered that normal refers to the coil *after* it has been de-energized (turned *OFF*), and the time set for the timer has elapsed. The other symbols used for OFF delay contacts are shown in Figure 11–7.

Figure 11–8 compares both types of symbols used for ON delay and OFF delay timer relays.

Reviewing the two types of symbols commonly used in motor control diagrams, an electrician or technician should have no trouble determining the type of timing relay (ON or OFF delay) used, or what is normal (open or closed) for the timed contacts.

The basic pneumatic timing relay is designed so that additional instantaneous contacts may be added. The instantaneous contacts operate when the coil is energized or de-energized independent

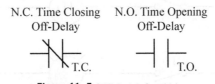

Figure 11–7 OFF Delay Symbols

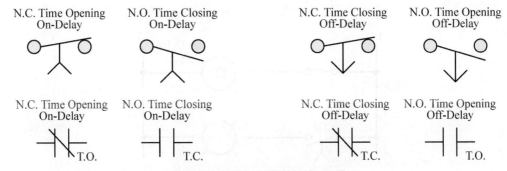

Figure 11–8 ON and OFF Delay Symbols

N.C. Instantaneous Contacts
Time Delay Relay

N.O. Instantaneous Contacts
Time Delay Relay

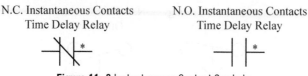

Figure 11–9 Instantaneous Contact Symbols

of the timing mechanism. Figure 11–9 shows the electrical symbol for contacts with an asterisk (*), which is sometimes used to indicate instantaneous contacts of a timing relay.

Figure 11–10a shows a simple light circuit controlled by an ON delay timer set for 5 seconds. The amount of delay is written near the timer coil on the diagram for understanding and troubleshooting. Figure 11–10b shows that when S^1 is closed, the coil of the pneumatic timer energizes, lifts the armature up and off the bellows, and the timing starts. Figure 11–10c shows the circuit after 3 seconds have elapsed (not enough time for the timer to time out) with the lamp circuit still open. After 5 seconds have elapsed (Figure 11–10d), the normally open time closing contacts close, and the lamp lights. As long as S^1 remains closed, the timer coil is energized, and the timed contacts stay closed. When S^1 is opened (Figure 11–10e), the coil circuit is broken, and the coil de-energizes. This causes the timed contacts to open, thereby turning *OFF* the lamp. The timed contacts will open the instant the coil de-energizes because they are timed only when power is applied to the coil.

Figure 11–11a shows the same circuit but with an OFF delay timer. When S^1 is closed (Figure 11–11b), the TD coil energizes, drawing the armature down and compressing the bellows. This causes the normally

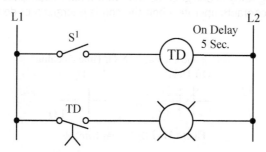

Figure 11–10a ON Delay Timer Circuit

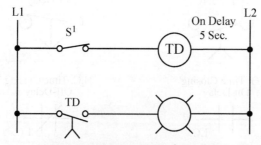

Figure 11–10b The Instant S^1 Is Closed

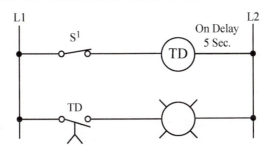

Figure 11–10c Three Seconds After S¹ Is Closed

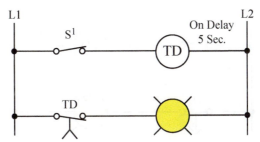

Figure 11–10d Five Seconds After S¹ Is Closed

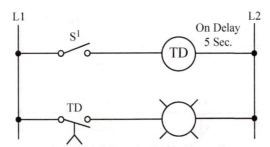

Figure 11–10e The Instant S¹ Is Opened

open OFF delay contacts to go closed instantly, and the lamp lights. When S¹ is opened (Figure 11–11c), the TD coil is de-energized, the spring-loaded armature is lifted up and off the bellows, and the 5-second timing begins. Figure 11–11d shows the circuit after 3 seconds have elapsed. The lamp remains energized until the full 5 seconds have elapsed, and the normally open contacts time out and open (Figure 11–11e).

Instead of a bellows assembly, like the pneumatic time delay relay, PLC timers use internal solid-state circuitry (clocks) for timing intervals or time base. The various PLC manufacturers use varying approaches for the actual programming of timers. The examples presented in this chapter will focus on timers in the Allen-Bradley Logix controllers. Although Allen-Bradley timers will be shown, the basic timing functions are the same in almost all PLCs.

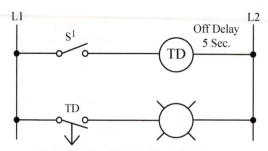

Figure 11–11a OFF Delay Timer Circuit

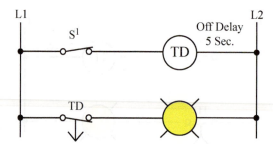

Figure 11–11b The Instant S^1 Is Closed

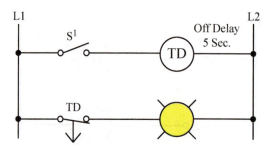

Figure 11–11c The Instant S^1 Is Opened

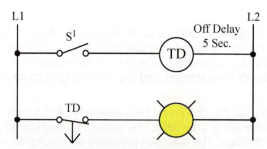

Figure 11–11d Three Seconds After S^1 Is Opened

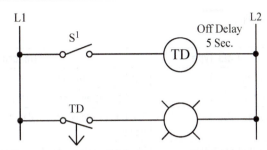

Figure 11–11e Five Seconds After S¹ Is Opened

Allen-Bradley Timers

Figure 11–12 shows the timer instruction format used in Allen-Bradley's Logix 5000 PLCs. The timer consists of a timing block containing the timer tag (timer address), the preset, and accumulated times. The time base is 1 millisecond for Logix 5000 timers. For example, for a 1-second timer, you would enter 1000 for the preset value. The preset time can be programmed with any value from 0–2,147,483,647 (DINT). The primary difference between the Logix 5000 timers and other PLC timers, is the fixed time base. The time base on some PLC timers is selectable, allowing you to select the time base that best fits your timing application.

The two lines to the right of the timing instruction are the enable (EN) and done (DN) outputs that indicate the current status of the timer instruction. The timer tag (address) uniquely identifies the timer and is used to store the status, present, and accumulated values. The structure of a timer tag is shown in Table 11–1.

The timer block is an output instruction and the enable (EN) bit is set to 1, or turned *ON*, when there is a true logic path to the timer block (rung goes true) and remains *ON* until the rung goes false. The enable (EN) bit can be used as an instantaneous contact within your ladder logic routine as needed.

The done (DN) bit is set to 1, or turned *ON*, when the accumulated value is equal to or greater than the preset value (accumulated value => preset value) and remains *ON* until the logic path to the timer block goes false or a reset instruction resets the timer. When the done (DN) bit is set to 1, or turned *ON*, this is often referred to as "the timer has timed out".

The timer timing (TT) bit is set to 1 (or true) whenever the timing operation is in progress and the accumulated value is less than the preset value. The timer timing (TT) bit could be used to control

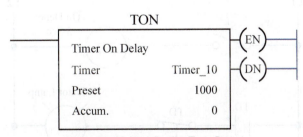

Figure 11–12 Allen-Bradley Timer Format

Table 11–1 Logix 5000 Timer Structure

Mnemonic	Data Type	Description
(tag name).EN	Boolean (BOOL)	The timer instruction enabled.
(tag name).TT	Boolean (BOOL)	The timer is enabled and timing (accumulated < preset).
(tag name).DN	Boolean (BOOL)	The timer is done, accumulated value is equal to the preset value.
(tag name).PRE	32-Bit Word (DINT)	The preset value of the timer in milliseconds. The value that the accumulated value must reach in order for the timer to be done (DN).
(tag name).ACC	32-Bit Word (DINT)	The current accumulated value of the timer. The number of milliseconds that has elapsed since the timer instruction was enabled (EN).

a timer timing light that is only *ON* when the timer is actually timing for example. Figure 11–13a shows how the TT bit is used to control an indicator light, and Figure 11–13b shows the equivalent circuit using a pneumatic timer.

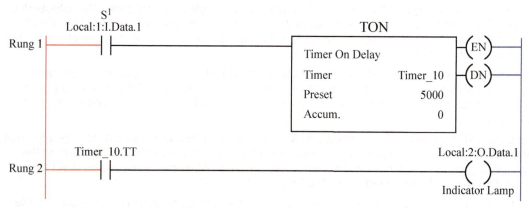

Figure 11–13a TT Bit Used to Control an Indicator Lamp

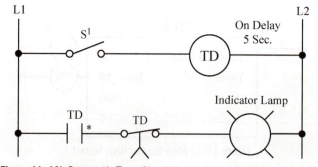

Figure 11–13b Pneumatic Timer Circuit Used to Control an Indicator Lamp

Note: The EN, TT, and DN bits can be used to control an output or other logic within a program routine as desired.

Figure 11–14 shows an ON delay timer instruction (TON) and how it is programmed to control outputs Local:2:O.Data.1, 2, 3, and 4.

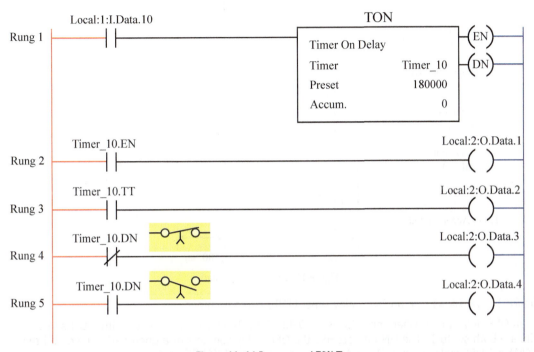

Figure 11–14 Programmed TON Timer

When input Data:1:I.Data.10 in Rung 1 is true, or set to 1, the processor starts timer "Timer_10" and sets the EN and TT bits to 1. This turns *ON* outputs Data:2:O.Data.1 and Data:2:O.Data.2 in Rungs 2 and 3. The accumulated value begins to increase in 1 millisecond intervals. The output in Rung 4, controlled by an EXAMINE OFF instruction, is true as long as the accumulated value is less than the preset value. The EXAMINE OFF instruction addressed with the timer DN bit acts like a normally closed time-opening contact and does not open until the accumulated value is equal to or greater than the preset value. The EXAMINE ON instruction in Rung 5 with the DN bit address acts like a normally open time-closing contact and does not close (or go true) until the accumulated value is equal to or greater than the preset. When the accumulated value does equal the preset value, the DN bit is set to 1, and output Local:2:O.Data.4 is turned *ON* and output Local:2:O.Data.3 is turned *OFF*. Once the timer instruction has completed timing, the TT bit is reset to 0 and output Local:2:O. Data.2 in Rung 3 is turned *OFF*. The timer instruction is reset when the input Local:1:I.Data.10 in Rung 1 is no longer true. Like a pneumatic ON delay timer, when power is removed, the timer instruction is reset (non-retentive).

Figure 11–15 shows a typical timing chart. Notice that when the rung condition is true (*ON*), the timer will time, but if the rung goes false (*OFF*), the timer resets to 0, as illustrated during the first 2 minutes of the timing diagram.

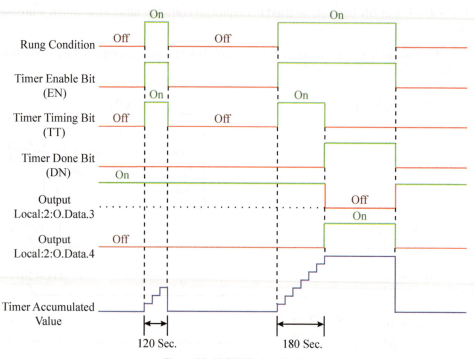

Figure 11–15 TON Timing Chart

Figure 11–16 shows how an OFF delay timer (TOF) is programmed.

In an OFF delay timer, when input Data:1:I.Data.10 in Rung 1 is set to 1, the timer DN and EN bits are both set to 1. The DN bit acts like the OFF delay contacts of a pneumatic timer, and the EXAMINE OFF instruction in Rung 4 goes false, while the EXAMINE ON instruction in Rung 5

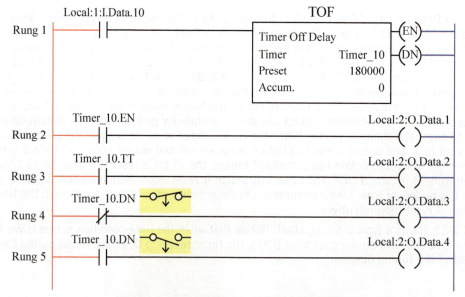

Figure 11–16 Programming an OFF Delay Timer (TOF)

goes true. When input Data:1:I.Data.10 is reset, or set to 0, Rung 1 goes false, and the timer starts to accumulate time in 1-millisecond increments as long as the rung remains false. When the accumulated value equals the preset value, the timer stops timing. The TT bit was set to 1 while the timer was timing and output Local:2:O.Data.2 in Rung 3 was *ON*. When the accumulated value equaled the preset value and the timer stopped timing, the TT bit was reset to 0 and output Local:2:O.Data.2 was turned *OFF*. When the TT bit is reset to 0, the DN bit is also set to 0, and output Local:2:O.Data.3 in Rung 4 is turned *ON* and output Local:2:O.Data.4 in Rung 5 is turned *OFF*. Figure 11-17 shows a typical timing chart for an OFF delay timer.

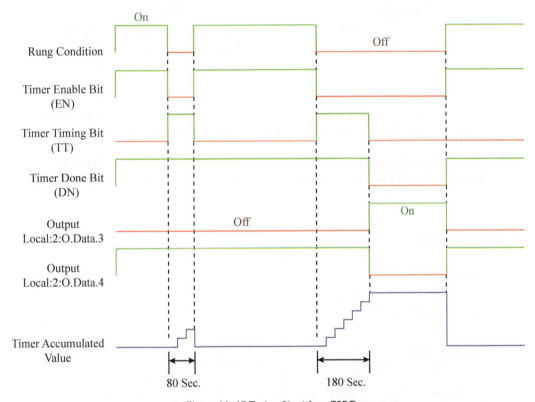

Figure 11–17 Timing Chart for a TOF Timer

During the first timing cycle, the timer was only *OFF* for 80 seconds. That was not long enough for the timer to time out, so the outputs controlled by the DN bit did not change state. During the second timing cycle, the timer was allowed to time out and the outputs changed state.

Retentive timers (RTO) can be programmed to replace motor-driven timers. The retentive timer lets the timer START and STOP without resetting the accumulated value to 0. The bits associated with the timer EN, TT, and DN function the same as the TON instruction. The RTO instruction begins timing when the rung goes true. As long as the rung remains true, the timer continues to time until the accumulated value reaches the preset value. If the timer rung goes false, the timer holds the accumulated time, rather than resetting the accumulated value to 0. When the timing rung goes true again, the count picks up from where it was, and continues to accumulate time. Once the accumulated time is equal to or greater than the preset time, the processor will set the DN bit to 1. The DN bit remains *ON* (or set to 1) as long as the accumulated value is equal to or greater than the preset value.

Because the retentive timer does not reset the accumulated value when the timer is de-energized, a reset (RES) instruction must be used. The reset instruction must be given the same tag address as the retentive timer it is intended to reset.

A common problem in programs that have retentive timers is that the timer can appear to not be accumulating time, even though the timer rung is true. The problem is that a corresponding RES instruction for the retentive timer is likely also true, which prevents the timer from timing by continually resetting the timer. Figure 11–18 shows how the RTO timer is programmed, including the reset instruction (Rung 3), and shows a typical retentive timer timing chart.

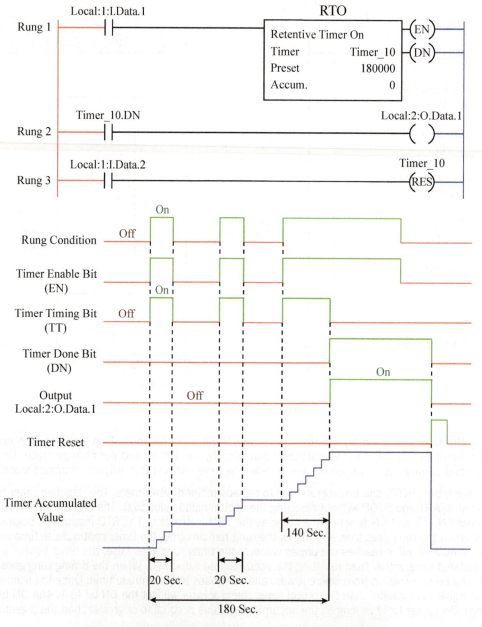

Figure 11–18 Programmed Retentive Timer (RTO) and Timing Chart

Programming Allen-Bradley Logix 5000 Timers

Figure 11–19 shows a TON timer programmed in a Logix 5000 PLC. Note that the *Timer* parameter contains the tag (address) for the timer. In this example, tag name "Flasher" was assigned to the timer. Figure 11–20 shows the new tag window. By default, when creating new tags, they will be configured as program-scoped tags.

After the tag has been created, the preset and accumulated values can be entered. The finished TON timer is shown in Figure 11–21.

The Logix 5000 controllers have three additional timer instructions—TONR, TOFR, and RTOR—that have a built-in reset function. These timers are only available with Function Block programming, which will be covered in Chapter 18.

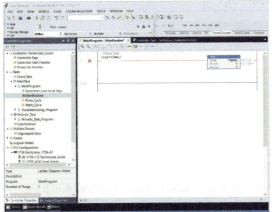

Figure 11–19 Logix 5000 TON Instruction

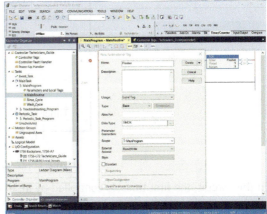

Figure 11–20 New Tag Dialog Box for Timer Tag "Flasher"

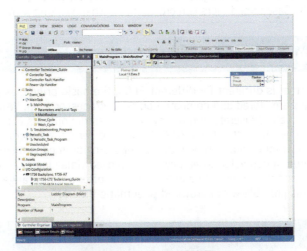

Figure 11–21 Finished TON Timer

Cascading Timers

When circuit requirements demand more time than is available from a single timer, two or more timers can be programmed together, as shown in Figure 11–22. Programming two or more timers together to extend the timing range is called **cascading**.

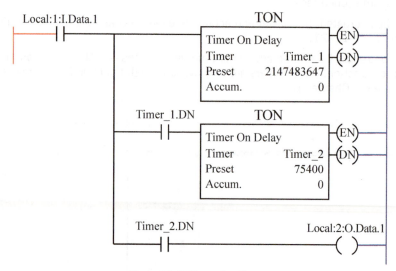

Figure 11–22 Cascading Timers

In this circuit, the first timer is controlled by input device Local:1:I.Data.1. When the device is true, the timer starts to time. When the accumulated time is equal to the preset time, the timer done bit is set to 1, or *ON*. When the done bit (DN) of timer "Time_1" is set to 1, the second timer (Timer_2) is enabled and begins timing. When the done bit of timer "Timer_2" is set to 1, output Local:2:O.Data.1 is turned *ON*. The total time to turn *ON* output Local:2:O.Data.1 was 2,147,559,047 milliseconds or approximately 35,792.6 minutes. As you can see, with a 32-bit word (DINT) the need to cascade timers is not very likely. Cascading timers is more common when working with 16-bit word controllers.

Chapter Summary

Although the timer format is different for different PLCs, the basic principles are the same. Preset and accumulated times are stored and compared on each processor scan. When the accumulated value equals the preset value, discrete output devices or internal outputs can be turned *ON* or *OFF*. Timers can be programmed for *ON delay* or *OFF delay*, or as *retentive* timers. The only limit to the number of timed and instantaneous contacts that can be programmed is memory size. Retentive timers retain their accumulated value when power is removed (false logic) and require the use of a reset instruction to reset the timers. Programmed timers offer a wider range of time settings and greater accuracy than is possible with hardwired pneumatic timers.

Key Terms

ON delay timer

retentive timers

OFF delay timer

cascading

Review Questions

1. Match the standard time delay symbols.

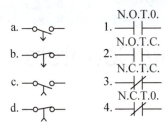

   a.
   b.
   c.
   d.

   N.O.T.O.
   1.
   N.O.T.C.
   2.
   N.C.T.C.
   3.
   N.C.T.O.
   4.

2. The amount of time for which a timer is programmed is called the:

   a. preset.

   b. set point.

   c. desired time (DT).

   d. all of the above.

3. As scan time increases, so does the accuracy of any programmed timers.

   T   F

4. When the timing of a device is not reset due to a loss of power, the timer is said to be:

   a. holding.

   b. secured.

   c. retentive.

   d. continuous.

5. When more time is needed than can be programmed with one timer, two or more timers can be programmed together. This programming technique is called:

   a. stacking.

   b. cascading.

   c. doubling.

   d. synchronizing.

6. When using Allen-Bradley timers, which mnemonic status bit will act as an instantaneous contact?

   a. DN

   b. TT

   c. EN

   d. IN

7. When the accumulated time is equal to the preset time, which mnemonic status bit in the Allen-Bradley timer will be true?

   a. DN

   b. TT

   c. EN

   d. IN

8. When programming a Logix 5000 timer, what preset value would be entered to create a 23-minute timer?

9. An Allen-Bradley off-delay timer (TOF) has been programmed for a delay time of 30 seconds. If the timer is enabled (true) for 15 seconds and then becomes disabled (false) for 45 seconds, show a timing chart for the enable (EN), timer timing (TT), and done (DN) status bits of the timer.

Chapter *12*

Programming Counters

Learning Objectives

After completing this chapter, you should have the knowledge to:

- Write a program using up and down counters.
- Define the terms *increment* and *decrement*.
- Define the terms *underflow* and *overflow*.

Programmed counters serve the same function as the mechanical counters used in the past. Programmed **counters** can count up, count down, or be combined to count up and down. Counters are similar to timers except they do not operate on an internal clock but instead are dependent on external or program sources for counting.

Allen-Bradley Counters

Allen-Bradley offers two types of counters, up counters (CTU) and down counters (CTD). Both counters are retentive until reset by a reset instruction (RES). Figure 12–1 shows a typical Allen-Bradley Logix 5000 up counter instruction.

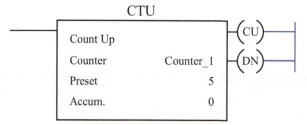

Figure 12–1 Allen-Bradley Counter Format

The Allen-Bradley counter instruction format is similar to the timer format. The counter instruction, CTU or CTD, consists of a counter block containing the counter tag name (address), the preset, and accumulated values. The counter's preset value can be set for any value between 0 and 2,147,483,647 (DINT). The two lines to the right of the counter instruction are the enable (CU or CD) and done (DN) output bits that indicate the status of the counter. The structure of a counter tag is shown in Table 12–1.

Table 12–1 Logix 5000 Counter Structure

Mnemonic	Data Type	Description
(tag name).CU	Boolean (BOOL)	The up counter instruction is enabled.
(tag name).CD	Boolean (BOOL)	The down counter instruction is enabled.
(tag name).DN	Boolean (BOOL)	The counter is done when the accumulated value is equal to or greater than the preset value.
(tag name).OV	Boolean (BOOL)	The accumulated counts exceeds the upper limit of 2,147,483,647 when counting up.
(tag name).UV	Boolean (BOOL)	The accumulated counts exceeds the lower limit of −2,147,483,648 when counting down.
(tag name).PRE	32-Bit Word (DINT)	The preset value of the counter. The value that the accumulated value must reach in order for the counter to be done (DN).
(tag name).ACC	32-Bit Word (DINT)	The current accumulated value of the counter.

The counter preset and accumulated values can be either positive or negative, −2,147,483,648 to +2,147,483,647. The positive numbers are stored in 32-bit binary, while the negative numbers are stored in 2s complement (2s complement was covered in Chapter 2). In almost all cases, counter preset values are entered as a positive number. Negative values are most often seen when the accumulated value is allowed to rollover the maximum (2,147,483,647) when counting up or under the minimum (0) when counting down.

The counter is an output instruction and the enable (CU or CD) bit is set to 1, or turned *ON*, when there is a true logic path to the counter instruction (rung goes true) and remains *ON* until the rung goes false. Since there are two counter instructions, *count up* (CTU) and *count down* (CTD), they each control their own enable status bit (CU or CD). Because down counters are rarely, if ever, programmed without a corresponding up counter, the same tag name (address) is used in both instruction blocks. This is why you see both enabled bits in the tag structure of a counter, as shown in Table 12–1. Counter instructions only count when there is a false-to-true transition of the instruction.

The counter done bit (DN) is *ON*, or set to 1, as long as the counter accumulated value is equal to or greater than the preset value. The DN bit is only reset to zero (0), turned *OFF*, when the accumulated count is less than the preset value.

In addition to the counter enable and done status bits, the counter instruction also has an *overflow* (OV) and *underflow* (UN) status bits. The counter OV bit is set by the processor to 1, or *ON*, when the accumulated count exceeds the upper limit of 2,147,483,647. When counting up and this limit is reached, the count wraps (rolls over) to −2,147,483,647 and the up-counter increments from there.

On the other hand, the counter UN bit is set by the processor to 1, or *ON*, when the accumulated count exceeds the lower limit of −2,147,483,648. When counting down and this limit is reached, the count wraps around to +2,147,483,647, and the countdown instruction counts down from there.

Figure 12–2 shows a CTU counter and how it is programmed to control outputs Local:2:O.Data.1, Local:2:O.Data.2, and Local:2:O.Data.3. Rung 5 is the reset rung that is used to reset the counter's accumulated value to zero. An output RES instruction is used to reset counters. The reset instruction must have the same tag name (address) as the counter to be reset.

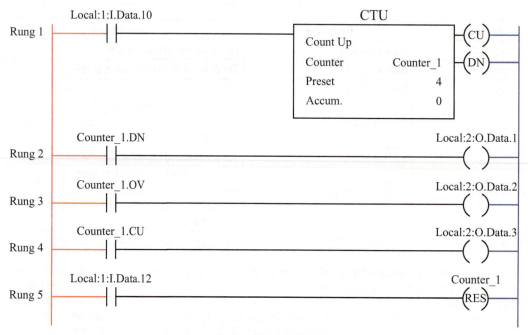

Figure 12–2 Programmed Up Counter

Each time input device Local:1:I.Data.10 in Rung 1 makes a transition from false to true, the counter **increments**, or counts up by 1. When the accumulated value (count) is equal to or greater than the preset, the done (DN) bit is set to 1 and Rung 2 becomes true, turning *ON* output Local:2:O.Data.1. Rung 3 is not true unless the count exceeds the counter's upper limit of +2,147,483,647. If the count exceeds the upper limit, output Local:2:O.Data.2 turns *ON* and remains *ON* until the counter is reset by closing input device Local:1:I.Data.12 in Rung 5. The count up bit (CU) can be programmed and used to indicate that the counter is enabled and that Rung 1 is true. The CU bit in Rung 4 is set to 1 by the processor any time that input device Local:1:I.Data.10 in Rung 1 is true and the counter is enabled. The CU bit is reset to 0 when Rung 1 goes false.

Figure 12–3 shows a typical counting chart for a CTU instruction.

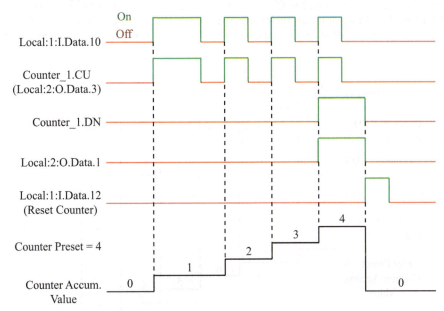

Figure 12–3 CTU Counting Chart

Figure 12–4 shows how a CTD is programmed.

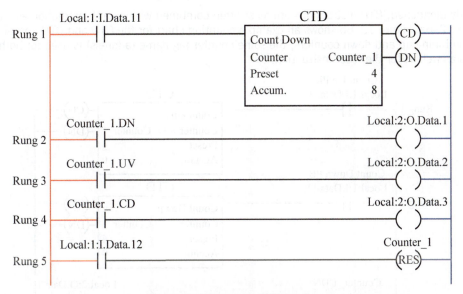

Figure 12–4 Programmed CTD Counter

The down counter counts down or **decrements** each time input device Local:1:I.Data.11 in Rung 1 goes from false to true. As long as the accumulated count is equal to or greater than the preset, the output in Rung 2 (Local:2:O.Data.1) remains *ON*. When the accumulated count falls below the preset of 4, output Local:2:O.Data.1 is set to 0, or *OFF*. Rung 3 contains the underflow bit, which is opposite the overflow bit used with the CTU counter and is only set to 1 when the count goes below −2,147,483,648. Rung 4 contains the CD bit and is *ON*, or true, any time the counter is enabled. The CD bit mirrors the status of input device Local:1:I.Data.11 in this example. Rung 5 is the reset rung and uses input device Local:1:I.Data.12 to reset the counter. Figure 12–5 shows the counting chart for a CTD instruction.

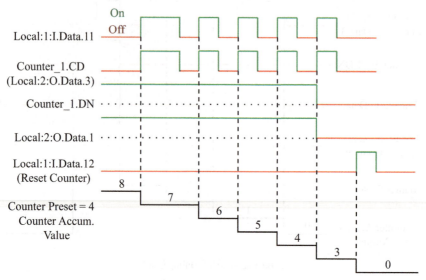

Figure 12–5 CTD Counting Chart

As already mentioned, CTD instructions are most often combined with a CTU instruction as shown in Figure 12–6a. Figure 12–6b shows an example counting chart for the CTD and CTU combination. When combining up and down counters, the same counter tag name (address) is used for both counters as well as for the reset instruction.

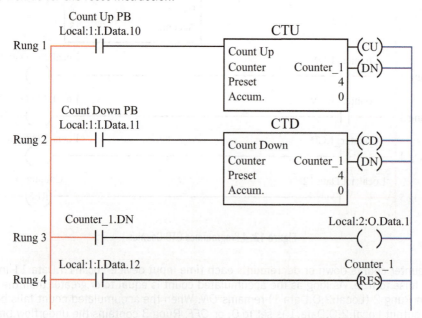

Figure 12–6a Combining CTU and CTD Instructions

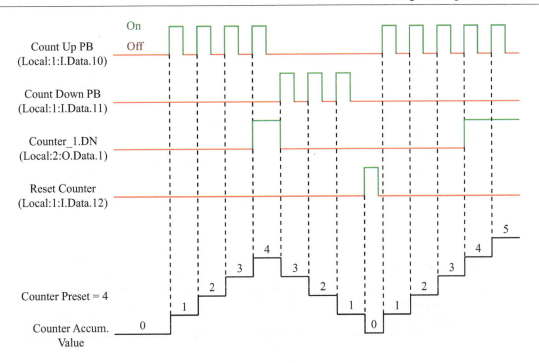

Figure 12–6b Combined CTU and CTD Counting Chart

Up and down counters can be programmed together as shown in Figures 12–7a and 12–7b to count products as they enter a conveyor line (count up) and as they leave the conveyor line (count down).

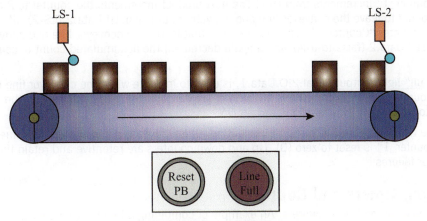

Figure 12–7a Applying Up and Down Counters

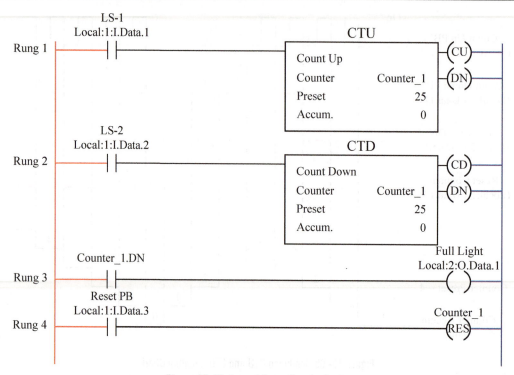

Figure 12–7b Up and Down Counter Logic

As a product enters the conveyor line, input Local:1:I.Data.1 (LS-1) is activated (false-to-true) and counter "Counter_1" increments from 0–1. The next product increments the counter to 2 and so on. The first product to leave the conveyor line and activate input Local:1:I.Data.2 (LS-2) will decrement the count in counter "Counter_1". The next product that leaves the conveyor line and activates input Local:1:I.Data.2 (false-to-true) would again decrement the accumulated count in counter "Counter_1".

The "Line Full" light, output Local:2:O.Data.1, is used to indicate when the conveyor line is full. When the counter accumulated value is equal to or greater than the preset value (25), the red "Line Full" light (output Local:2:O.Data.1) is turned *ON*.

Input device Local:1:I.Data.3 is a reset push button. When pushed, the accumulated value in counter "Counter_1" is reset to zero (0). Up and down counters are retentive and retain their values during power failures.

Combining Timers and Counters

Timers can be combined with counters. An example of combining a timer with a counter is shown in Figure 12–8.

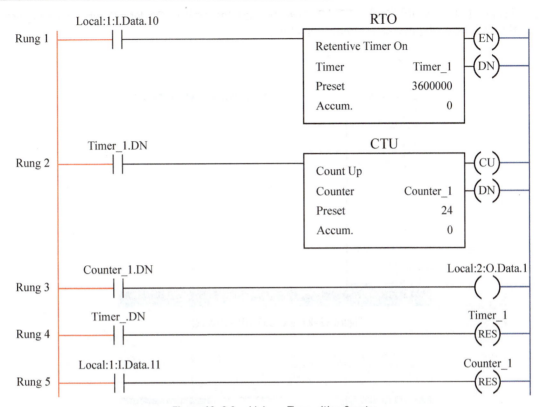

Figure 12-8 Combining a Timer with a Counter

The timer ("Timer_1") has a preset value of 3,600,000. The 3,600,000-millisecond preset value is equal to 1 hour. When input device Local:1:I.Data.10 is closed, the timer starts to time in 1-millisecond increments. When the accumulated time is equal to the preset value, the timer DN bit is set to 1 and the CTU counter counts, or increments, by one. The timer DN bit also resets the retentive timer and the timer starts to accumulate time again. When the accumulated time on the timer has reached 3,600,000 milliseconds, the timer increments counter "Counter_1" again and resets itself. The counter continues to count each time the timer DN bit makes a false-to-true transition (every 3,600,000 milliseconds, or 1 hour) until the accumulated count equals the preset value of 24. When the counter has counted to 24 (24 hours), the counter DN bit is set to 1 and output Local:2:O.Data.1 is turned *ON.* Input Local:1:I.Data.11 is used to reset the counter.

Note: Remember that the length of the program affects scan time, which in turn affects timer accuracy and total time. The actual time it takes for the counter to count to 24 hours may be slightly more than 24 hours.

Programming Allen-Bradley Logix 5000 Counters

Figure 12-9 shows a CTU counter being programmed in a Logix 5000 controller. Note that the *Counter* parameter contains the tag (address) for the counter. In this example, tag name "Counter1" was assigned to the counter. Figure 12–10 shows the new tag window. By default, when creating new tags, they will be configured as program-scoped tags.

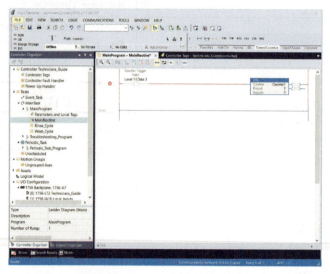

Figure 12–9 Logix 5000 CTU Instruction

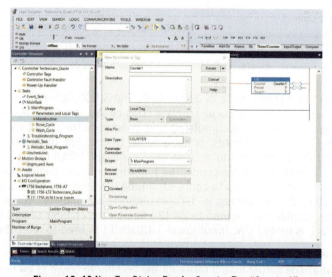

Figure 12–10 New Tag Dialog Box for Counter Tag "Counter1"

After the tag has been created, the preset and accumulated values can be entered. The finished CTU counter is shown in Figure 12–11. The tag structure is shown in Figure 12–12 for the tag "Counter1".

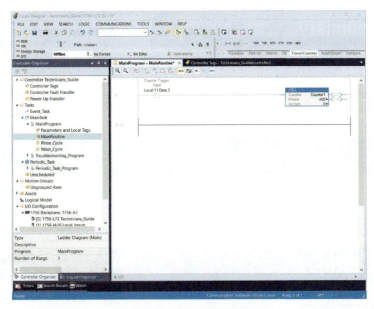

Figure 12–11 Finished CTU Counter

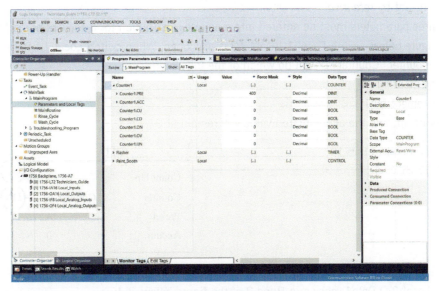

Figure 12–12 "Counter1" Tag Elements

The Logix 5000 controllers have one additional counter instruction, CTUD, that is both an up and down counter in one instruction with a built-in reset function. This counter is only available with Function Block programming, which will be covered in Chapter 18.

Chapter Summary

Programmed counters give added flexibility and control to process equipment and/or driven machinery. Like timers, counters store values in binary format for the preset and accumulated counts. The processor compares the preset and accumulated values on each scan of the rung and updates the counter's status bits as appropriate.

Key Terms

counters   increments

decrements

Review Questions

1. Define the term *increment*.
2. Define the term *decrement*.
3. What is meant by *retentive*?
4. What three parameters must be entered into a counter instruction for it to work properly?
5. In the figure below, switch Local:1:I.Data.10 is now open. When switch Local:1:I.Data.10 is closed, counter "Counter_1" will do which of the following?
   a.   increment by 1
   b.   decrement by 1
   c.   not count, and the accumulated value will remain at 0

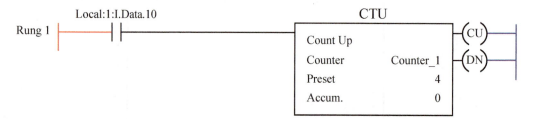

6. Output Local:2:O.Data1, shown in Rung 2 of the figure below, is true:
   a.   only when the count is equal to the preset value
   b.   when the count is equal to or greater than the preset value
   c.   when the count is less than the preset value
   d.   when the accumulated value reaches +2,147,483,647 and overflows
   e.   when the count goes to 11

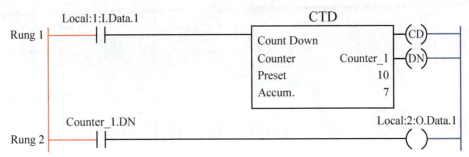

7. When the accumulated count is equal to or greater than the preset count, which counter status bit will be true?
   a. CU   b. CD
   c. OV   d. DN

8. When is the OV status bit set on an up counter?

9. Up and down counters can be programmed together to count up and down.
   T   F

10. The reset rung shown in the figure below resets counter "Counter_1":
    a.   automatically when the count reaches 10
    b.   automatically when the count reaches 11
    c.   only when the count reaches 2,147,483,647
    d.   only when switch Local:1:I.Data.12 is closed
    e.   only when switch Local:1:I.Data.12 is closed and then opened

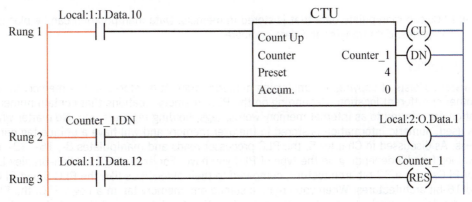

11. Define the term *overflow*.

12. Define the term *underflow*.

13. When an up counter accumulated value equals the preset value, the counter will:
    a.   reset itself
    b.   stop counting
    c.   continue to count
    d.   continue to count but go into an overflow condition as soon as the accumulated value exceeds the preset value

Chapter 13

Data Manipulation

Learning Objectives

After completing this chapter, you should have the knowledge to:

- Explain what data transfer is.
- Define the term *writing over*.
- Write a rung of logic that transfers data from one memory location to another.
- Identify the standard data compare instructions.
- Write logic that compares data to control an output.

Nearly all PLCs can manipulate data that is stored in memory. Data manipulation can be placed in two broad categories: data transfer and data compare.

Data Transfer

Data transfer consists of copying or transferring numeric information stored in one memory location into another or different location. Depending on the PLC, memory locations that contain numeric information are referred to as internal memory words, tags, holding registers, etc. No matter what name is used, numeric information is stored in the user memory and will have a unique tag name or address. As discussed in Chapter 5, the PLC processor reads and manipulates 8-, 16-, 32-, or 64-bit values (words) depending on the type of PLC you have. For example, the Allen-Bradley Logix family of PLCs have a 32-bit architecture, compared to their older PLCs like the PLC-5 and SLC-500 that had 16-bit architectures. When you create a standalone memory tag in a Logix PLC, the PLC allocates 4 bytes (32-bits) which is the minimum allocation for data, even if the data to be stored requires less than 4 bytes.

Figures 13–1a and 13–1b illustrate the concept of copying numerical data from one word location to another word location. Figure 13–1a shows the numeric (binary) data stored in tag "Memory_1" and "Memory_2".

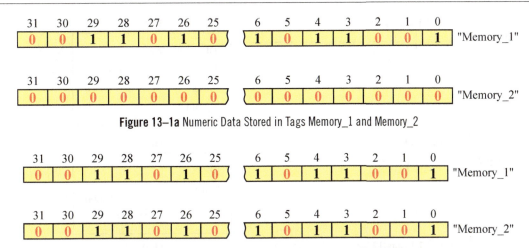

Figure 13–1a Numeric Data Stored in Tags Memory_1 and Memory_2

Figure 13–1b Data Transferred from Tag Memory_1 into Tag Memory_2

After the data transfer (Figure 13–1b), tag "Memory_2" now holds the exact or duplicate information that was in tag "Memory_1". If tag "Memory_2" had information already stored, rather than all zeros (0), the information would have been replaced. When new data replaces existing data in a word after a transfer, it is referred to as **writing over** the existing data.

Allen-Bradley Logix Data Transfer Instructions

The Logix family of PLCs use a move (MOV) instruction for moving data from one memory location to another. MOV is an output instruction that copies a value from one word (source tag) to another word (destination tag). When the rung that holds the MOV instruction is true, the instruction moves or copies data from the source tag into the destination tag on each processor scan that the rung is true. Figure 13–2 shows the MOV instruction format.

When input device Local:1:I.Data.1 is closed and the rung is true, the MOV instruction reads the data from the source tag and copies it into the destination tag leaving the source unchanged. In this case, the source and destination tags are DINT words. As long as the rung is true, the value stored in tag "Memory_1" will be copied into tag "Memory_2" each time the rung is scanned by the processor.

To illustrate how a MOV instruction could be used, consider a machine that produces two types of products. Product A requires a time delay of 10 seconds and product B requires a 20-second delay. The ON delay (TON) timer would be programmed as shown in Figure 13–3.

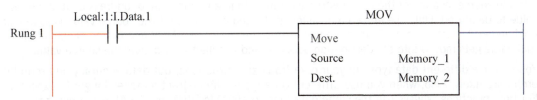

Figure 13–2 Move Instruction (MOV) Format

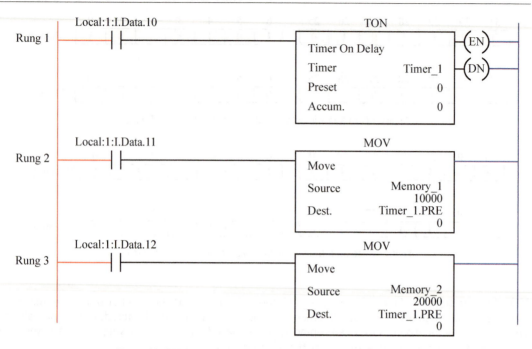

Figure 13–3 MOV Instruction Used to Change Timer Preset Values

The ON delay timer (Timer_1) is programmed initially with no preset value as shown in Rung 1. In Rung 2, when input device Local:1:I.Data.11 is true or *ON*, the move (MOV) instruction is programmed to copy the numeric value found in tag "Memory_1" (Source) into tag "Timer_1.PRE" (Destination). The destination word is the location that holds the preset value for Timer_1. Rung 3 is programmed to copy the value in tag "Memory_2" (Source) into the timer preset of "Timer_1" (Destination) when the rung is true.

When product A is being processed, input devices Local:1:I.Data.10 and Local:1:I.Data.11 are activated or true, and the numeric value of 10,000 in tag "Memory_1" is copied into the preset word of ON delay timer "Timer_1". This gives timer "Timer_1" a preset value of 10 seconds. When product B is to be processed, input device Local:1:I.Data.11 is opened and input device Local:1:I. Data.12 is closed or true (Rung 3), and the value in tag "Memory_2" (20000) is copied into the preset word of ON delay timer "Timer_1". The value 20,000 in tag "Memory_2" overwrites the previous value in the preset word for "Timer_1", which was 10,000.

In the MOV instruction, immediate values (constants) such as 250, 400, 5.5, etc., can be entered into the source location of the MOV instruction rather than a memory tag or address containing the value to be moved. Only use immediate values in the source location if the source value is to remain unchanged. In fact, entering immediate values into the source of a MOV instruction ensures that only that value gets moved into the destination address, and is called *hard coding* the source value.

You can use different data types in your data transfer instructions, but data accuracy and rounding errors are likely. Also, when you use different data types, the instruction takes longer to execute. For best practices, always use the same data type, preferable DINT and REAL when using the Allen-Bradley Logix instructions.

The move instruction (MOV) can be used to change preset values of timers or counters, as well as for transferring data between any two tags (words) to meet program requirements. An example of how an

MOV instruction can be used to change the preset value of a counter is in a program that counts boards in a saw mill. When the mill is producing 2x4s, it wants 400 boards in a stack. However, when the mill is producing 2x6s, only 250 boards are needed for a full stack. Figures 13–4a and 13–4b show how to change the preset value of an up counter for each lumber size by using push buttons.

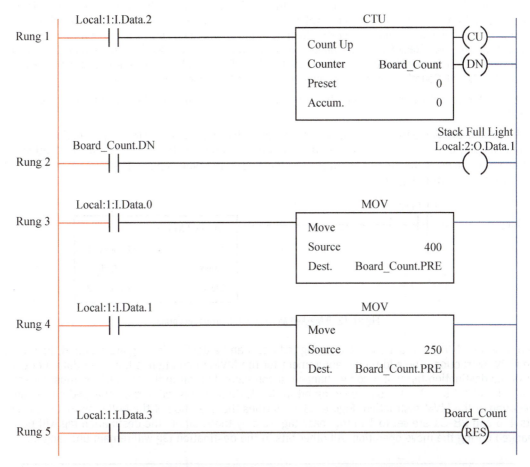

Figure 13–4a Changing Preset Values with an MOV Instruction

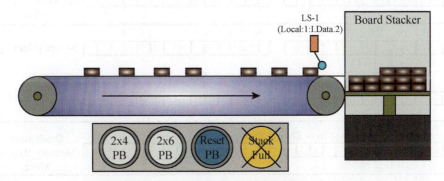

Figure 13–4b Pushbuttons Used to Change Preset Values

Up counter "Board_Count" is initially programmed with no preset value (0). The preset of the counter is determined by which pushbutton is pressed, Local:1:I.Data.0 or Local:1:I.Data.1. If 2x4s are to be counted, pushbutton Local:1:I.Data.0 is pushed, enabling the MOV instruction in Rung 3. When Rung 3 is enabled, or true, it tells the processor to copy (move) the value 400 into the preset word of counter "Board_Count". A single false-to-true transition of the MOV instruction is all that is required to enable the move. Each false-to-true transition of limit switch input Local:1:I.Data.2 will increment the board counter. When the board counter is equal to or greater than 400, the counter done bit "Board_Count.DN" will be set to 1, or true, and the "Stack Full" light will be turned *ON* in Rung 2. To reset the counter and start counting a new stack of boards, pushbutton Local:1:I.Data.3 in Rung 5 is used to enable a reset instruction (RES) for counter "Board_Count", causing the counter accumulated value to be reset to zero (0).

To change the preset value of counter "Board_Count" from 400 to 250, the pushbutton for 2x6s is pressed.

Another Allen-Bradley data manipulation instruction is the masked move (MVM) instruction. The MVM is an output instruction that copies a value from a source location to a destination location but allows portions of the data to be **masked** or blocked from being copied. The format for the MVM instruction is shown in Figure 13–5.

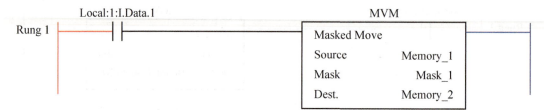

Figure 13–5 Masked Move (MVM) Instruction Format

To program an MVM instruction, a source tag (address) and a destination tag are required, just as in the MOV instruction. The additional requirement for the MVM instruction is the mask data. For each bit in the destination tag that is to be changed, a corresponding bit in the "Mask" tag must be set to a 1. If not set to a 1, the corresponding bit in the destination tag will not be changed during an execution of the MVM instruction. Figure 13–6 clarifies the operation of the MVM instruction. Only bits 4–6 and 28–31 are set to 1 in the mask tag, so only those bits in the destination tag will be changed during the move operation. All other bits in the destination tag will remain unchanged.

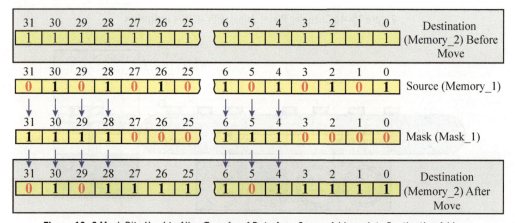

Figure 13–6 Mask Bits Used to Allow Transfer of Data from Source Address into Destination Address

The status of the bits (0 or 1) in the mask tag must be manually entered or changed in logic. The Mask can also be a decimal value that corresponds to the bits in the destination tag to be changed by the source. This is referred to as entering an immediate mask value. If you prefer to enter the mask using another data format, precede the value with the correct data type prefix. For example, if you want to enter the mask in hexadecimal, use the prefix "16#" followed by the hexadecimal value that corresponds to the mask bits to be set to 1. For binary data type use the prefix "2#". You can also enter the source in the MVM instruction as an immediate value if desired based on your application needs.

Data Compare

Data compare opens up a new realm of programming possibilities and demonstrates why PLCs are rapidly replacing most, if not all, hardwired control systems.

Data compare instructions, as the name implies, compare the data stored in two or more tags (words) and makes a true or false decision based on that comparison. Numeric values can be compared for *less than* (<), *equal to* (=), *greater than* (>), *less than or equal to* (≤), *greater than or equal to* (≥), and *not equal to* (≠) conditions, depending on the PLC. Data compare instructions are input instructions just like XIC and XIO instructions, except they compare two values for equality based on the instruction type.

The concept of comparing two values was in play when timers and counters were discussed previously. For example, the ON delay timer turns *ON* an output when the accumulated value equals the preset value (ACC = PRE). What happens is that the accumulated value is compared to the preset value on each scan the rung is true, and when the accumulated value is equal to the preset value (ACC = PRE), the output of the timer is turned *ON*.

Allen-Bradley Logix Data Compare Instructions

The Allen-Bradley family of PLCs have a set of data compare instructions that include *equal* (EQU), *greater than or equal* (GEQ), *greater than* (GRT), *less than or equal* (LEQ), *less than* (LES), *not equal* (NEQ), and *mask equal* (MEQ). Figure 13–7 shows how the EQU instruction is programmed.

The EQU instruction is true and turns *ON* output Local:2:O.Data.1, when the value in Source A is equal to the value stored in Source B. Sources A and B can be either numeric values or memory tags (addresses) that contain values. In Figure 13–7, the value in Source A is the accumulated value in timer "Timer_1", whereas the value in Source B is the numeric value 200. To use either the accumulated or preset values of timers and counters, a period is entered after the timer or counter's tag name followed by ACC or PRE, respectively.

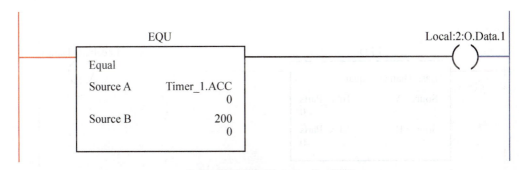

Figure 13–7 Equal To Instruction (EQU)

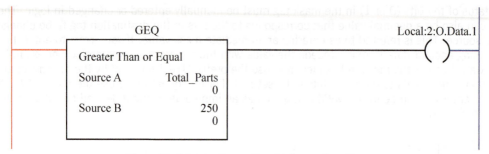

Figure 13–8 Greater Than or Equal To Instruction (GEQ)

Figure 13–8 illustrates how the GEQ instruction operates.

This instruction becomes true and turns output Local:2:O.Data.1 *ON* when the value in Source A is greater than or equal to the value in Source B. Again, the value that is in Source A or B can be numeric values or memory tags (addresses) that contain values. In this illustration, the value in Source A is the value stored in tag "Total_Parts". The value in Source B is the numeric value of 250.

The GRT (greater than) instruction is programmed as shown in Figure 13–9. This instruction is true when the value in Source A is greater than the value in Source B. In Figure 13–9, output Local:2:O. Data.1 is turned *ON* whenever the accumulated value in counter "Parts_A" is greater than the accumulated value in counter "Parts_B".

The less than or equal instruction (LEQ) is programmed as shown in Figure 13–10. This instruction is true whenever the value in Source A is less than or equal to the value stored in Source B.

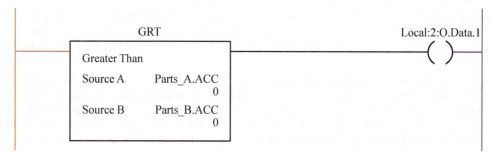

Figure 13–9 Greater Than Instruction (GRT)

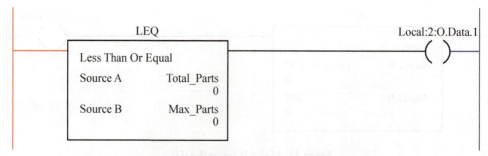

Figure 13–10 Less Than or Equal To Instruction (LEQ)

The LEQ instruction is true as long as the value in tag "Total_Parts" is less than or equal to the value in tag "Max_Parts". When the value in Source A is less than or equal to the value in Source B, output Local:2:O.Data.1 is turned *ON*.

The LES (less than) instruction is logically true when the value in Source A is less than the value in Source B. Figure 13–11 shows an LES instruction.

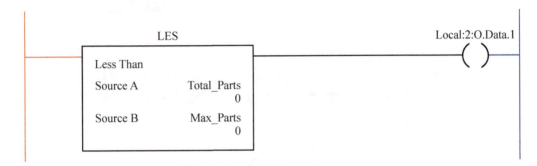

Figure 13–11 Less Than Instruction (LES)

In Figure 13–11, output Local:2:O.Data.1 is *ON* whenever the value in tag "Total_Parts" is less than the value in tag "Max_Parts" (Source B). When the value in Source A is equal to or greater than the value in Source B, the instruction is not logically true, and output Local:2:O.Data.1 is set to 0, or *OFF*.

The not equal instruction (NEQ) is programmed as shown in Figure 13–12. This instruction is true whenever the value in Source A is not equal to the value stored in Source B.

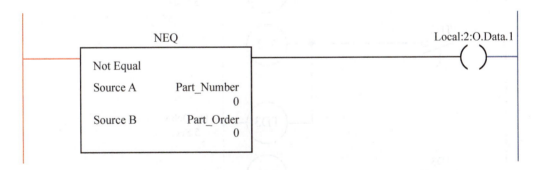

Figure 13–12 Not Equal To Instruction (NEQ)

The NEQ instruction will be logically true any time the value in tag "Part_Number" (Source A) is not equal to the value in tag "Part_Order" (Source B). If the values are not equal, output Local:2:O. Data.1 will be turned *ON*. This instruction is logically false only when the value in Source A is equal to the value in Source B.

To graphically demonstrate how data compare instructions can be used, consider the hardwired circuit in Figure 13–13. This circuit uses three pneumatic time delay relays to start a 4-motor conveyor system in inverse order (4–3–2–1).

The same circuit can be programmed with a PLC using only one timer and two data compare instructions, as shown in Figure 13–14.

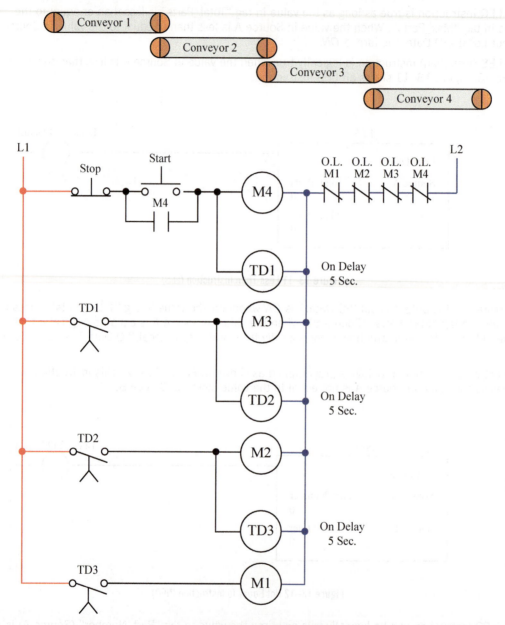

Figure 13–13 Hardwired Conveyor Control System

Assume that the STOP button (Local:1:I.Data.0) in Rung 1 is closed or true. When the START button (Local:1:I.Data.5) is pushed, output M4 (Local:2:O.Data.4) energizes and holding contacts Local:2:O.Data.4 close and hold the circuit in. When motor M4 starts, timer "Conv_Delay" is also enabled. The ON delay timer is preset to 15 seconds (15,000) and the timer's accumulated value is stored in tag "Conv_Delay.ACC".

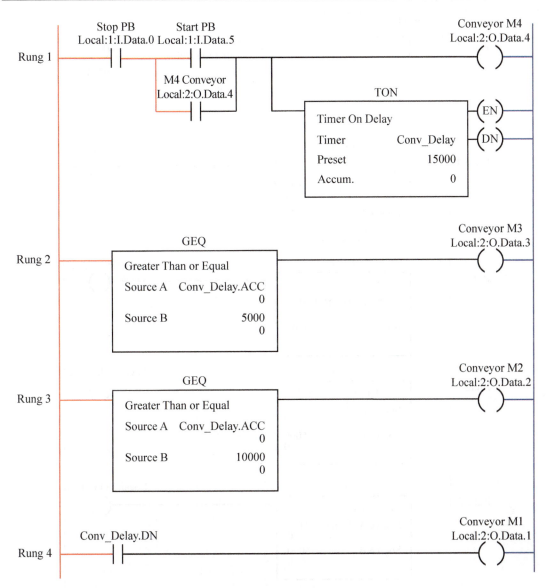

Figure 13–14 Data Compare Logic

The GEQ instruction in Rung 2 controls motor M3 (Local:2:O.Data.3) and is logically true when the accumulated value of timer "Conv_Delay" (Source A) is greater than or equal to 5000 or 5 seconds (Source B). Similarly, when the accumulated value reaches 10,000 or 10 seconds, motor M2 (Local:2:O.Data.2) is turned *ON*. When the accumulated value of the timer reaches 15,000 or 15 seconds, the done bit (DN) of timer "Conv_Delay" will be set to 1 and the last motor, M1 (Local:2:O.Data.1), will be turned *ON*.

To further illustrate how the data compare instructions work, consider the program in Figure 13–15 and the timing chart for the data comparisons in Figure 13–16.

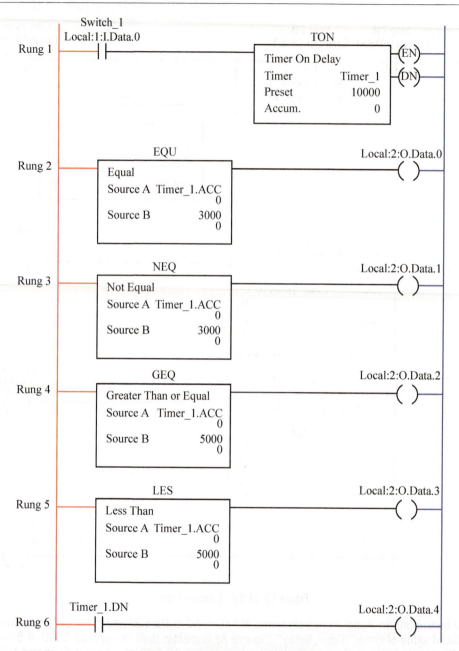

Figure 13–15 Data Compare Instructions

When power is applied, but before Switch_1 (Local:1:I.Data.0) is closed to enable the timer, outputs Local:2:O.Data.1 (Rung 3) and Local:2:O.Data.3 (Rung_5) are *ON* or energized. The NEQ instruction in Rung 3 is true since the value in Source A is not equal (≠) to 3000 and the LES instruction in Rung 5 is true since the value in Source A is less than (<) 5000. With Switch_1 open, the timer "Timer_1" is not enabled and its accumulated value is zero (0).

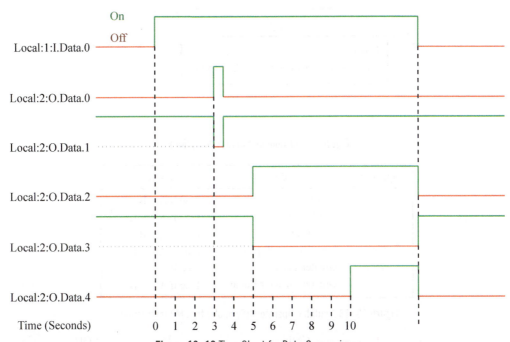

Figure 13–16 Time Chart for Data Comparisons

When Switch_1 closes, the timer "Timer_1" is enabled and starts to time. When the timer accumulated value reaches 3000, the EQU instruction in Rung 2 becomes true and output Local:2:O.Data.0 is turned *ON* and the NEQ instruction in Rung 3 becomes false, turning *OFF* output Local:2:O.Data.1. During the next increment of the timer (1-millisecond), the accumulated value goes to 3001 and the output in Rung 2 turns *OFF* and the output in Rung 3 turns *ON*.

When the accumulated time reaches 5000, the GEQ (≥) instruction in Rung 4 becomes true and output Local:2:O.Data.2 turns *ON*. The output in Rung 5 (Local:2:O.Data.3) turns *OFF* since the value in Source A is no longer less than (<) 5000. Output Local:2:O.Data.2 will remain *ON* and output Local:2:O.Data.3 will remain *OFF* until the timer instruction is no longer true and the accumulated value resets to zero (0).

When the timer is done, accumulated value equals the preset (10,000), the timer done bit (DN) in Rung 6 becomes true and output Local:2:O.Data.4 turns *ON*. The timing chart in Figure 13–16 illustrates the *ON* and *OFF* states of the outputs in relation to time and the data compare instructions.

The Allen-Bradley compare instruction (CMP) is an input instruction that can perform complex expressions in one instruction. All the comparison types just discussed can be used in the CMP instruction along with math operators (compare & math). Figure 13–17 shows a CMP instruction and a very simple "equal" expression.

The CMP instruction in Figure 13–17 is true when the accumulated value in counter "Counter_1" is equal to the value stored in tag "Target".

The table in Figure 13–18 shows the different compare operators (symbols) that the CMP instruction uses. Because standard computer keyboards do not have keys for not equal, less than or equal to, or greater than or equal to, the CMP instruction uses variations, as shown in Figure 13–19.

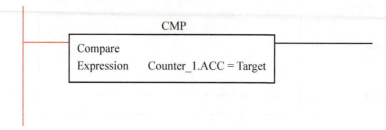

Figure 13–17 Compare Instruction (CMP)

Operator	Description	Example
=	Equal	True if A = B
< >	Not Equal	True if A < > B
<	Less Than	True if A < B
< =	Less Than Or Equal	True if A < = B
>	Greater Than	True if A > B
> =	Greater Than Or Equal	True if A > = B

Figure 13–18 Available Operators (Symbols) for CMP Instruction

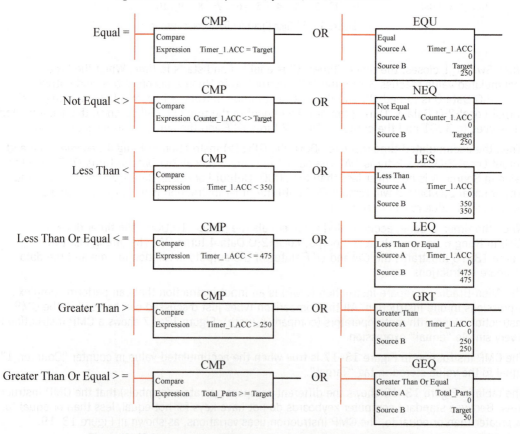

Figure 13–19 Comparing the CMP Instruction to Data Compare Instructions

Note: While the CMP instruction duplicates the other data compare instructions, the execution time for the CMP instruction is longer than the execution time for equivalent comparison instructions (for example, GRT, LEQ, etc.). A CMP instruction also uses more words per instruction than the equivalent comparison instructions. The advantage, however, is that the CMP instruction can also perform math operations as part of the compare as well as multiple compare operations. Use parentheses ( ) to define more complex expressions. Figure 13–20 shows an example of a CMP instruction with embedded math operation.

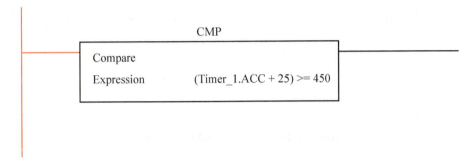

Figure 13–20 CMP with Embedded Math Operation

The limit test instruction (LIM) is an input instruction used to test for values inside or outside a specific range. The instruction is false until it detects that the test value is within certain limits, then goes true. When the instruction detects that the test value has again gone outside the prescribed limits, the instruction goes false. This instruction is perfect for monitoring analog signals and making program decisions based on a range. Figure 13–21 shows the format for an LIM instruction.

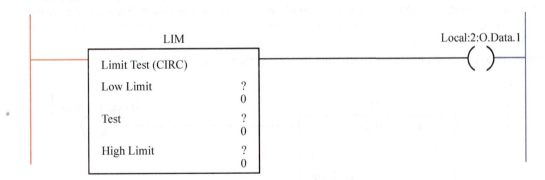

Figure 13–21 Limit Test Instruction Format (LIM)

To program an LIM instruction, the following information must be provided:

Low Limit – The value in the Low Limit operand determines the lower limit of the test range and can be either an immediate value or a memory tag that contains a value.

Test – The Test is a memory tag that contains a value to be examined to determine if it is inside or outside the specified range as determined by the low and high limit values. The test can be either an immediate value or memory tag that contains the value to be tested. Most often, the test is a memory tag (address) such as an analog value.

High Limit – The value in the High Limit operand determines the upper limit of the test range and can be either an immediate value or a memory tag that contains a value.

An example of how an LIM instruction is used is shown in Figure 13–22. In this example, the LIM instruction is used to turn *ON* an output whenever the accumulated value of timer "Timer_1" is between the values stored in tags "Timer_1_Low" and "Timer_1_High".

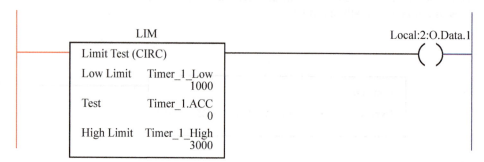

Figure 13–22 Limit Test Instruction Example

If the lower limit is set to 1000 (value stored in tag "Timer_1_Low") and the upper limit is set for 3000 (value stored in tag "Timer_1_High"), the instruction will be false as long as the accumulated value in timer "Timer_1" is less than 1000 or greater than 3000. When the accumulated value in timer "Timer_1" reaches 1000 and becomes equal to or greater than the low limit, the instruction becomes true and output Local:2:O.Data.1 is turned *ON*. If the accumulated value in timer "Timer_1" becomes greater than 3000, the instruction again goes false and the output is turned *OFF*. The instruction remains *OFF* as long as the accumulated value in timer "Timer_1" is outside the test range (1000–3000) that was established for the LIM instruction.

The values that are used for the low limit and high limit can be entered as numeric values when the instruction is being programmed if there are no plans to change these values. For example, instead of referencing tag "Timer_1_High" (which held a value of 3000), a value of 3000 could have been entered at the high limit prompt.

Another example of the limit test instruction (LIM) is shown in Figure 13–23.

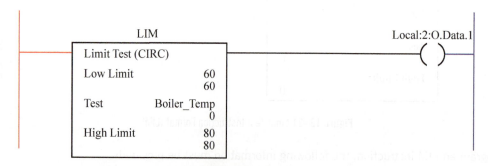

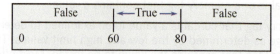

Figure 13–23 LIM Instruction with Low and High Immediate Values

The low limit is an immediate value (constant) of 60 whereas the high limit has an immediate value of 80. The test address is tag "Boiler_Temp". This could be the numeric value from an analog input device such as a thermocouple. Programmed this way, the instruction is true, and output device Local:2:O.Data.1 is ON, if the test value is equal to or greater than the low limit of 60 and less than or equal to the high limit of 80. Any time the value stored in tag "Boiler_Temp" is below 60 or greater than 80, the instruction is false and output device Local:2:O.Data.1 is turned OFF. The output device could be an indicator lamp that is ON if the temperature measured by the thermocouple is within the specified range of 60–80°.

If an LIM instruction is programmed with the low limit value higher than the high limit value, as shown in Figure 13–24, the logic of the instruction is reversed.

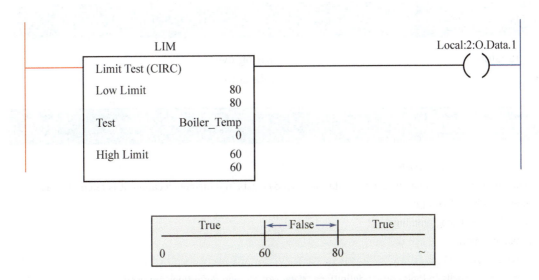

Figure 13–24 LIM Instruction with the Low Limit Value Higher Than the High Limit Value

In this illustration, the low limit has been set at 80, and the high limit has been set at 60. The test value is still the value stored in tag "Boiler_Temp". Programmed this way, the logic of the LIM instruction is true when the value stored in tag "Boiler_Temp" is equal to or greater than 80 (low limit) and also true when the value is less than or equal to 60 (high limit). Output device Local:2:O. Data.1 will be ON whenever the value stored in tag "Boiler_Temp" is equal to or less than 60 and equal to or greater than 80.

All compare instructions will execute faster and with the lowest memory requirements if all the operands (parameters) of the instruction use the same data type, typically DINT or REAL. To prevent unexpected results, it is highly recommended that you use the same data types. For example, if you have a tag called "Boiler_Temp" that is of the REAL data type and you are making a comparison to another internal memory tag, make that tag also a REAL data type.

Once the technician becomes familiar with a specific PLC, the many applications and advantages of using the various data compare instructions become evident.

Chapter Summary

Although formats and instructions vary with each PLC manufacturer, the concepts of data manipulation are the same. Data manipulation enables an operator to transfer data from one memory (word) location to another, whereas data comparison allows a value in one memory location to be compared to another memory location or immediate value.

Both data transfer and data comparison instructions give new dimension and flexibility to process control and motor-control circuits, and the application of either is only limited by programmer imagination.

Key Terms

data transfer

mask

writing over

data compare

Review Questions

1. Define the term *data transfer*.
2. When numerical information replaces data that already exists in a memory location, it is referred to as:
   a. exchanging info (data)
   b. replacement programming
   c. blanket move
   d. writing over
3. Match the symbols to their correct definitions. ***Note:*** *Not all nine definitions are used.*
   a. > _____ 1. less than 1
   b. < _____ 2. less than
   c. = _____ 3. less than or equal to
   d. ≥ _____ 4. greater than
   e. ≠ _____ 5. greater than or equal to
   f. <> _____ 6. equal to
   g. >= _____ 7. not equal to
   h. <= _____ 8. not equal to 1
   i. ≤ _____ 9. greater than 1
4. Define the term *data compare*.
5. Write a program that compares the accumulated value of timer "Timer_2" to a constant of 2500. The instruction is to be true when the accumulated value of timer "Timer_2" is greater than 2500.
6. Define the term *mask*.
7. In a mask move instruction (MVM), how would you enter the mask value in hexadecimal for a binary value of 0000100101101110111010101001011111?
8. Show how the LIM instruction would be programmed to turn ON a light when the value in tag "Temp" is within the range of 120–180.

Chapter *14*

Math Functions

Learning Objectives

After completing this chapter, you should have the knowledge to:

- List the four standard math functions available with most PLCs.
- Describe how math functions are used.
- Program the logic to solve a right triangle problem.

Using Math Functions

The four basic math, or arithmetic, functions available in almost all PLCs are addition (+), subtraction (-), multiplication (×), and division (÷). In addition to the four basic math functions, many PLCs have advanced functions such as trigonometric functions (cosine, sine, tangent, arc sine, etc.), log functions, square root, X to the power of Y, and so on.

The math functions can be used with immediate values (constants) or values stored in user memory. You can mix data types, but there may be a loss of accuracy and in most cases the instruction will take longer to execute. To avoid these problems, use the same optimal data type, typically DINT or REAL. It is recommended that when performing complex math operations that you use REAL data types to avoid rounding, compounding errors, and loss of accuracy.

A typical application of a math function might be a chemical batch plant where a 2:1 ratio of two chemicals (A and B) is required. Using analog input devices (discussed in Chapter 3) to measure the actual weight of chemicals A and B, along with math functions, we are able to provide the proper ratio. For a 2:1 mix, only half as much of chemical B (by weight) is needed. The weight of chemical A can be divided by 2 to determine the amount of chemical B that is needed for a proper 2:1 mix.

Allen-Bradley LOGIX Math Instructions

Figure 14–1 illustrates a divide (DIV) instruction showing how the value of chemical A (Source A) can be divided by 2 (Source B), and the result placed into the destination tag "Chem_B_Target". A data comparison can then be made to limit the amount of chemical B to one-half the amount of chemical A.

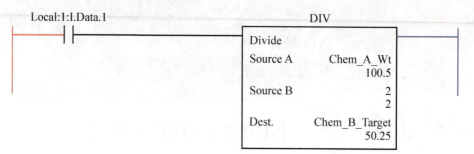

Figure 14–1 Divide (DIV) Instruction

Once the total weight of chemical A has been determined, input device Local:1:I.Data.1 becomes true and the weight of chemical A is divided by 2, and the result stored in destination memory tag "Chem_B_Target". A data comparison instruction can then be used to control the amount of chemical B to be added to chemical A as shown in Figure 14–2.

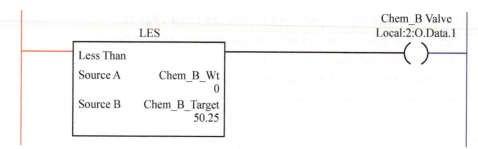

Figure 14–2 Data Compare Instruction to Control Amount of Chemical B

The control valve for chemical B (Local:2:O.Data.1) is true, or *ON*, as long as the value in Source A is less than Source B. Source A (Chem_B_Wt) is the current weight of chemical B and Source B (Chem_B_Target) is the desired or target weight. When the weight of chemical B is equal to or greater than one-half the weight of chemical A (Chem_B_Target), the compare instruction goes false and chemical B valve closes.

Math functions can also be used with timer and/or counter values, accumulated or preset, to change a given process or machine operation under varying conditions.

The DIV instruction divides the value in Source A by the value in Source B and stores the result in the Destination. Figure 14–3 shows another example of the DIV instruction.

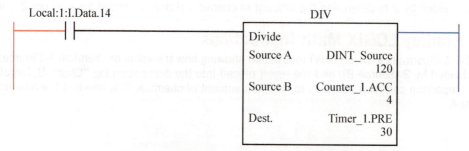

Figure 14–3 Divide (DIV) Instruction Used to Enter Preset

When input device Local:1:I.Data.14 is true, the value in Source A (DINT_Source) is divided by the value in Source B which is the accumulated value of counter "Counter_1", and the result is placed in the preset of timer "Timer_1".

Figure 14–4 shows the ADD instruction. This instruction adds the value in Source A to the value in Source B and places the result in the destination address. In Figure 14–4, Source A is a double integer data type, Source B is a floating point data type, and the Destination is a double integer data type. When input address Local:1:I.Data.10 is true, the value in Source A (20) is added to the value in Source B (40.52) and the result (61) is placed in the destination tag "Dest_DINT". Notice that the result in the destination tag is not 60.52. This is because the destination tag is an integer data type and the processor rounded the result. To prevent a rounding error, make the destination tag a REAL data type (floating point).

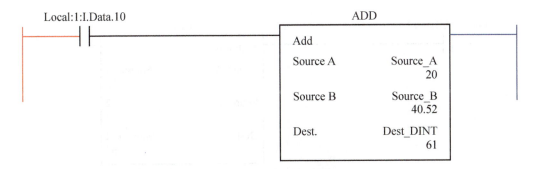

Figure 14–4 Addition (ADD) Instruction

Sources A and B can be memory locations (tags) that contain values or immediate values (constants). If the result of the addition is greater than the maximum value for the destination data type, the processor sets an internal overflow arithmetic status flag. The overflow arithmetic status flag is also set if the value to be placed in the destination tag is less than the minimum value for the destination data type. You can use arithmetic status flags in your program logic as desired. Figure 14–5 shows the arithmetic status flags for the Allen-Bradley Logix controllers.

Status Flag	Controller Action
S:V (Overflow)	Set to 1 if the value being stored cannot fit into the destination data type.
S:Z (Zero)	Set to 1 if the instruction's destination value is zero (0).
S:N (Sign)	Set to 1 if the instruction's destination value is negative.
S:C (Carry)	The Carry flag is not actually a part of the data type. The Carry Flag represents the bit that would be in the data type if it were stored to a larger data type.

Figure 14–5 Allen-Bradley Logix Arithmetic Status Flags

The subtract (SUB) instruction is shown in Figure 14–6. The value in Source B is subtracted from the value in Source A, and the result is placed in the destination memory tag specified.

When input device Local:1:I.Data.11 is true, the value in Source B (15) is subtracted from the value in Source A (20) and the result (5) is stored in destination tag "Sub_Dest".

The multiply (MUL) instruction is shown in Figure 14–7.

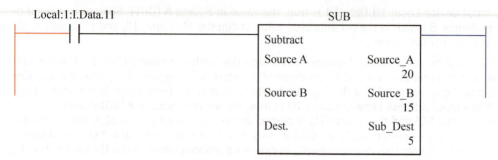

Figure 14–6 Subtract (SUB) Instruction

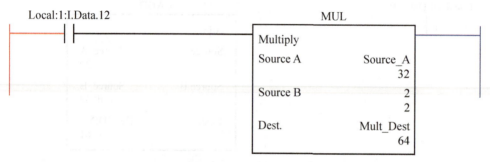

Figure 14–7 Multiply (MUL) Instruction

Like previous instructions, the MUL instruction multiplies the value in Source A by the value in Source B and places the result into the destination memory tag specified. In Figure 14–7, when input device Local:1:I.Data.12 is true, the value stored in Source A (32) is multiplied by the constant 2 and the result (64) is placed into destination tag "Mult_Dest". As with previous math instructions, Sources A and B can be immediate values (constants) or memory tags (addresses) that contain values.

In many of the Allen-Bradley PLCs, a more powerful math instruction called the Compute (CPT) is available. The CPT instruction is an output instruction that performs the operations defined in the expression and places the result into the destination tag specified. Figure 14–8 shows the format of the CPT instruction.

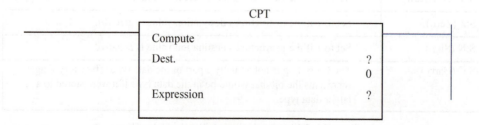

Figure 14–8 Compute (CPT) Instruction

All of the basic, advanced, and trigonometric math functions (operators) can be used in the expression field. The instruction evaluates the operations in the expressions using the standard order of operations for calculations that involve more than one arithmetic operation. You can use parentheses ( ) to force the instruction to perform the operation within the parentheses ahead of other operations. Operations of equal order are performed from left to right. Refer to the Logix general instruction manuals for the specific order of operations for the CPT instruction.

Figure 14–9 shows how the CPT instruction might be programmed.

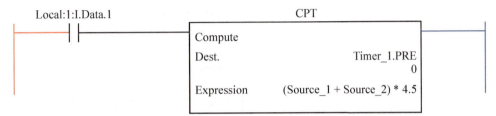

Figure 14–9 Programmed Compute (CPT) Instruction

When programming the CPT instruction, enter the destination tag address first followed by the expression. The destination address in Figure 14–9 is tag "Timer_1.PRE". The expression states that the value in tag "Source_1" is to be added to the value in tag "Source_2" and the sum multiplied by 4.5 when the instruction is true. The result of the calculation is placed in the destination tag "Timer_1.PRE".

The clear instruction (CLR), as the name implies, clears the destination memory address to zero (0). Figure 14–10 shows the CLR instruction. When input device Local:1:I.Data.1 is true, tag "Total_ Parts" is set to zero (0) or cleared. This instruction should be used over a MOV instruction to clear a memory address because it executes faster.

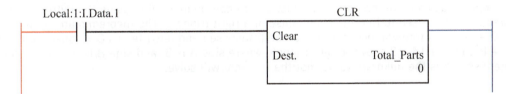

Figure 14–10 Clear (CLR) Instruction

The square root instruction (SQR) is shown in Figure 14–11.

When the logic that precedes the SQR instruction is true, the instruction calculates the square root of the value stored in the Source and places the result into the Destination location. In the example of Figure 14–11, the square root of the value stored in tag "Flow" (64), which is 8, is placed into the destination tag "Flow_SQR".

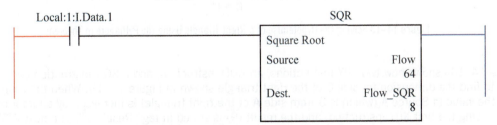

Figure 14–11 Square Root (SQR) Instruction

The XPY instruction raises a number (Source A) to the power specified in Source B (X to the Power of Y). This instruction is shown in Figure 14–12.

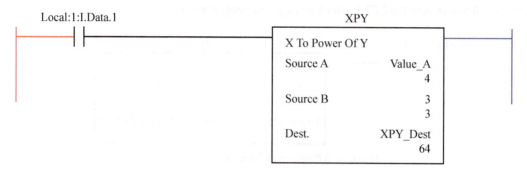

Figure 14–12 XPY Instruction

If you wanted to raise the value stored in a memory tag by a power of 3, which is the same as saying X^3 or $X \bullet X \bullet X$, you could use the XPY instruction. If the value of Source A is 4 and the value of Source B is 3, the Destination would hold the result of raising 4 by a power of 3, which is 64. Raising 4 by a power of 3 could be written as 4^3 or could be written as $4 \bullet 4 \bullet 4$. In either case, the answer is 64 ($4 \bullet 4 \bullet 4 = 64$).

Combining Math Functions

An example of using more than one math instruction can be illustrated using the Pythagorean Theorem, which is used to find the hypotenuse of a right triangle. The theorem states that $C^2 = A^2 + B^2$. This formula, or theorem, can be transposed and rewritten as C = square root ($\sqrt{}$) of $A^2 + B^2$. Figure 14–13 shows a right triangle where side A is 3″ and side B is 4″. Side C the hypotenuse and is the unknown value that the theorem will solve.

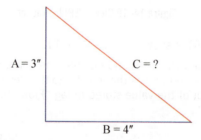

Figure 14–13 Solving the Hypotenuse of a Right Triangle Using the Pythagorean Theorem

Figure 14–14a shows how two XPY instructions, an ADD instruction, and a SQR instruction can used to find the dimension for side C of the right triangle shown in Figure 14–13. When the rung is true, the value in Source A (which is 3 from side A of the right triangle) is increased by a factor of 2 (3^2) using the first XPY instruction, and the result (9) is stored in tag "Result_1". The next XPY instruction takes the dimension of side B, which is 4, and raises it by a factor of 2 (4^2) and places the result (16) into tag "Result_2". The ADD instruction then takes the XPY instruction results and adds them together to get 25, as shown in tag "Result_3". The last instruction (SQR) finds the square root of the value stored in tag "Result_3", which is 5, and places the result into tag "Side_C". We all remember from our high-school geometry class, if side A is equal to 3, and side B is equal to 4, then side C must be equal to 5. This is often referred to as a 3, 4, 5 triangle.

We could have solved the problem using a single compute (CPT) instruction as shown in Figure 14–14b. With the CPT instruction, a formula or expression is written that will solve the problem and the result is placed into the destination tag "Side_C".

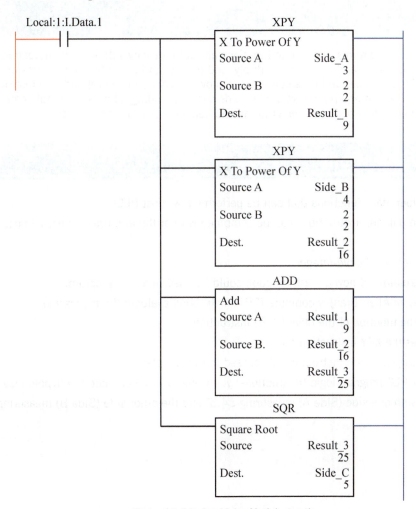

Figure 14–14a Combining Math Instructions

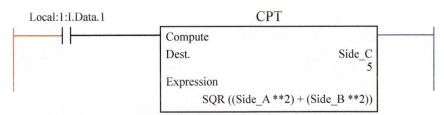

Figure 14–14b Using the CPT Instruction to Solve the Hypotenuse of a Right Triangle

The limits and limitations of the different arithmetic functions vary from manufacturer to manufacturer, but the concepts are basically the same. The values stored in memory address (tags), or immediate values (constants), are combined arithmetically, and the results are stored in memory tags. The only way to learn how to apply the arithmetic functions on a given PLC is to read the manufacturer's literature and work with that PLC.

Chapter Summary

As with most other features, arithmetic functions and formats vary with each manufacturer. Arithmetic functions of add, subtract, multiply, and divide can be combined with data manipulation instructions (data transfer and data compare) to provide expanded control and information for and from process or driven equipment. Memory words such as holding, storage, and data can be used with the arithmetic functions as well as words and constants, or just constants.

Review Questions

1. List the four math functions that can be performed by most PLCs.
2. Data manipulation instructions can be combined with arithmetic (math) instructions.

   T   F
3. Define the term *double-integer*.
4. Give an example of how a math function could be used in a PLC program.
5. How does an Allen-Bradley compute (CPT) instruction evaluate the expression?
6. What is the function of the clear (CLR) instruction?
7. What does the XPY instruction do?
8. List the trigonometric instructions discussed in this chapter.
9. Show the PLC program logic (instructions) you would use to calculate the hypotenuse of a right triangle with one side (Side A) measuring 34.5″ and the other side (Side B) measuring 96″.

Chapter *15*

Array Instructions

Learning Objectives

After completing this chapter, you should have the knowledge to:

- Describe the function of a synchronous shift register.
- Describe the operation of *element-to-array*, *array-to-element*, and *array-to-array*.
- Describe the difference between an *asynchronous shift register (FIFO)* and an *element-to-array* move.
- Describe the operation of a FAL instruction.

This chapter will cover array type instructions. Although Allen-Bradley array instructions will be covered, the general concept is the same in other PLCs with similar instructions. The primary difference is in the names and definitions used.

Before array instructions are discussed, the technician should understand the definitions used to address data and how user memory is created and organized. Before proceeding, let us take a few minutes to review several of the definitions that were covered previously in Chapter 4.

Words, or registers as they are often called, are locations in memory that can be used to store different kinds of information. Typically, a word or register can store the status of inputs and outputs, hold numerical values used for math functions and other numerical data used for timers, counters, etc. The Allen-Bradley Logix family of PLCs use 32-bit words, whereas many of the older PLCs used 16-bit words.

Tags are alphanumeric names that are given to memory locations. When you create a standalone tag that stores data, the minimum memory allocation for data is 4 bytes (one 32-bit word). For example, if you create an integer (INT) data type tag, only the first 16 bits are used in the 32-bit word. If you create a Bool tag, only the first bit of the 32-bit word is used. Tags represent a memory location of a specific data type or structure. When tags are created, the processor allocates the necessary memory, in 32-bit word increments, based on the type of data or structure it represents. Tag names must be unique, no two tags can have the same alphanumeric name.

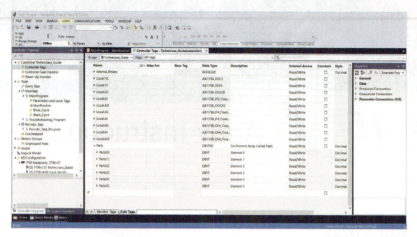

Figure 15–1 Six Element Single-Dimension Array

Structures store a group of data and each data in the structure can be of a different type. Each individual data type within a structure is called a member. Like tags, members have a name and data type.

An **element** is a single addressable unit of an array or structure. *Arrays* are groups of data of the same data type under a common name. Logix controllers use arrays to organize data and are a key element when programming array type instructions. If you recall from Chapter 4, an array is a tag that contains a block of multiple pieces of data and can be one-, two-, or three-dimensional. An array is similar to a data table in that it contains individual pieces of data called *elements*. Each element uses the same data type, such as DINT. The elements of an array tag are arranged in a contiguous block of memory in the controller with each element in sequence. An example of a single-dimension array called "Parts" is shown in Figure 15–1. In this example, the array has been configured for six DINT elements. A subscript identifies each individual element within the array. A subscript starts at [0] and extends to the number of elements in the array.

Only single-dimensional arrays will be covered in this chapter. If you want an instruction to access different elements in an array, you must use a tag in the subscript [tag] of the array. This subscript tag is called an indirect address. By changing the value of the subscript tag, you are changing the location (element) within the array that your logic would be referencing. Using indirect addresses can take longer to execute than directly referencing the element in the array, therefore use indirect addresses only when required. Figure 15–2 shows how an indirect address is used in the subscript of an array tag.

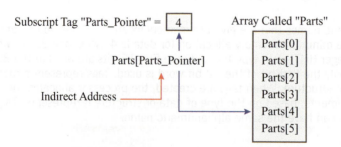

Figure 15–2 Subscript Indirect Addressing of Arrays

Synchronous Bit Shift Arrays

When information is shifted, one bit at a time, within an array, it is often called a **synchronous shift register** or synchronous bit shift. The bits within the array may be shifted forward (left) or reverse (right). Figure 15–3 shows a 50-bit array used as a forward (left) synchronous bit shift. Figure 15–3a shows the bit status of the array prior to the forward shift, while Figure 15–3b shows the same array after the bits have been shift one place to the left or forward. Notice that when the bit array is shifted, the information (1 or 0) in bit position 50 is shifted out, and is lost. If the bit array is continually shifted with a zero (0) in bit position 1, all of the 1s are shifted left (forward) until only zeros (0) remain.

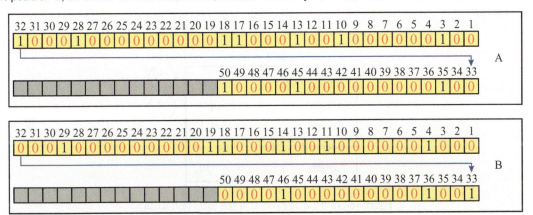

Figure 15–3 Bit Shift Left Array Before and After Execution

In a forward shift array, bit position 1 of the array is used to enter data (1 or 0) into the array. The data in the array is first shifted forward, one bit at a time. Figure 15–4 shows a 64-bit (2-word) forward shift array. In this case, data is entered at bit position 1 and shifted one bit at a time to the left. With a 64-bit or 2-word shift array, the information (1 or 0) in bit position 32 of word one (bit-31) is not shifted out and lost but is shifted into bit 0 of word two or bit array position 33.

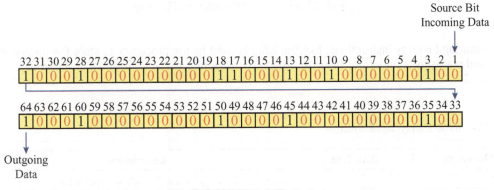

Figure 15–4 Two Word (64-bit) Array

Allen-Bradley Bit Shift Instructions

The Bit Shift Left (BSL) instruction is an output instruction that shifts all bits one position to the left with each false-to-true transition of the instruction. Figure 15–5 shows an Allen-Bradley BSL instruction.

When programming the BSL instruction, the *Array* parameter is the array tag of the first element of the array. In our example we specified element [0] in the array called "Parts", with an array length of 50 bits. The bits will shift to the left starting at bit array position 1, bit 0 of element [0], and end at bit array position 50 or bit 17 of element [1], the second word in the array. The *Control* parameter in Figure 15–5 is a user-created tag that is defined as a control type tag. You must create a unique control tag for each type of array instruction you program. The *Source* bit is a bit tag in memory containing the status to be moved into the array each time the rung goes from false to true. In our example, the source is digital input Local:1:I.Data.0 (limit switch). Any unused bits in the array are invalidated and cannot be used. Figure 15–6 shows the 50-bit array and the 2 elements that make up the array.

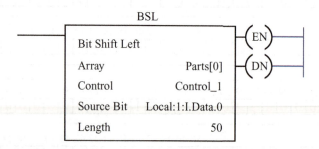

Figure 15–5 Allen-Bradley Bit Shift Left (BSL) Instruction

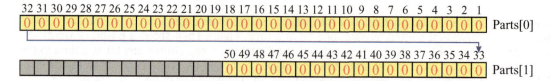

Figure 15–6 Two Element 50-Bit Shift Left (BSL) Array

The control tag for array instructions, like the BSL, is used by the processor to store the instructions length and status information for the instruction. The structure of the control tag is shown in Table 15–1.

Table 15–1 Bit Shift Array Control Structure

Mnemonic	Data Type	Description
(Control Tag).**EN**	Boolean (Bool)	Enable bit, set to 1 when the instruction is true or enabled.
(Control Tag).**DN**	Boolean (Bool)	Done bit, set to 1 when all bits have shifted one position to the left.
(Control Tag).**UL**	Boolean (Bool)	Stores the status, 0 or 1, of the last bit in the array that was shifted out, or exits, on each false-to-true rung transition.
(Control Tag).**ER**	Boolean (Bool)	Error bit, set to 1 if the instruction's length was entered as a negative number (<0).
(Control Tag).**LEN**	DINT	The length of the number of array bits to shift.

Note: The status bit ".UL" stores the status of the bit that was shifted out of the instruction (last bit position) when the shift was executed.

An example of a practical application for a bit shift array is the overhead parts conveyor in Figure 15–7, which is used to transport parts into a paint booth for painting. When a part enters the paint booth, limit switch 1 (LS-1) is activated. Limit switch 2 (LS-2) is activated each time a hook on the conveyor system passes the limit switch, even if no part is present.

Both limit switches (LS-1 and LS-2) are wired to a digital input module and the solenoid that operates the paint spray nozzle is wired to a digital output module, as shown in Figure 15–8.

Figure 15–9 shows the program logic and how the BSL instruction, digital inputs, and output are programmed.

When the conveyor is operating, limit switch LS-2 (Local:1:I.Data.2) will be triggered (*ON* and *OFF*) each time a hook on the conveyor system passes the limit switch, whether a part is present or not. Limit switch LS-2 is used in Rung 1 to provide the false-to-true transition needed for the BSL instruction.

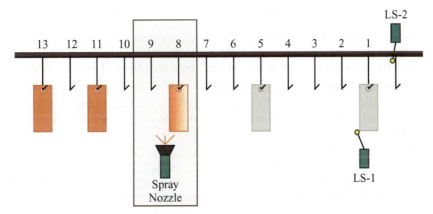

Figure 15–7 Applying a Bit Shift Array

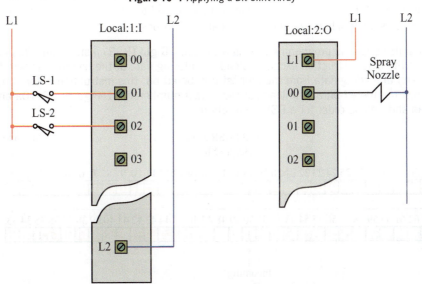

Figure 15–8 Input and Output Devices Wired to I/O Modules

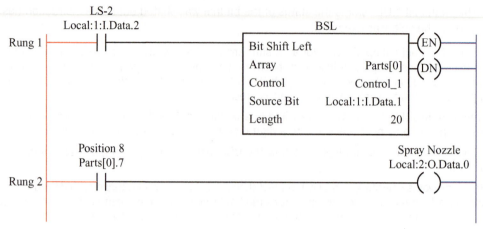

Figure 15–9 Programming a Forward Bit Shift Array Using BSL Instruction

In our example, each bit in the array represents a hook position within the paint booth starting at limit switch LS-1, as shown in Figure 15–7. As long as the conveyor system is operating, the BSL instruction will shift the bits in the array one position to the left each time limit switch LS-2 is activated or true.

When a part triggers limit switch LS-1 (Local:1:I.Data.1), a 1 is placed into the first location of the BSL array, indicating that a part is present at that location. As the BSL instruction continues to shift the bits one position to the left, the 1 that was placed into the array is also shifted one position to the left in the array.

When the bit representing the part is shifted into the array location that corresponds to the location of the paint sprayer, output (Local:2:O.Data.0) is activated (Rung 2) and the part is sprayed. Because each bit location in the array corresponds to a physical location beginning at limit switch LS-1, it becomes easy to track the parts as they travel through the spray booth.

Note: This example is only representative of how a bit shift instruction could be used. By no means does it represent all of the control logic that would be required to ensure reliable and safe operations. This is only used as a simplified example for educational purposes.

A reverse bit shift register is programmed using a Bit Shift Right (BSR) instruction. This instruction is the same as the BSL instruction except instead of entering data at the lowest numbered bit in the bit array and unloading data from the highest numbered bit, this instruction enters (loads) data into the highest numbered bit and unloads at the lowest numbered bit. Figures 15–10a and 15–10b show the load and unload order for a BSR instruction.

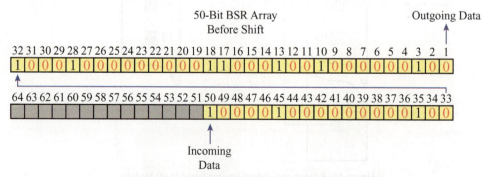

Figure 15–10a Load and Unload Bits for BSR Instruction prior to Shift

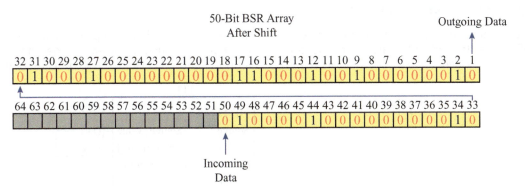

Figure 15–10b Load and Unload Bits for BSR Instruction after Shift

You can use a bit shift left and right instruction together to create a forward and reverse bit shift array. This can be useful when a conveyor can be operated in both directions (forward and reverse).

Asynchronous Shift Arrays (FIFO)

In an **asynchronous shift array**, instead of shifting bits of information within a bit array like the bit shift instructions just discussed, a value (data) is shifted into an array or stack as it is sometimes called. There are similarities between the two types of shift arrays, but there is also one major difference. In the asynchronous shift array, the data from a word is shifted into the last unused word location of the array with each implementation, or indexing, of the instruction. This difference is why the asynchronous shift array is often referred to as a FIFO stack (first-in first-out). Figure 15–11 shows an asynchronous shift array before and after loading data into the array.

Asynchronous shift arrays are very useful when there is a need to store data values sequentially in a table (array) and then retrieve those values in the same order they were placed, first-in first-out. When using an asynchronous shift array to store and retrieve data, it requires two separate instructions, one to load and one to unload the data. In asynchronous shift arrays, you create an array of the data type desired, usually DINT or REAL, and of the size needed for your application. For example, if you needed to store the hourly temperature for a 24-hour period and then later retrieve those temperatures, you would create a 24-element array of data type REAL.

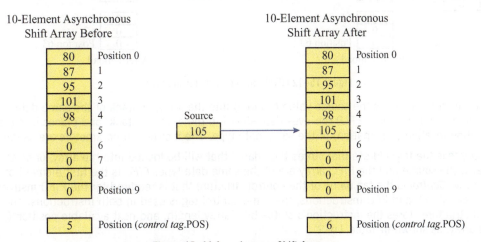

Figure 15–11 Asynchronous Shift Array

Allen-Bradley FIFO Instructions

Allen-Bradley uses a pair of output instructions to store and retrieve data in a prescribed order, or FIFO (First-In First-Out). Values are loaded into an array and unloaded in the same order in which they were loaded. Figure 15–12 shows the FIFO Load (FFL) and FIFO Unload (FFU) instructions.

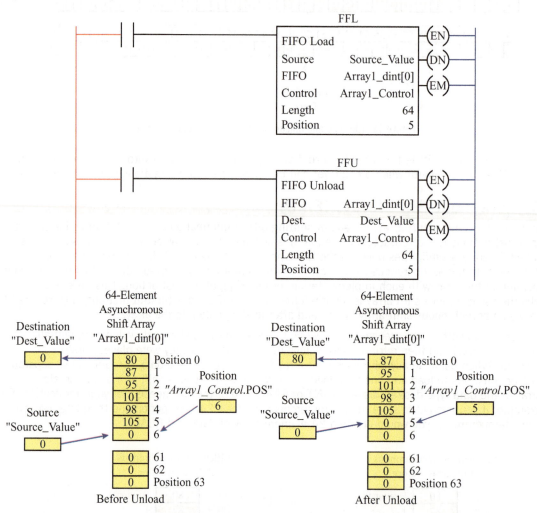

Figure 15–12 FIFO Load and Unload Instructions

The FFL instruction loads the source value (Source) into the array, or stack of the same data type. The FIFO instruction (FFU) retrieves data stored in the array, or stack, and places it into a destination location (Destination) of the same data type. The instruction parameters are as follows:

The *Source* is the tag address that stores the "data" that will be loaded into the array, or stack. Typically, the source and the FIFO array are of the same data type. *FIFO* is the first element of the FIFO array. *Control* is the tag name of the control structure that is required for the FIFO instruction. When pairing FFL and FFU instructions, the same control tag is used in both instructions. The control structure stores the instruction's status bits, array length, and next available position (pointer) in the array, as shown in Table 15–2.

Table 15–2 FIFO Array Control Structure

Mnemonic	Data Type	Description
(Control Tag).**EN**	Boolean (Bool)	Enable bit, set to 1 when the instruction is true or enabled.
(Control Tag).**DN**	Boolean (Bool)	Done bit, set to 1 when the FIFO is full (.POS = .LEN).
(Control Tag).**UM**	Boolean (Bool)	Empty bit, set to 1 when the FIFO is empty.
(Control Tag).**LEN**	DINT	The maximum length of elements in the FIFO.
(Control Tag).**POS**	DINT	Current position (location) in the FIFO where the instruction will load the next value.

In Figure 15–12, the FIFO is array "Array1_dint[0]" for both the FIFO Load and FIFO Unload instructions. Tag address "Array1_dint[0]" is the first element of the 64-element array, and tag address "Array1_dint[63]" is the last element in the array. Position 0 is the first element of the array, or stack, and position 63 would be the last element of the 64-element array. With each false-to-true transition of the FFL instruction, data from the source is loaded into the array (FIFO) and the position indicator will advance, or increment, by one.

When the rung that contains the FFL instruction goes true, the processor sets the EN control bit to *ON*, and loads the source data from tag "Source_Value" into the next available position in the array. The processor loads data from the source into the array with each false-to-true transition. When the array is full, the processor sets the DN control bit to 1. Once the array is full, no additional data is loaded into the array.

When the rung that contains the FFU instruction goes from false to true, the processor sets the EU control bit to 1 and unloads data from the first element of the array into the destination tag "Dest_Value". As the data is shifted out, the processor shifts the remaining data in the array *up* one position toward the first element of the array. The processor continues to unload the array each time the rung goes from false to true until it empties the FIFO array.

Allen-Bradley Last-In First-Out (LIFO) Instructions

The Last-In First-Out Load (LFL) instruction loads data into an array or stack. Then, using a LFU (LIFO Unload) instruction, it retrieves them in inverse order. In other words, the instruction removes the last value that was entered into the array when the rung that has the LFU instruction makes a false-to-true transition. Figure 15–13 shows an LFL and an LFU instruction and illustrates how the values are entered and retrieved.

When the rung condition makes a false-to-true transition, the LFL enable bit (EN) will be set to 1 and the LFL instruction will load the contents (data) from the source tag "Source_Value" into the array, and the position value will be incremented by 1. The LFL instruction will continue to load data into the array, or stack, on each false-to-true transition until the array is full (64 elements). When the array is full, the processor will set the done (DN) bit to 1. This prevents any additional information from being loaded into the array.

When the LFU rung makes a false-to-true transition, the LFU unload bit (EU) is set to 1 and the LFU instruction will unload the last value that was loaded into the array and place it into the destination tag "Dest_Value". The array data will continue to be unloaded from the array each time the rung makes a false-to-true transition. Each time a value is removed from the array, or stack, the position indicator will decrement (decrease in value) by 1.

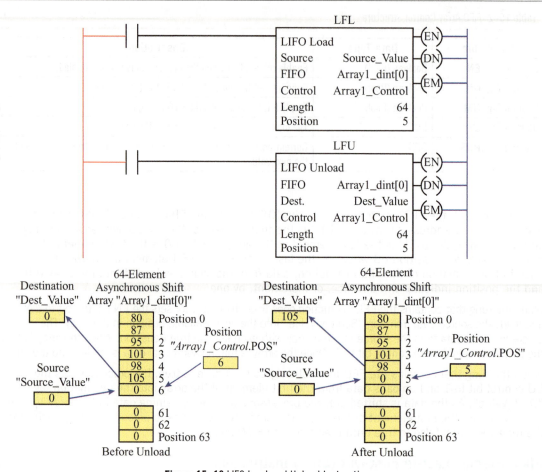

Figure 15–13 LIFO Load and Unload Instructions

Miscellaneous Array Instructions

In addition to the array shift instructions, the Allen-Bradley Logix family of controllers also has these additional array type instructions:

FAL – File Arithmetic and Logic
FSC – File Search and Compare
COP – File Copy
FLL – File Fill
AVE – File Average
SRT – File Sort
STD – Standard Deviation (not covered)
SIZE – Size in Elements (not covered)

Note: In older Allen-Bradley PLCs, data was grouped into files, whereas the Logix PLCs group consecutive data types into arrays. When you see the term "file" used, we are talking about an array. Both files and arrays are consecutive blocks of data of the same type. Do not be confused when you see the two terms used interchangeably.

Try to avoid mixing data types when using array instructions. For example, if you create a DINT type array, then use the same data type for the incoming source and outgoing destination tags. As mentioned earlier in this chapter, DINT and REAL data types are the preferred data types to use in array instructions.

Before discussing the FAL and FSC array type instructions, there is an additional parameter that is unique to just these two instruction called the "Mode" parameter. The mode parameter tells the PLC how to distribute the array operation. When you program the instruction, you must select one of three possible modes of operation, *all*, *numerical*, and *incremental*.

All Mode – All mode tells the PLC to operate on all of the specified elements in the array before continuing on to the next instruction or rung. When the instruction goes from false-to-true, the elements in the array are executed one at a time until the position (.POS) value equals the length (.LEN) value. When you set the all mode, you are in essence telling the PLC to perform the array operation requested on all elements within the array before executing the next instruction or rung in the program. The all mode is the most common mode used with FAL and FSC instructions.

Numerical Mode – Numerical mode tells the PLC to distribute the array operation over a number of scans. You enter the number of elements to operate on each time the PLC scans the instruction. For example, if you entered 5 in the mode parameter, the PLC would only perform 5 element operations per scan until finished (POS=LEN). Once the instruction is triggered (enabled), it will continue to perform the number of operations each scan until finished, regardless of the rung condition. Numerical mode is primarily used when you are working on large amounts of data that is non-time-critical. This helps to spread the operation over multiple scans improving overall scan times.

Incremental Mode – When you select incremental mode, the PLC will operate on one element of the array each time the instruction's goes from false to true. This is an important difference from the numerical mode. For example, if you had an array of five elements, it would take five false to true transitions of the instruction to perform the requested operation on all five elements of the array. This mode allows the rung logic proceeding the instruction to control when each operation is performed on the array elements.

File Arithmetic and Logic (FAL) Instruction

The File Arithmetic and Logic (FAL) instruction is an output instruction. The instruction performs copy, arithmetic, logic, and function operations on the data stored in an array. This instruction operates on an expression that you enter into the instruction. The same operations that you can perform with the CPT instruction can also be performed with the FAL instruction, except you are working on an array of data. This instruction is very powerful and versatile based on how you enter the expression and destination parameters. Figure 15–14 shows the FAL instruction.

Some of the instruction parameters are similar to those used in the array shift instructions covered earlier in this chapter. The *Control* parameter requires that you enter a unique control tag for each FAL instruction programmed. The *Length* parameter is the number of elements in the array to be manipulated. For example, if you have 20 elements in an array called "TempsC[0]" that you wish to manipulate, you would simply enter a length of 20. The *Position* parameter is the position of the current element in the array that the instruction is accessing. You typically always enter a value of 0 when first programming the instruction. The position value can become an important value when programming and using the FAL instruction. Remember, the position value can be accessed by referencing the control structure element "(control tag).POS". The *Mode* parameter, as already discussed, determines the mode of operation. Enter *ALL*, *INC*, or a value for numerical mode. The *Destination* parameter is a tag location to store the result of the expression. The destination can be a single element or a number of elements (array). If you enter a single element tag such as

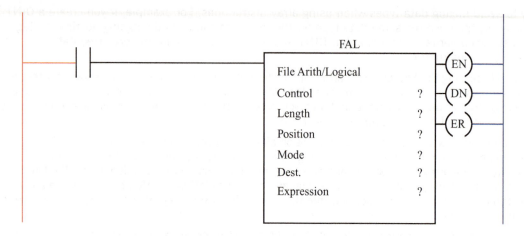

Figure 15–14 File Arithmetic and Logic (FAL) Instruction

"Dest_Tag", you are creating what is called an array-to-element operation. If you enter an array tag with the subscript pointing to the position value, you would be creating either an element-to-array operation or array-to-array operation.

For example, an array tag called "Array1[*control tag*.POS]" would tell the controller to place the result of the expression into an array called "Array1" at the element location as determined by the current value in the position parameter of the instruction's control tag.

The expression can consist of tags and/or immediate values separated by operators. Tags can be single element tags or array tags. If you use the instructions position in the subscript of an array tag, it would be called an array-to-array, array-to-element, or element-to-array operation.

Figures 15–15 shows how to use the (control tag).POS value to step through the array(s).

The example in Figure 15–16 shows how you could use the FAL instruction to convert twenty Celsius temperature readings in an array called "TempsC[0]" to Fahrenheit and place the results into an array called "TempsF[0]".

File Search and Compare Instruction

The File Search and Compare (FSC) instruction is an output instruction that compares values in an array, element-by-element, for the logical operations you specify in the expression. You can use the same operators (=, <>, <=, >=, <, >), including math operators, as you could with the CMP instruction. When the FSC instruction finds a true comparison, it sets the instruction's found (FD) and inhibit (IN) status bits to *ON* or 1, and stops execution at that position. The instruction's POS displays the array location where the instruction found the true comparison. To continue searching the remainder of the array, clear the IN bit. Figure 15–17 shows the FSC instruction.

The FD, IN, and POS elements of the control structure are key in using this instruction. When the FSC instruction is enabled, your program should monitor the FD and DN status bits. If the FD bit is set, you can use it and/or the POS value to make logical decisions in your program. If you want to continue searching, simply reset the IN bit by using an unlatch instruction. If not, you can reset the instruction by enabling a reset (RES) instruction using the FSC instruction's control tag. An RES can be used with timers, counters, and control structures.

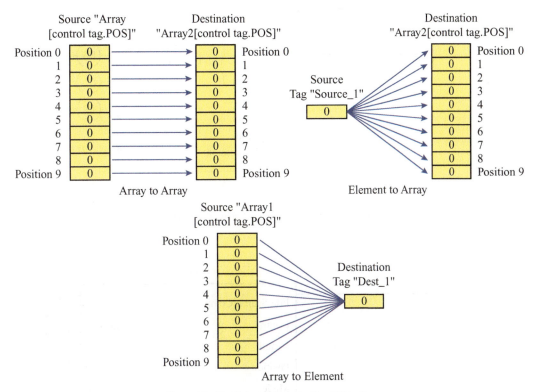

Figure 15–15 FAL Instruction Position Utilization

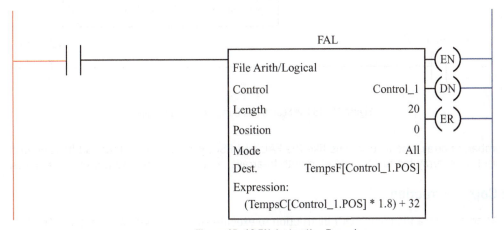

Figure 15–16 FAL Instruction Example

An example where you might use this instruction is checking for valid part numbers in a lookup table. Say that an array called "valid_parts[0]" has been preloaded with all the valid part numbers for that day's run. As parts travel down the conveyor system their part number is scanned and compared against the valid part numbers in the array using the FSC instruction. If a valid part number is found, the conveyor system allows the part to continue to the packaging machine. Figure 15–18 shows how the FSC would be configured.

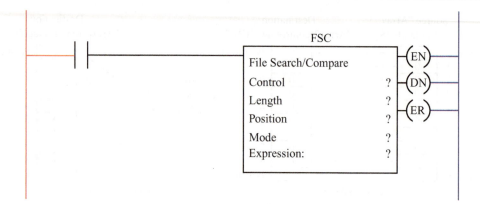

Figure 15–17 File Search and Compare (FSC) Instruction

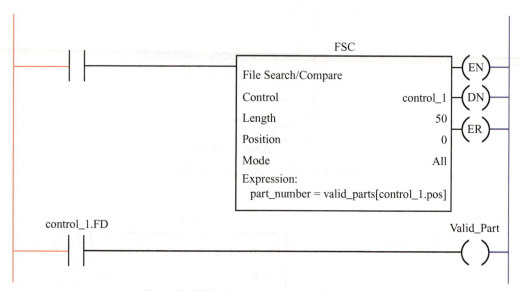

Figure 15–18 File Search and Compare (FSC) Example

Remember, in array type instructions like the FAL and FSC, you can use indirect addressing in the subscript of an array tag to compare elements-to-arrays, arrays-to-arrays, and arrays-to-elements.

File Copy Instruction

Allen-Bradley uses a file copy (COP) instruction to execute simple array-to-array moves. The file copy instruction is an output instruction and is programmed as shown in Figure 15–19.

You specify the source and destination arrays along with the length of elements to copy. There is no control structure required for the COP instruction. When the instruction has a true logic path and is enabled, it will copy the number of elements specified in the length, from the source array to the destination array each time the rung is scanned. The data in the source array remains unchanged. The COP instruction is executed faster than the FAL and should be used when you only want to copy data from one array to another.

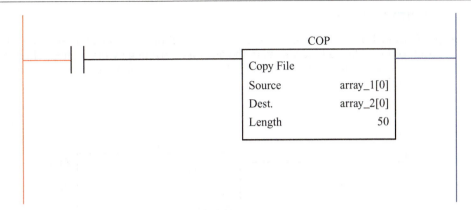

Figure 15–19 File Copy (COP) Instruction

Another version of the COP instruction is the Synchronous Copy File (CPS) instruction. The only difference between the COP and CPS instructions is that the CPS instruction prevents data from changing during the copy operation. This instruction is useful when buffering I/O data as was discussed in Chapter 9. When the CPS instruction is enabled, the controller takes an image of the data in the source array and copies that image into the destination array. During the copy operation, no I/O updates can change the data.

File Fill (FLL) Instruction

The file fill (FLL) instruction fills the elements of an array with a source value. The source can either be an immediate value or a tag that contains a value. The destination is the array to be filled and the Length is the number of elements in the array to be filled. Figure 15–20 shows the FLL instruction.

When using the file COP and FLL instructions, the source and destination should be of the same data type. If the length is greater than the total number of elements in the array, the instruction stops at the end of the array boundary. This is not the case when using the FAL and FSC instructions, the array(s) size should be equal to or greater than the length specified in the instruction, otherwise unexpected results or a major fault could occur.

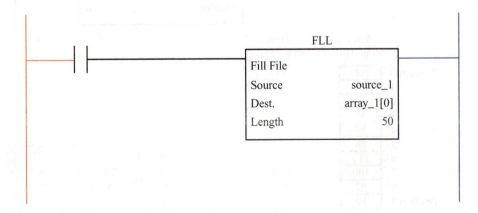

Figure 15–20 File Fill (FLL) Instruction

File Average (AVE) Instruction

The file average (AVE) instruction calculates the average of a set of values in an array. The instruction calculates the average by adding all the values in the array and dividing by the number of elements in the array (Length). The AVE instruction requires a false-to-true transition to execute. Figure 15–21 shows the AVE instruction.

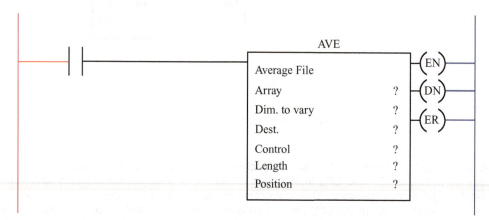

Figure 15–21 File Average (AVE) Instruction

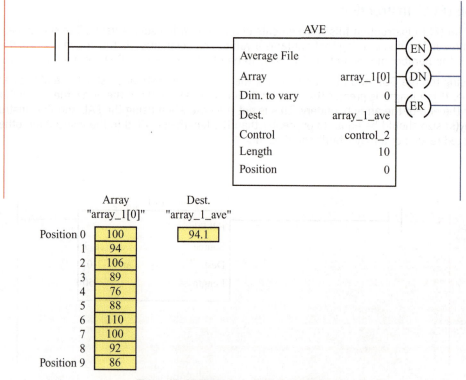

Figure 15–22 File Average (AVE) Example

The *Array* parameter is the tag name of the array to be averaged and you specify the first element in the array. The *Dim to vary* parameter specifies which dimension (0, 1, or 2) to use depending on the number of dimensions in the array. If you are averaging just a single dimensional array, the parameter value entered would be 0. The *Destination* parameter is the tag name where the result of the operation is stored. It is recommended that this tag be of the REAL data type for maximum accuracy. The *Control* parameter is the control tag that you assign when entering the instruction. The *Length* parameter is the number of elements in the array to average. Make sure that the length value does not exceed the number of elements in the array. In the *Position* parameter, enter a value of 0, which is the initial value. Figure 15–22 shows the AVE instruction used to average the values in an array called "array_1" and the result is stored in a tag called "array_1_ave".

File Sort (SRT) Instruction

The file sort (SRT) instruction is similar to the AVE instruction just discussed. Instead of averaging a group of elements in an array, the SRT instruction sorts the values within the array into ascending order. The SRT instruction requires a false-to-true transition to execute. Figure 15–23 shows the SRT instruction.

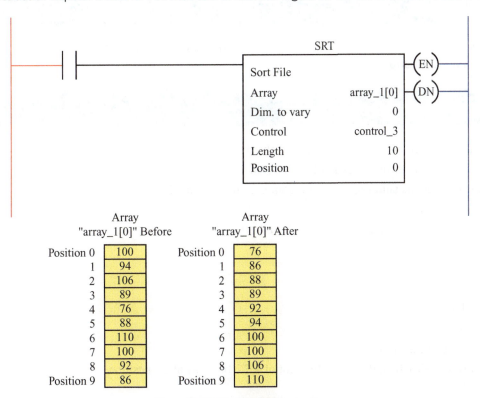

Figure 15–23 File Sort (SRT) Instruction

Array instructions bring another level of programming capability to PLCs. When you have a need to work with and manipulate large amounts of data, array instructions, like the ones just discussed, can make that process very easy and efficient. When programming array instructions, take the time to fully understand how the instructions work and are executed by the controller. When first learning array instructions, it is best to start with a simple one-dimension array of the data type DINT or REAL in a test program until you become familiar with the instructions.

Chapter Summary

Although the instructions vary with each PLC manufacturer, the principles of element (word) and array (file) moves are the same. The synchronous shift register shifts bits of information left or right (forward and reverse) within an array. The asynchronous shift register is referred to as FIFO, or first-in first-out, and transfers data from a source into the array at the next position in the array. The data is retrieved in the order it enters the array (first-in first-out). The array can also be programmed to retrieve the data that was last in to be the first out. This convention is referred to as a LIFO stack. In addition, other array instructions are available for managing large amounts of data such as the File Arithmetic and Logic (FAL), File Search and Compare (FSC), File Copy (COP), File Fill (FLL), File Average (AVE), and File Sort (SRT). When using the FAL instruction, the instruction can be configured to manipulate or move data from elements-to-arrays, arrays-to-elements, or arrays-to-arrays.

Key Terms

element   synchronous shift register
asynchronous shift array

Review Questions

1. Define the term *element* as used in this chapter.
2. Define the term *array* as used in this chapter.
3. The synchronous shift register shifts data in a forward direction only.

   T   F

4. In a BSL instruction, data enters the 32-bit array at bit position:
   a. 1
   b. 2
   c. 8
   d. 16
   e. 32
   f. none of the above

5. In a BSR instruction, data is shifted out of a 32-bit array at bit position:
   a. 1
   b. 2
   c. 16
   d. 32
   e. 8
   f. none of the above

6. Define the term *FIFO*.

7. Define the term *LIFO*.

8. Briefly describe the difference between *synchronous* and *asynchronous shift registers*.

9. When using the FFL instruction, which bit is set to 1 when the array or stack is full?

10. When using the FFU instruction, which bit is set to 1 when the array is empty?

11. List two other terms that could be used to refer to an array.

12. What is a FAL instruction?

13. What is the purpose of the COP instruction?

14. Which of the following group of elements could *not* be an array?

    a.   50, 51, 52

    b.   50, 51, 52, 53

    c.   100, 101, 102, 103

    d.   100, 101, 102, 103, 105

15. Briefly describe the function of an element-to-array move.

16. Briefly describe the function of an array-to-element move.

17. Briefly describe the function of an array-to-array move.

18. Which instruction is also known as a FIFO?

    a.   synchronous shift register

    b.   word-to-file move

    c.   file-to-word move

    d.   asynchronous shift register

    e.   file-to-file move

19. When data is transferred into an array using an element-to-array move, the data is entered at the last unused element of the array.

    T   F

Chapter 16

Sequencers

Learning Objectives

After completing this chapter, you should have the knowledge to:

- Describe what a sequencer output instruction does.
- Describe the basics sequencer output operation.
- Define the term mask.

The sequencer instruction transfers information from elements within an array into a destination tag (word). Sequencer instructions are typically used to control automatic assembly machines that have consistent and repeatable operations.

Sequencer

A programmed **sequencer** replaces the old mechanical drum sequencer that was used before PLCs. On the mechanical sequencer, when the drum cylinder rotated, contacts opened and closed mechanically to control output devices. Figure 16–1 shows a mechanical drum cylinder with pegs placed at varying horizontal positions for step 1 of the sequence. When the cylinder rotated, contacts that aligned with the pegs closed, and contacts where no pegs existed remained open. In this example, the presence of a peg should be thought of as a 1, or *ON*, and the absence of a peg as a 0, or *OFF*.

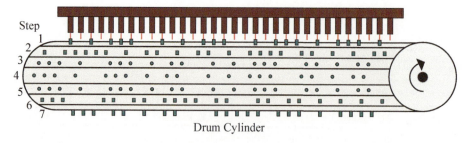

Figure 16–1 Mechanical Drum Cylinder

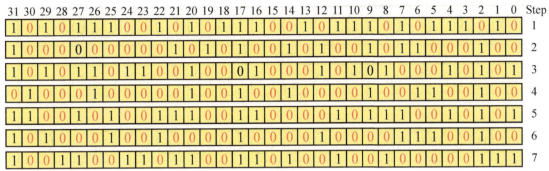

31	30	29	28	27	26	25	24	23	22	21	20	19	18	17	16	15	14	13	12	11	10	9	8	7	6	5	4	3	2	1	0	Step
1	0	1	0	1	1	1	0	0	1	0	1	0	1	1	1	0	0	1	0	1	1	1	0	1	0	1	1	1	0	1	0	1
1	0	0	0	0	0	0	0	0	0	1	0	1	0	1	0	0	1	0	1	0	0	1	0	1	1	0	0	0	1	0	0	2
1	0	1	0	1	1	0	1	1	0	0	1	0	0	0	1	0	0	0	1	0	1	0	1	0	0	0	1	0	1	0	1	3
0	1	0	0	0	1	0	0	0	0	0	1	0	0	1	0	0	1	0	0	0	0	1	0	0	1	1	0	0	1	0	0	4
1	1	0	0	1	0	1	0	0	1	1	1	0	0	1	1	0	0	0	0	1	0	1	1	1	0	0	1	0	1	0	1	5
1	0	1	0	0	0	1	0	0	1	0	0	0	0	0	1	0	0	0	1	0	0	0	0	1	1	1	0	0	1	0	0	6
1	0	0	1	1	0	0	1	1	0	1	1	0	0	1	1	0	1	0	0	1	0	0	1	0	0	1	0	0	1	1	1	7

Equivalent Sequencer Table

Figure 16–2 Sequencer Array

To program a sequencer, binary information is entered into a series of consecutive memory elements or array. Information from the elements in the array are transferred sequentially to the destination tag (word) to control digital outputs.

If the first seven steps on the drum cylinder in Figure 16–1 are removed and flattened out, they might appear as illustrated in Figure 16–2.

For step 1, each horizontal location where a peg was located is now represented by a 1 (*ON*), and the positions where there were no pegs are represented by a 0 (*OFF*).

The seven steps could also be viewed as a 7-element array with each 32-bit element (word) representing a sequencer step. If one enters different binary information (1s and 0s) into each element of the array, the array replaces the rotating drum cylinder.

To illustrate how this works, 16 lamps are used for outputs as shown in Figure 16–3.

Each lamp represents one bit address (0 through 15) of tag "digital_out". Assume, for the sake of discussion, that the operator wants to light the lamps in the 4-step sequence as shown in Figures 16–4.

The bit addresses of the lamps that are to be lit in each step of the sequence are written down to assist in entering data into the elements of the sequencer array.

Step 1. 0, 1, 2, 3

Step 2. 0, 1, 2, 3—12, 13, 14, 15

Step 3. 0, 1, 2, 3, 4—8—12, 13, 14, 15

Step 4. 0, 1, 2, 3, 4—7, 8—11, 12, 13, 14, 15

Tag "digital_out"

Figure 16–3 Output Lamps

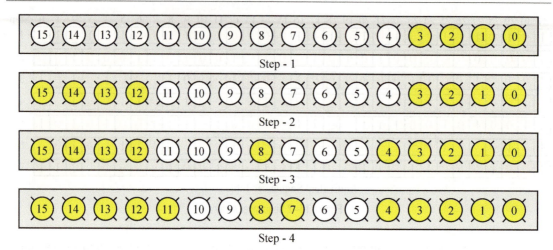

Figure 16–4 Output Lamp Sequence

The next step is to define an array to store the binary data required for each step of the sequencer. Elements 1, 2, 3, and 4 are used for the 4-element array. Using the programming software, one enters binary information (1s and 0s) into each element of the array to reflect the desired lamp sequence (Figure 16–5).

Note: Some PLCs allow the data to be entered using BCD, which speeds the entry process. To use this feature, the required binary information for each sequencer step is converted to BCD. The information is then entered using a programming device into the array with 4 key strokes for each element, rather than 16.

Once the sequencer has been programmed and the data entered into the array, the sequencer is ready to control the lamps. When the sequencer is activated and advanced to step 1, the binary information in element 1 (Figure 16–5) is transferred into tag "digital_out" and the lamps light in the pattern shown in Figure 16–4. Advancing the sequencer to step 2 transfers the data from element 2 into tag "digital_out" for the next light sequence shown in Figure 16–4. Step 3 transfers the data from array element 3 into tag "digital_out", and step 4 transfers information from element 4 into tag "digital_out". When the last step is reached, the sequencer can be reset and sequenced again.

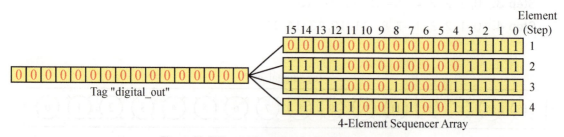

Figure 16–5 Binary Information for Each Sequencer Step

Depending on the PLC, sequencers can be programmed from a few steps up to hundreds of steps, and can control one output or several.

Mask

When a sequencer operates on a destination tag (word), there may be outputs or bits associated with the destination tag that are not to be controlled or altered by the sequencer. To prevent the sequencer from altering these bits in the destination tag (word), a **mask** or immediate value is used. Figure 16–6 shows how a mask tag works.

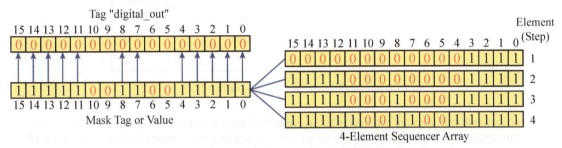

Figure 16–6 Using a Mask Tag (Word)

The mask or immediate value is a means of selectively screening out (blocking) data from the sequencer array to the destination tag (word). For each bit in the destination tag "digital_out" that is to be changed, the corresponding bit in the mask tag must be set to 1.

In Figure 16–4, bits 5, 6, 9, and 10 are not used. Setting bits 5, 6, 9, and 10 of the mask to zero (0) means that these bits in the destination tag will not be altered by the sequencer.

In Figure 16–6, mask bits 0, 1, 2, 3, 4, 7, 8, 11, 12, 13, 14, and 15 are set to 1 corresponding to the destination bits to be changed by the sequencer array. The sequencer works much like the array-to-element move discussed in Chapter 15. For PLCs that don't have a dedicated sequencer instruction, an array-to-element move instruction, if available, could be used.

Allen-Bradley Sequencer Instructions

The Allen-Bradley Sequencer Output Instruction (SQO) creates a sequencer array of information that is used to control various output devices. When the rung that contains the SQO instruction makes a false-to-true transition, the instruction increments to the next element in the sequencer array and copies the bit status from the array element into the destination tag (word) through the mask. Figure 16–7 shows an SQO instruction.

The *array* parameter of the instruction is the tag name of the sequencer array. You must specify the first element of the sequencer array in the subscript of the array tag name. In this illustration, the sequencer array is identified as "seq1_array[0]".

The *mask*, as explained earlier, is a filter through which all data from the sequencer array must pass before being placed into the destination, or output tag (word). The mask can either be a memory tag or an immediate value. If an immediate value is used, it must be entered as a decimal value or use the correct prefix (16# hex, 8# octal, 2# binary). A 1 must be placed in the mask bit location for information to be passed through to the destination.

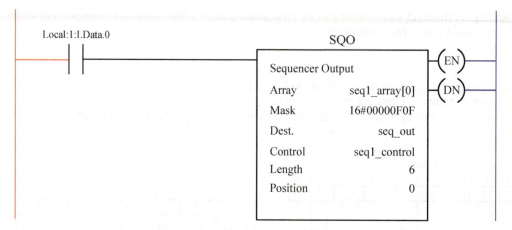

Figure 16–7 Sequencer Output Instruction (SQO)

The *destination* is the memory tag that is to be controlled by the sequencer instruction. In Figure 16–7, the destination tag is "seq_out". You can use this tag in your program to control the logic of the outputs or more commonly, directly to a digital output module tag (word) if desired.

The control tag structure for all sequencer type instructions is shown in Table 16–1.

Table 16–1 Sequencer Control Structure

Mnemonic	Data Type	Description
(Control Tag).**EN**	Boolean (Bool)	Enable bit, set to 1 when the instruction is true or enabled.
(Control Tag).**DN**	Boolean (Bool)	Done bit, set to 1 when all the specified elements have been moved to the destination.
(Control Tag).**ER**	Boolean (Bool)	Error Bit, set to 1 if the instruction's length is equal to or less than 0, or the position is less than 0 or greater than length.
(Control Tag).**LEN**	DINT	Length, is the number of steps in the sequencer array.
(Control Tag).**POS**	DINT	Position, is the current element in the sequencer array that the controller is manipulating.

The *length* value in the instruction is the number of steps in the sequencer array starting with step 1. Step 0 is the startup, or default value of the sequencer on PLC program startup (program-to-run). On the first false-to-true transition and after the last step of the sequencer, the sequencer will reset to step 1 not step 0.

The *position* indicates the step, or element, where the sequencer is currently positioned. The position number will increment with each false-to-true transition of the instruction.

Figure 16–8 shows a programmed SQO instruction with the sequencer array, mask, destination tag, and the status of the external output devices when the sequencer instruction is on step 4.

As programmed in Figure 16–8, the sequencer instruction transfers data from the sequencer array each time that input device Local:1:I.Data.0 closes. Even though the data in element 4 (step 4) has a 1 set

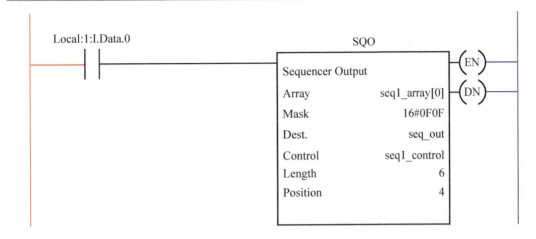

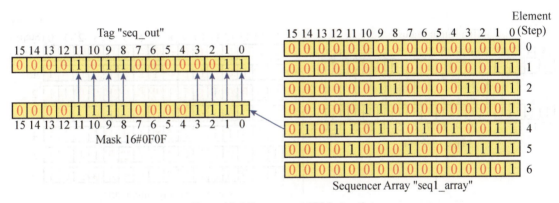

Figure 16–8 Programmed SQO Instruction

for bits 0, 1, 4, 6, 8, 9, 11, 12, and 14, the destination tag "seq_out" only shows that bits 0, 1, 8, 9, and 11 are *ON*. The reason outputs 4, 6, 12, and 14 are not *ON* is that these bits are masked out.

The mask in Figure 16–8 has been entered as an immediate value in hexadecimal (16#0F0F). Notice that we used the hexadecimal prefix 16#.

Another Allen-Bradley sequencer instruction is the Sequencer Input Instruction (SQI). This instruction is an input instruction that compares all of the masked bits of a source element to an array element (sequence step) for equality. If all of the bits *match*, the instruction becomes true. This instruction can be used to compare the status of machine/equipment input devices with what is expected during operation. The SQI instruction is normally paired with a Sequencer Output Instruction (SQO), and is used to detect when a step is complete in a sequence. The SQI instruction can operate independently, but requires additional logic to work. This instruction is also a great way to do machine diagnostics. Figure 16–9 shows an SQI instruction.

As programmed in Figure 16–9, the sequencer array is "input_array[0]". The mask has been set to all 1's by entering the hexadecimal number 16#FFFF. The source element is tag "input_status". This is the element that holds the status of the input devices for the process equipment. The control tag is the same control tag used in the SQO when the instructions are paired. If you are not

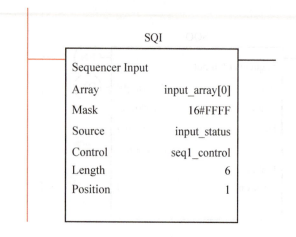

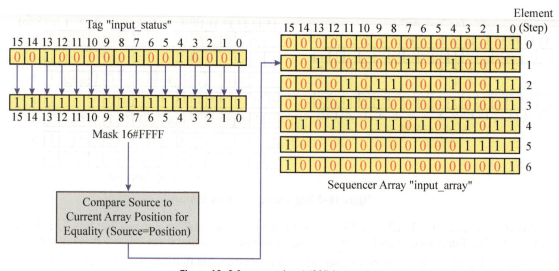

Figure 16–9 Sequencer Input (SQI) Instruction

pairing the SQI with an SQO instruction, then enter a unique control tag for the instruction, and manipulate the *.POS* element in your logic. The length is the number of steps in the sequencer array, and is the same as the SQO when paired. Position displays the current step the instruction(s) is on. Remember, the initial value for the position is typically zero (0) when first programming the sequencer instruction(s) and is called the initialization step. After the initialization step, the position value will be 1 to length. For example, when the position value equals the length value in a SQO instruction, on the next iteration it wraps to position 1, not 0. Your sequencer arrays should have an element length that is equal to or greater than the length plus one (1). When the controller goes from program-to-run mode, the controller clears (initializes) the position value for all SQO instructions in the program. This initialization position (0) is called the startup or safe position, and you should set the bits in the SQO array element for position 0 to all zeros to ensure a safe machine startup. On the first false-to-true transition of the SQO instruction, the position will advance to position 1 and on the last step it will wrap back around to position 1. You can use an RES instruction to reset the control structure of a sequencer instruction, but it will reinitialize the position to 0. Figure 16–10 shows the SQI and SQO instructions paired together. Notice how each instruction is using the same control tag so that they work in unison.

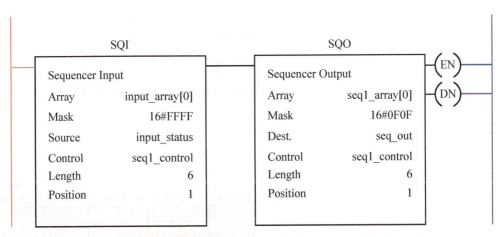

Figure 16–10 Sequencer Input and Output Instructions Paired

As mentioned, the SQI instruction can be used as a powerful diagnostic tool to determine correct machine operation. This instruction can be used to compare input elements as well as output elements that represent input and output devices used for the manufacturing process. You are only constrained by your imagination.

The Sequencer Load (SQL) instruction allows data to be loaded into a sequencer array from a source tag (word) on each false-to-true transition. Figure 16–11 shows the sequencer load instruction, the source tag, and the six-element sequencer array.

On each false-to-true transition of input device Local.1:I.Data.0, the current status (1 or 0) of each device represented in tag "digital_status" will be transferred (copied) into the sequencer array. Figure 16–11 shows the SQL instruction on step 3. The information shown in the source tag "digital_status" has been copied into step 3 of the sequencer array. On each false-to-true transition, the instruction will increment the position and the current status of tag "digital_status" will be copied into the array "seq1_array[0]". When the instruction has reached step 6, the done bit (.DN) will be set to 1. On the next false-to-true transition of the instruction, the sequencer load instruction will wrap around to step 1 and the status of tag "digital_status" will be copied into the array, overwriting any previous information that had been loaded into step 1 of the array.

The SQL instruction is typically used to load reference conditions into a sequencer array. This instruction is like an element-to-array move and could also be used to store numeric data into an array.

When using sequencer instructions, make sure you are using the same data type (INT, DINT, etc.) and your array sizes are long enough to prevent boundary faults (length > size of array).

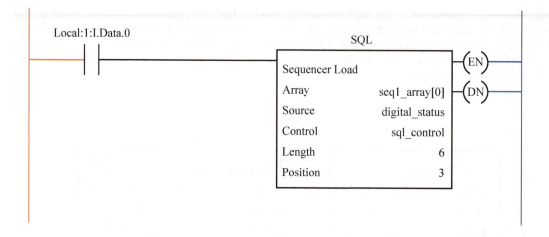

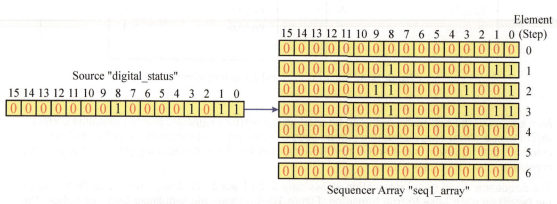

Figure 16–11 Sequencer Load (SQL) Instruction

Chapter Summary

Although sequencers, like other data manipulation and arithmetic instructions, are programmed differently with each PLC, the concepts are the same. Data is entered into an array for each sequence step, and, as the sequencer advances through the steps, binary information is transferred sequentially from the array to the destination tag (word). Destination bits not to be altered can be masked out so they can operate independently of the sequencer. The SQI and SQL instructions are typically used with the sequencer output instruction to create a very efficient machine control program where machine operation is repeatable (cyclic).

Key Terms

sequencer   mask

Review Questions

1. Briefly describe a *sequencer*.
2. A series of consecutive elements is referred to as a:
   a. deck
   b. group
   c. array
   d. chain
3. What is the purpose of a *mask* in a sequencer?
4. What device is commonly replaced by a PLC sequencer instruction?
5. Set up the array in the following figure so the sequencer will operate the motors as shown for steps 1, 2, 3, and 4. Configure the mask so motors 5, 6, 7, 14, and 16 cannot be energized.

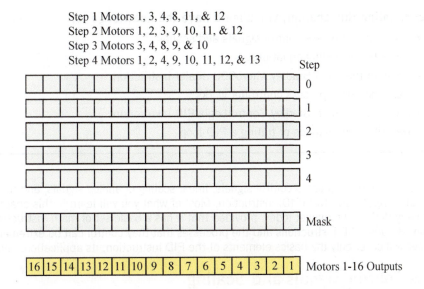

6. Which Allen-Bradley sequencer instruction is like an *array-to-element* move?
7. Which Allen-Bradley sequencer instruction is like an *element-to-array* move?
8. What would be the mask value in hexadecimal if only the first 5 bits (0–4) were to be controlled by an SQO instruction?
9. How is an SQC instruction used?

Chapter 17

Process Control Signals, Scaling, and PID Instructions

Learning Objectives

After completing this chapter, you should have the knowledge to:

- Understand process control signals and scaling.
- Apply linear scaling equations.
- Program the Allen-Bradley Micro800 Scaler instruction.
- Describe how a process controller works.
- Program the Allen-Bradley ControlLogix PID instruction.
- Describe two methods of tuning a PID loop.

This chapter covers basic process control signals, linear scaling of analog signals, and the Allen-Bradley Proportional Integral Derivative (PID) instruction. Most of what you will learn in this chapter can be applied to any PLC on the market today, provided that it has a basic set of math instructions and a PID instruction. Because PID instructions and the processes that they control can be extremely complex, this chapter will cover only the basics elements of the PID instruction, its application, and tuning.

Process Control Signals and Scaling

The ability to monitor and/or control a process depends on having accurate and meaningful information about the process. This information can include such things as temperature, pressure, level, flow, weight, etc., and is commonly referred to as the **process variable (PV)**. Electrical and pneumatic signals are used to represent the PVs and are typically generated by field-mounted devices such as transducers and transmitters. These process signals interface to PLCs, loop controllers, displays, and other devices using one of the following standard analog type signals:

- DC electrical current (4–20 mA, 0–20 mA, etc.)
- DC electrical voltage (0–10V DC, 1–5V DC, +/−10V DC, etc.)
- Air pressure (3–15 psig)

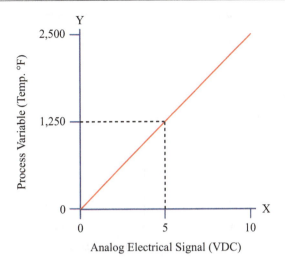

Figure 17–1 Relationship between Analog Signal and
Process Measurement

An analog signal has a value that uniquely corresponds to the PV being measured. That is, as the measured PV changes, so does the analog signal representing it.

Figure 17–1 shows an xy graph of an electrical analog signal compared to the PV being measured. The PV is shown along the vertical or *y*-axis and is in degrees Fahrenheit. The electrical analog signal is shown along the horizontal or *x*-axis and is represented as a low-level DC voltage. This low-level DC voltage is the analog signal that is sent to a PLC or other device to represent the measurement of the PV, temperature in this example.

As you study the graph in Figure 17–1, you will notice that the PV has a range of 0–2500°F and the corresponding analog signal representing that PV has a range of 0–10V DC. There is a linear relationship that exists between these two signals, meaning that when the temperature is at 50% of its range (1250°F), the electrical analog signal will be at 50% of its range (5V DC). Simply put, for every change in the PV there is an equally proportional change in the electrical signal representing that PV.

Before an electrical analog signal can be used by the PLC or other digital device, it must first be converted into a corresponding digital value. This conversion from analog to digital is accomplished by an Analog-to-Digital (A/D) converter that is an integral part of the hardware of the device. The digital range that the electrical signal is converted into depends on the electrical range and digital resolution of the A/D converter. No matter what that range is, there is again a linear relationship between the electrical signal and the corresponding digital value, just as there was with the PV and the electrical analog signal described above. Figure 17–2 shows the relationship between the electrical signal and the corresponding digital range of the A/D converter.

In Figure 17–2 the electrical signal is shown on the horizontal or *x*-axis and has a range of 0–10V DC. The corresponding digital range is shown on the vertical axis or *y*-axis and has a range of 0–32,767. When the analog signal is at 50% of its range (5V DC), the corresponding digital value will be 16,384 or 50% of the digital range.

In order for process information to be transmitted and made available for display and/or control, there will always be some type of signal conversion that takes place, and in most cases, more than

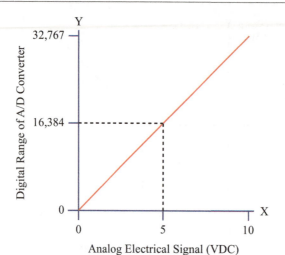

Figure 17–2 Relationship between Analog Signal and Digital Range

one conversion. Shown in Figure 17–3 is an example of a complete PV signal from the output of the transmitter in the field to a PLC memory location that stores the current PV measurement in engineering units.

As you can see from Figure 17–3, there are three signal conversions taking place. The first conversion converts the physical measurement into an electrical analog signal by the transducer/transmitter. The second conversion converts the electrical analog signal into a digital value by the A/D converter. The third conversion is required to convert the digital output of the A/D converter into a corresponding scaled value representative of the engineering units being measured, pressure in this example.

To maintain measurement accuracy of PV signals, every signal conversion required must be done correctly and accurately. In most cases, this conversion is done by solid-state and digital solid-state

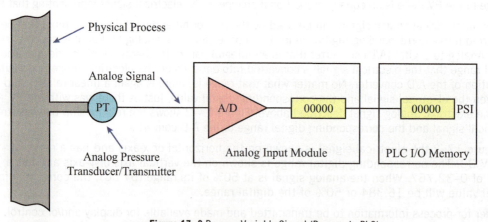

Figure 17–3 Process Variable Signal (Process to PLC)

devices like transducers, transmitters, and A/D converters. With these devices we merely need to make sure they are programmed and/or calibrated correctly. On the other hand, converting the digital output of an A/D converter into a value that represents the engineering units being measured may require the need to calculate the linear conversion between the two ranges using linear interpolation or scaling formulas in the PLC program. Many PLCs today, like Allen-Bradley, have analog I/O modules that can perform the scaling conversions within the module itself. Whether or not the I/O hardware can perform the scaling function, having an understanding of how this is done can be important when working with PLCs and process instrumentation. Let us begin by calculating the linear relationship between the output of an A/D converter and the corresponding scaled value in engineering units using linear interpolation or scaling formulas.

The following mathematical equations can be used to express the linear relationship between an input value and the resulting scaled value if the minimum and maximum ranges of both are known:

$$\text{Scaled Value} = (\text{input value} \times \text{slope}) + \text{offset}$$

The *slope* (sometimes called *rate*) and *offset* values in the above equation can be calculated from the following:

$$\text{Slope} = (\text{scaled maximum} - \text{scaled minimum}) / (\text{input maximum} - \text{input minimum})$$
$$\text{Offset} = \text{scaled minimum} - (\text{input minimum} \times \text{slope})$$

Note: The offset value will always be equal to what the scaled value would be when the input value is at zero.

Example 1 In this example, we will use the graph data shown in Figure 17–4 to calculate a scaled value in psi. The digital output range of the A/D converter is shown on the horizontal or x-axis and has a range of 0–32,767. The corresponding scaled range for the PV is shown on the vertical or y-axis and has a range of 0–250 psi.

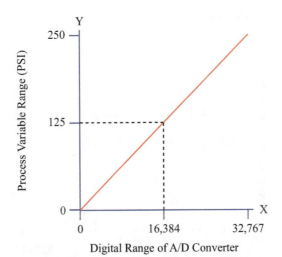

Figure 17–4 Relationship between Process Variable and Digital Range

Let us assume in our example that the current digital output value from the A/D converter is 18,325. What would be the corresponding scaled value in psi?

Solution:

First we must calculate the *slope* (*rate*) using the following equation:

Slope = (scaled maximum − scaled minimum) / (input maximum − input minimum)

Slope = (250 − 0) / (32,767 − 0)

Slope = **0.0076296**

Next we must determine what the offset value should be. The offset can be calculated from the following equation:

Offset = scaled minimum − (input minimum × slope)

Offset = 0 − (0 × 0.0076296)

Offset = **0**

Remember that the offset value is always equal to what the scaled value would be when the input is at zero. In this case, we know that when the input is zero the scaled value will also be zero. Since there is no offset in this example, we could choose to omit the offset in the scaling equation.

The last step is to calculate the scaled value for the digital input value given (18,325) as follows:

Scaled Value = (input value × slope) + offset

Scaled Value = (18,325 × 0.0076296) + 0

Scaled Value = **139.8 psi**

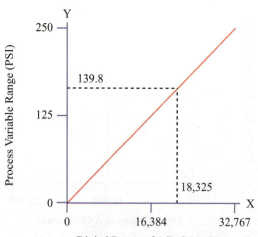

Figure 17–5 Intercept of Two Ranges

Figure 17–5 shows the same graph as in Figure 17–4, but this time with dashed lines showing the intercept of the two ranges for the above example.

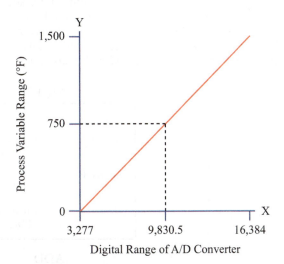

Figure 17–6 Relationship between Process Variable and Digital Range

Example 2 This example uses the range data shown in Figure 17–6 to calculate a scaled value. The digital input range is again shown on the horizontal or x-axis and has a range of 3,277–16,384. The corresponding scaled range is shown on the vertical or y-axis and has a range of 0–1500°F.

Let us assume that the current digital output value from the A/D converter is 10,121. What would be the corresponding scaled value in °F for this value?

Solution:

Slope = (scaled maximum − scaled minimum) / (input maximum − input minimum)

Slope = (1,500 − 0) / (16,384 − 3,277)

Slope = **0.1144426**

Offset = scaled minimum − (input minimum × slope)

Offset = 0 − (3,277 × 0.1144426)

Offset = **−375.0284**

Scaled Value = (input value × slope) + offset

Scaled Value = (10,121 × 0.1144426) + −375.0284

Scaled Value = **783.25°F**

After working through the previous two examples you can see that once the *slope* (*rate*) and *offset* values have been determined, you can calculate a corresponding scaled value for any input value given. To further illustrate this, let us take the slope and offset values just calculated and use them in a PLC program so that we will have a scaled value representing °F for any digital output value given by the A/D converter. Figure 17–7a shows the PLC logic using basic math instructions. The output of the A/D converter is stored in PLC memory tag "Local:3:I.Ch0Data" and the corresponding scaled value will be stored in a REAL data type tag called "scaled_input1". Figure 17–7b shows the same math operation using the Allen-Bradley Compute (CMP) instruction.

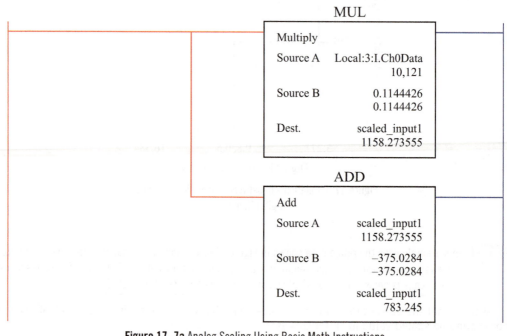

Figure 17–7a Analog Scaling Using Basic Math Instructions

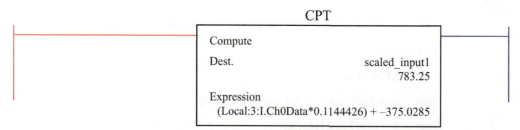

Figure 17–7b Analog Scaling Using Compute Instruction

Analog signals from process measurement devices are not the only type of analog signals found in industry. There are analog signals used to control devices like valves, variable speed drives, actuators, analog displays, etc. An analog signal that is used to control a device that has a direct influence on the process being controlled is called the **control variable (CV)**, and the field device is called the final control element. A PLC analog output used to operate a final control element is referred to as a PLC CV signal.

The same analog signal conversions described previously for PV signals also applies to CV signals, but only in reverse. This time a digital value is converted to an electrical analog signal by means of a Digital-to-Analog (D/A) converter. The electrical analog signal is then converted into physical motion by the final control element. Figure 17–8 shows an example of a complete CV signal from a PLC memory location to the final control element that is controlling the process.

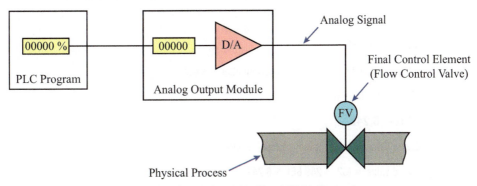

Figure 17–8 Control Variable Signal (PLC to Process)

When working with analog output signals in a PLC, you may be required to convert a decimal value in engineering units, such as valve position in % or speed in ft/min, into a digital value required by the D/A converter to output the proper electrical signal. The same linear interpolation or scaling equation and math instructions used previously can also be used here. The following example will help to illustrate this.

Example 3   In this example, an internal PLC memory word contains the desired position of a control valve (0–100%) that is controlled by an electrical analog output signal. We have determined that the analog electrical signal to the valve is a 4–20 mA current signal and the corresponding digital range required by the D/A converter to produce such a signal is 6,242–31,208. We will again use a graph to help illustrate the two data ranges we will be working with, as shown in Figure 17–9. The scaled range of the valve is shown on the horizontal or x-axis and has a range of 0–100%. The digital input range to the D/A converter to produce the 4–20 mA signal is shown on the vertical or y-axis and has a range of 6,242–31,208. This will also be our scaled range in the scaling equation.

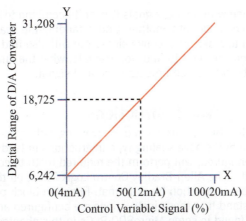

Figure 17–9 Relationship between Scaled Range and Digital Range

If we assume that the current position of valve needs to be 32%, what would be the corresponding digital value required by the D/A converter to produce the required analog current signal?

Solution:

Slope = (scaled maximum − scaled minimum) / (input maximum − input minimum)

Slope = (31,208 − 6,242) / (100 − 0)

Slope = **249.66**

Offset = scaled minimum − (input minimum × slope)

Offset = 6,242 − (0 × 249.66)

Offset = **6,242**

Scaled Value = (input value × slope) + offset

Scaled Value = (32 × 249.66) + 6,242

Scaled Value = **14,231**

The value of 14,231 would be placed into the PLC memory location that is used by the D/A converter to output the desired analog signal.

In the above example, what would be the analog current signal, in milliamps, to the control valve?

Solution:

Milliamps (mA) = (((20 − 4) / (31,208 − 6,242)) ∗ (14,231 − 6,242)) + 4

Milliamps = **9.12 mA**

or

Milliamps (mA) = ((20 − 4) ∗ .32) + 4

Milliamps = **9.12 mA**

You will find that when working with analog signals it is quite common to be required to convert analog signals from one form or value into another. Understanding the scaling formulas and examples presented here will allow you to make such conversions. You will also find that the preceding equations will be helpful when you want to find, for example, what the analog electrical signal would or should be on a digital meter for a given process or control signal.

Allen-Bradley Scaling Type Instructions

Depending on the Allen-Bradley PLC controller you are working with, some controllers like the Micro800's and the older SCL500's have scaling type instructions in ladder logic format. These output instructions, when configured, will perform the required math to convert an unscaled value to a value in engineering units. The Allen-Bradley ControlLogix PLCs, on the other hand, only have a scale-type instruction available in function block format. Function block programming is covered in Chapter 18. To better understand how these instructions are configured and used, we will cover only the Scaler (SCL) instruction found in some Micro800 PLCs in this chapter.

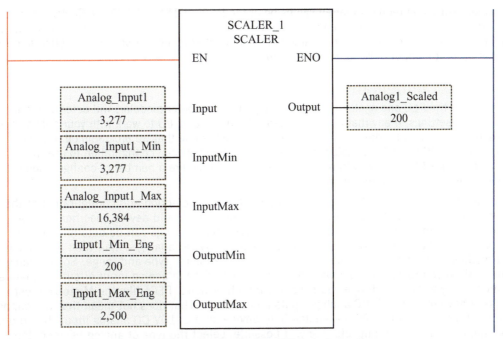

Figure 17–10 Allen-Bradley Micro800 Scaler Instruction

Micro800 Scaler Instruction (SCL)

The Scaler (scale) instruction is an output-type instruction used to produce a scaled value that has a linear relationship between an input value and the scaling parameters entered into the instruction. The instruction takes a source value and, based on the minimum and maximum ranges entered into the instruction, makes a linear conversion and stores the result in an output tag (word). You simply enter into the instruction the source location, the minimum and maximum ranges of both the input and output, and the output location. When the instruction is enabled, it will perform the scaling operation every time it is scanned. Figure 17–10 shows the Micro800 Scaler instruction programmed for the example in Figure 17–11.

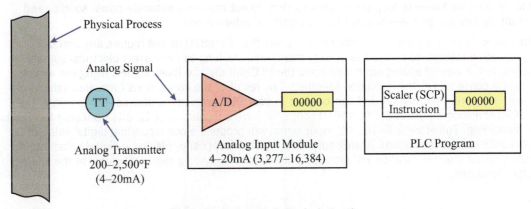

Figure 17–11 Example for Scaler Instruction

The Scaler instruction requires that all values be of the REAL data type. With the Scaler instruction, the scaled output value is not clamped between the minimum and maximum values. You should take additional steps to condition the input value before it is entered into the Scaler instruction to ensure that it is within the input minimum and maximum values.

Analog I/O Configuration

Depending on the PLC and analog I/O interface type (embedded or modular) will determine the configuration settings, if any that must be entered for the analog I/O to work with your field devices. For example, the Allen-Bradley Micro810 controller, like the one that we programmed in Chapter 8, has four 0–10V DC embedded analog inputs and requires no special configuration to use them, whereas the Logix 5000 PLCs with modular analog I/O will require a number of configuration settings in order to use them with your field devices.

The analog field device(s) will drive to a greater extent the selection and configuration of the analog I/O requirements for the PLC system. Once the analog field device specifications are known, the type of analog I/O module(s) and configuration settings can be determined. For example, if you have a field device that produces a 0–10V DC analog signal, then the analog input module and settings must be compatible with that signal type and range. Some analog I/O modules allow you to select the analog type and range for each analog input or output channel to match the type of field device you have. On the other hand, PLCs like the Micro810 just discussed have only 0–10V DC analog inputs and outputs. That does not mean that you cannot use a 0–5V DC or even a 4–20 mA analog field device on a 0–10V DC analog input. If you do, understand the resolution impacts if any. If possible, select the type of analog I/O module(s) that are fully compatible with your analog field devices. Many of the analog I/O modules on the market today are universal in that you configure each analog channel for the type and range of signal that matches your field device.

Once you know the analog field device's signal type and range, review the analog module's specifications. The specifications that are important to the technician are signal type, range, and digital resolution. The resolution of an analog channel is the corresponding digital range that the electrical signal is converted into or from. Resolution is most often denoted as the number of bits of resolution, such as 10-bit, 12-bit, 16-bit, etc. Knowing the digital resolution of an analog signal plays an important role in scaling, calibration, and troubleshooting. Analog signals normally fall into one of two types: voltage or current. The most common DC voltage ranges are 0–5, 0–10, and +/−10V DC, whereas with current signals they are 0–20 and 4–20 mA, with 4–20 mA being the most common of all analog signals found in industrial controls. That is because many 4–20 mA field devices are two-wire loop power devices that do not require a separate power source, and current signals are also less susceptible to electrical interference.

Many older PLCs and some of the smaller PLCs like the Micro810 do not require any special configuration settings to be used as already mentioned. As long as the correct electrical signals are applied to the correct analog input terminals, the PLC will digitize those signals using the analog-to-digital converters at the resolution specified. The result of the conversions (unscaled values) are placed into controller memory and can be used in your PLC program as desired. For example, the 0–10V DC analog inputs of the Micro810 controller have a resolution of 12 bits truncated to 10 bits for smoothing. Therefore, a 0–10V DC input signal will produce a corresponding digital value of 0 and 1,023 (10-bit resolution). Analog outputs are just the opposite, placing a digital value into an analog output memory location will produce a corresponding analog electrical signal at the analog output terminals.

The term "channel" is used to denote the analog I/O location on the module, and they are typically designated as channels 00, 01, 02, 03, etc. An 8-point analog input module would have analog input channels 00–07.

When analog modules require configuration settings to work correctly with your field devices, some of those settings may include:

- Wiring Method (single-ended or differential)
- Data Format (integer, floating point, etc.)
- Signal Type (voltage, current, etc.)
- Signal Range (0–10V DC, 4–20 mA, etc.)
- Scaling (corresponding engineering units)
- Sensor Offset
- Digital Filtering
- Sampling Rate
- Alarming
- Loss of Communications or Program Mode Response

This partial list shows that careful planning and consideration must be given when configuring analog modules in order for them to work correctly with your field devices and control strategy. In many cases, not all of the settings may be required or even desired. You can configure each analog channel based on your application needs.

ControlLogix Analog I/O Module Configurations

To demonstrate the typical settings required to configure analog I/O modules, the Allen-Bradley ControlLogix 1756-IF8 Analog Input and 1756-OF4 Analog Output modules will be shown.

The first task is adding the analog modules to the I/O configuration folder in the Controller Organizer window in the Logix Designer software for your project. Right-click on the chassis that the analog module will be located in and select "Add Module", as shown in Figure 17–12. After selecting "Add Module",

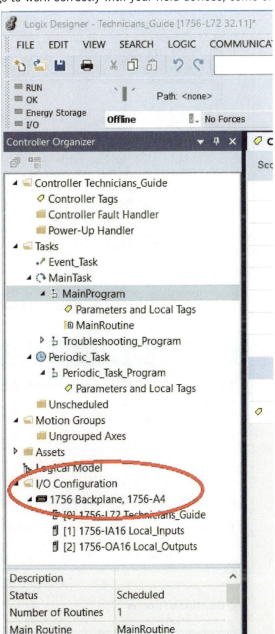

Figure 17–12 Adding Analog I/O Module to Project

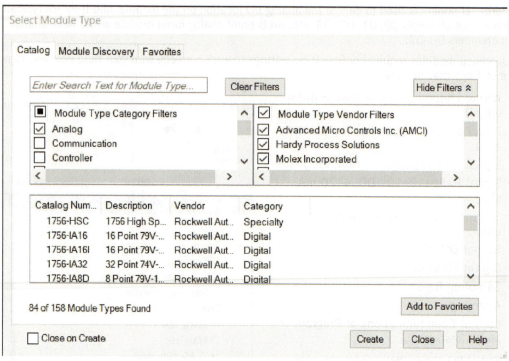

Figure 17–13 Selecting Module Type

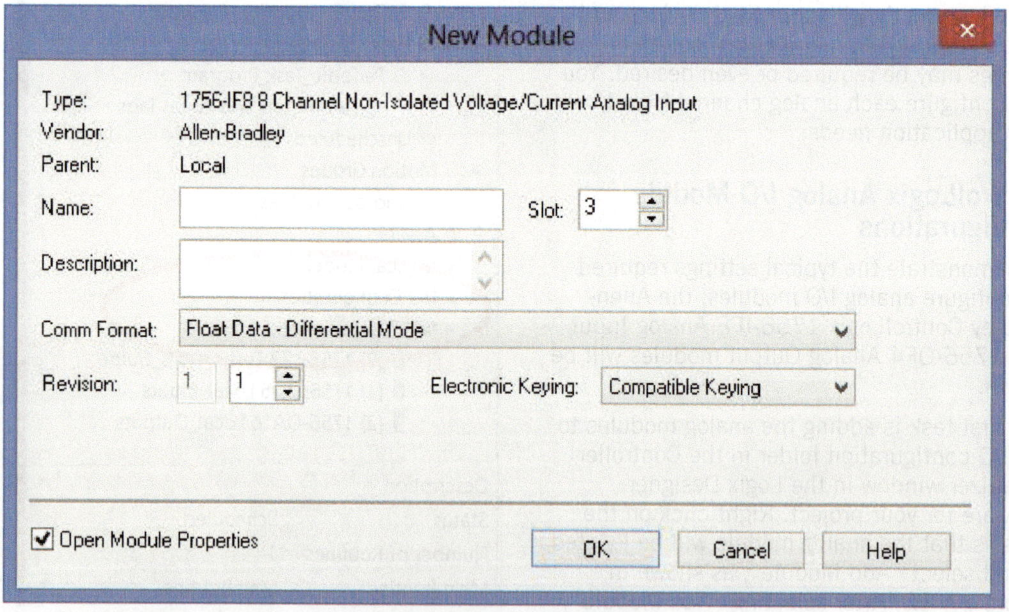

Figure 17–14 New Module Configuration

the Select Module Type window will appear as shown in Figure 17–13. Find and select the analog module type from the available list and then click on "Create". In our example, we have selected the 1756-IF8 Analog Input module first. After clicking on the "Create" button, the New Module window will appear as shown in Figure 17–14.

In the New Module window, select a name and description for the analog module if desired. Next, select the chassis slot number that the module will reside in from the dropdown list. Select the module revision number and electronic keying. Electronic keying is optional. The "Comm Format" selection is unique to analog and some other modules. The selection you make here cannot be changed later unless you delete the module and recreate the module. The dropdown list will only show those communication formats applicable to the type of analog module you are configuring. The communication format determines what type of configuration options are made available, the type of data transferred between the module and controller, and the tags that are created when the configuration is complete. For analog input modules there are over 30 formats to select from, and for analog output modules there are 8. The primary consideration when selecting a communication format is the data type, integer, or floating point. Floating point data format is recommended unless your application requires an integer data format. With some analog input modules like the 1756-IA8, you must also select the type of wiring method you have chosen to use, single-ended or differential. Single-ended wiring compares one side of the signal to signal ground. All analog input devices on the module are tied to a common ground. Single-ended wiring maximizes the number of usable channels on the module. For example, if you select differential mode for an eight (8) channel module like the 1756-IA8, only half of the analog inputs will be available, four (4) instead of the full eight. The differential wiring method is recommended for applications that can have separate signal pairs or a common ground is not available. Differential wiring is recommended for environments where improved noise immunity is required. One thing to keep in mind with differential mode is that the channels are not totally isolated from each other. Choose the wiring method that best fits your application. Refer to the ControlLogix Analog I/O Modules User Manual for additional information on communication formats and wiring methods.

In this demonstration, we have chosen the floating-point data format and differential wiring method. After selecting the communications format, slot, revision, name, description, and electronic keying, click on the "OK" button. If you left the box checked for "Open Module Properties" the module's properties window will appear as shown in Figure 17–15. You can also open the module properties window anytime from Controller Organizer window by double clicking on the module.

With the module properties window open as shown in Figure 17–15, you can choose from several property areas to view and configure. We have opened the channel 0 (Ch00) configuration window. Did you notice there are only four analog channels shown for the 1756-IA8 module? That is because we selected differential mode, not single-ended, and that limits the available analog channels to just four. The following is a brief description of each property area.

General – The General property area provides information on the type and vendor of analog module, name, slot, description, revision, electronic keying, and some module status information.

Connection – The Connection property area lets you enter a requested packet interval (RPI), inhibit the module, and set a connection fault when the module is in Run mode. The RPI provides a defined, maximum period of time when data is transferred to the controller and defaults to 100 ms. Most often you can leave this area in the default selections.

Module Info – The Module Info property area displays information about the analog module as well as additional status information.

Channels – The Channels property area contains two configuration areas for each available channel, general configuration, and alarming. This area is the most important to the technician and requires

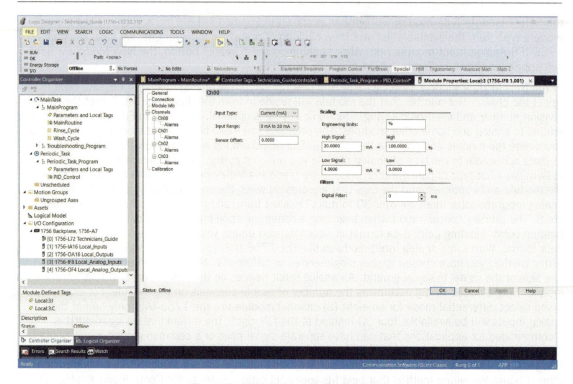

Figure 17–15 Analog Input Module Properties

proper configuration for correct analog operation. The first selection is the input type (current, voltage, etc.). Next is the electrical input range (+/− 10V, 4–20 mA, etc.) for the channel. Select the input range that matches your field device using the dropdown list of available ranges. This will determine the minimum and maximum signal that can be detected by the module. The Sensor Offset parameter allows you to enter a value to compensate for any sensor offset errors, the default is zero. Digital filter is a value in milliseconds that specifies the time constant for a first-order lag filter on the input, the default is zero. Scaling allows you to enter the high and low signals and their corresponding engineering values along with engineering units. For example, if you have a sensor that produces a 4–20 mA signal that represents 0–500 psi, enter these values in the scaling parameters. The analog value sent to the controller will be in engineering units (psi). Scaling is only available if you have selected the floating-point data format. See Figure 17–14.

The alarming area for each analog channel allows you to set alarm limits that, when triggered, will set corresponding alarm bits in the controller memory area for that module. You can monitor these alarm flags (bits) as desired in your PLC program or through an operator interface device. The process alarms you can configure are Low/Low, Low, High, High/High, and Rate Alarm for each analog channel, as shown in Figure 17–16. If you do not want alarming, set the alarm values to zero (default).

Calibration – The Calibration area lets you recalibrate the default factory calibrations, if necessary. Calibration corrects any hardware inaccuracies on a particular analog channel. Recalibration should rarely be required and if needed should only be done by qualified instrumentation technicians.

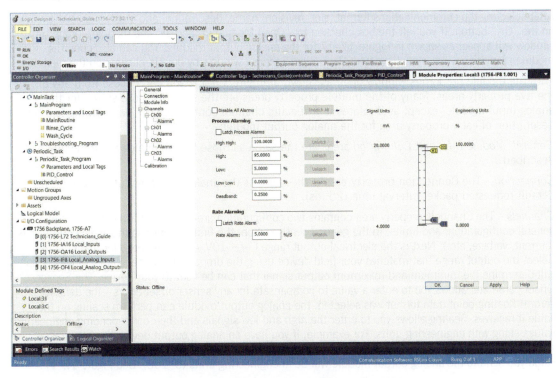

Figure 17–16 Analog Input Module Alarming Properties

Next, we will configure a 1756-OF4 Analog Output module to demonstrate the different configuration settings unique to analog output modules.

Add the analog output module to the I/O configuration folder in the Controller Organizer window by right-clicking on the chassis that the analog module will be located in and select "Add Module". Then, select the analog output module from the available list and click on "Create". After clicking on the "Create" button, the New Module window will appear as shown in Figure 17–14, but this time it will display the 1756-OF4 module.

As with the analog input module, select a name and description for the new module if desired. Next, select the chassis slot number that the module will reside in and the module's revision number. Configure electronic keying if desired. As with the analog input module, the "Comm Format" selection cannot be changed later unless you delete the module and recreate the module. The dropdown list will only show those communication formats applicable to the analog output module. The communication format determines what type of configuration options are made available, the type of data transferred between the module and controller, and the tags that are created when the configuration is complete. For analog output modules, there are only eight choices. The primary consideration when selecting a communication format for analog output modules is the data type, integer, or floating point. Floating point data format is recommended when sending analog output values to the module in engineering units.

After selecting the communications format, slot, revision, name, description, and electronic keying, click on the "OK" button. If you left the box checked for "Open Module Properties" the module's properties window will appear as shown in Figure 17–17 for the analog output module. You can also open the module properties window anytime from Controller Organizer window by double-clicking on the module.

With the module properties window open as shown in Figure 17–17, you can choose from any of the five property areas. Many of the module property areas are the same as those shown for the analog input module, except as it relates to a specific channel configuration. The following is a brief description of each property area for the analog output module.

General, *Module Info*, and *Calibration* areas are the same as those for the analog input module just described.

Connection – The Connection property area is the same as the analog input module except for the default requested packet interval time (25 ms).

Channels – The Channels property area contains two configuration areas for each available channel, general configuration, and limits. In the general configuration, the first selection is the analog output type (current, voltage, etc.). Next is the electrical output range (+/– 10V, 4–20 mA, etc.) for that channel. Select the output range that matches your field device using the dropdown list of available ranges. This will determine the minimum and maximum output signal that can be sent to your device. The Sensor Offset parameter allows you to enter a value to compensate for any sensor offset errors—the default is zero. If floating-point data format was selected, the analog output module can perform scaling on the value it receives. Scaling allows you to enter the high and low signals and their corresponding engineering values along with engineering units. For example, if you have an analog output device that requires a 4–20 mA signal that corresponds to 0–100%, enter these values in the scaling parameters. The analog value you send to the analog output module will be in engineering units (0–100). See Figure 17–17.

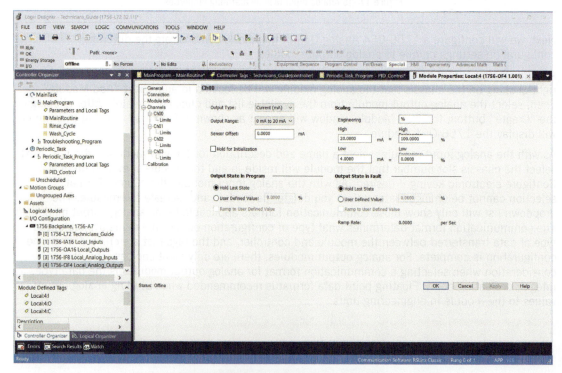

Figure 17–17 Analog Output Module Properties

With analog output modules, you also select how each analog output channel reacts when the PLC is in program mode or a fault is detected. The two choices are "Hold Last State" and "User Defined Value". If the User Define Value is selected, choose the value and whether to ramp to that value and at what rate. You should carefully consider these settings for your analog outputs to ensure failsafe operations.

The Limits area allows you to select output clamping, clamp alarming, and output ramping for each analog output channel; see Figure 17–18.

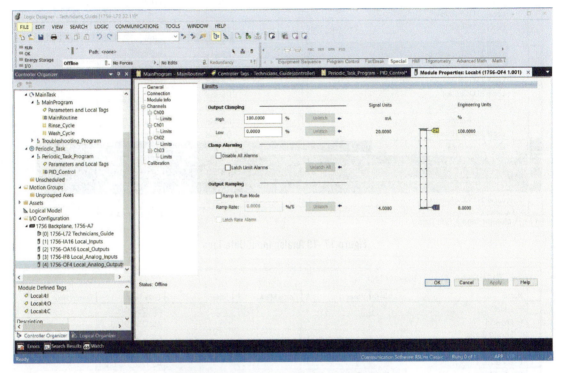

Figure 17–18 Analog Output Module Limit Properties

Once you have installed, wired, added, and configured the appropriate settings for your analog modules, they are ready to be used with your analog devices and user program. Depending on the type of analog module and how it was configured will determine the type of information created in PLC memory for each module. Figure 17–19 shows the analog input data words for the four analog input channels we configured for the 1756-IF8 module located in slot 3 of the local chassis. If everything was configured correctly and your analog devices are functioning, you should see these values change as your process changes. Figure 17–20 shows the analog output data words for the four analog output channels we configured for the 1756-OF4 module located in slot 4 of the local chassis. By placing a value in these memory words, the analog output module will produce a corresponding electrical signal at the module's output terminals.

As you can see with analog I/O, there can be a fair amount of additional configuration required in order for them to work for your application as compared to digital I/O. The examples presented in this chapter only demonstrated two types of analog modules and the steps required to configure them. When working with analog modules, take the time to become familiar with each module by reading the analog module's user manual and studying the configuration settings. Troubleshooting analog I/O will be covered in Chapter 21.

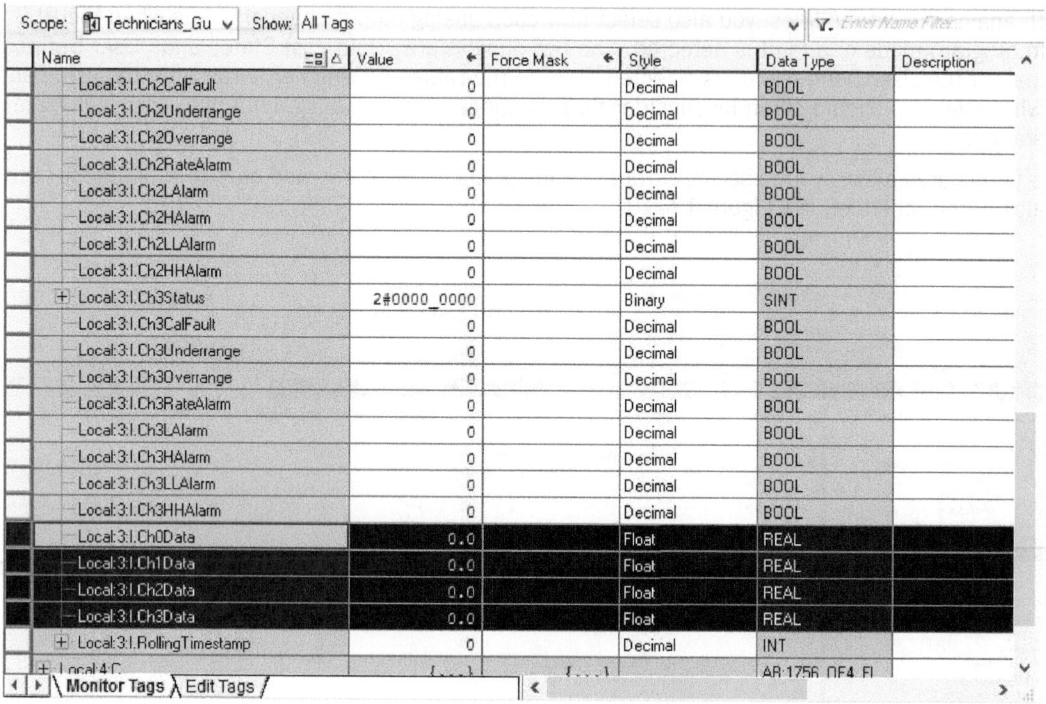

Scope: Technicians_Gu ∨ Show: All Tags ∨ ⟨Y. Enter Name Filter⟩

Name		Value	Force Mask	Style	Data Type	Description	
Local:3:I.Ch2CalFault		0		Decimal	BOOL		
Local:3:I.Ch2Underrange		0		Decimal	BOOL		
Local:3:I.Ch2Overrange		0		Decimal	BOOL		
Local:3:I.Ch2RateAlarm		0		Decimal	BOOL		
Local:3:I.Ch2LAlarm		0		Decimal	BOOL		
Local:3:I.Ch2HAlarm		0		Decimal	BOOL		
Local:3:I.Ch2LLAlarm		0		Decimal	BOOL		
Local:3:I.Ch2HHAlarm		0		Decimal	BOOL		
+ Local:3:I.Ch3Status		2#0000_0000		Binary	SINT		
Local:3:I.Ch3CalFault		0		Decimal	BOOL		
Local:3:I.Ch3Underrange		0		Decimal	BOOL		
Local:3:I.Ch3Overrange		0		Decimal	BOOL		
Local:3:I.Ch3RateAlarm		0		Decimal	BOOL		
Local:3:I.Ch3LAlarm		0		Decimal	BOOL		
Local:3:I.Ch3HAlarm		0		Decimal	BOOL		
Local:3:I.Ch3LLAlarm		0		Decimal	BOOL		
Local:3:I.Ch3HHAlarm		0		Decimal	BOOL		
Local:3:I.Ch0Data		0.0		Float	REAL		
Local:3:I.Ch1Data		0.0		Float	REAL		
Local:3:I.Ch2Data		0.0		Float	REAL		
Local:3:I.Ch3Data		0.0		Float	REAL		
+ Local:3:I.RollingTimestamp		0		Decimal	INT		
+ Local:4:C		{...}	{...}		AB:1756_OF4_FI...		

\ **Monitor Tags** ⟨ Edit Tags /

Figure 17–19 Analog Input Data Tags

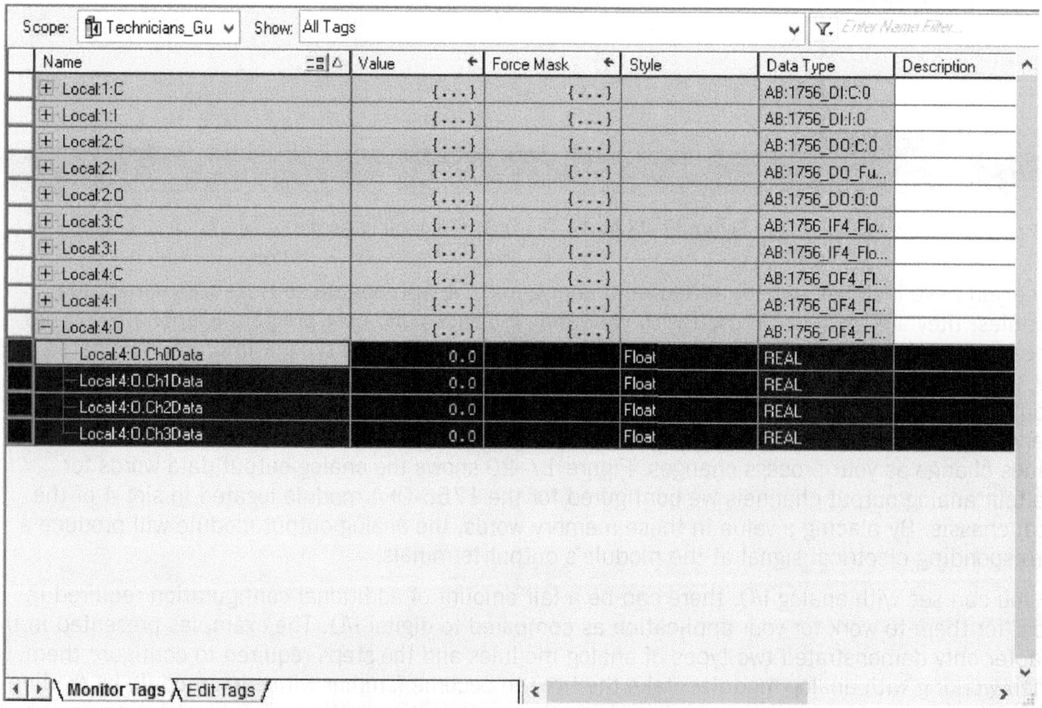

Scope: Technicians_Gu ∨ Show: All Tags ∨ ⟨Y. Enter Name Filter⟩

Name		Value	Force Mask	Style	Data Type	Description	
+ Local:1:C		{...}	{...}		AB:1756_DI:C:0		
+ Local:1:I		{...}	{...}		AB:1756_DI:I:0		
+ Local:2:C		{...}	{...}		AB:1756_DO:C:0		
+ Local:2:I		{...}	{...}		AB:1756_DO_Fu...		
+ Local:2:O		{...}	{...}		AB:1756_DO:O:0		
+ Local:3:C		{...}	{...}		AB:1756_IF4_Flo...		
+ Local:3:I		{...}	{...}		AB:1756_IF4_Flo...		
+ Local:4:C		{...}	{...}		AB:1756_OF4_FI...		
+ Local:4:I		{...}	{...}		AB:1756_OF4_FI...		
− Local:4:O		{...}	{...}		AB:1756_OF4_FI...		
Local:4:O.Ch0Data		0.0		Float	REAL		
Local:4:O.Ch1Data		0.0		Float	REAL		
Local:4:O.Ch2Data		0.0		Float	REAL		
Local:4:O.Ch3Data		0.0		Float	REAL		

\ **Monitor Tags** ⟨ Edit Tags /

Figure 17–20 Analog Output Data Tags

Allen-Bradley Logix 5000 PID Instruction

Before we cover the PID instruction, let us first look at what the elements of a process control system are, what a process control loop is, and how we might use the PID instruction.

Process control can be defined as a means by which we regulate a process. The heating/cooling system that maintains or regulates your home's temperature can be thought of as a process control system. The internal temperature of your home is the process under control and the means to regulate or control that temperature is your home's furnace or air conditioner. The hot water heater in your home is also a process system. Every day our lives are affected by some means of process control. Figure 17–21 illustrates the basic elements of a process control system.

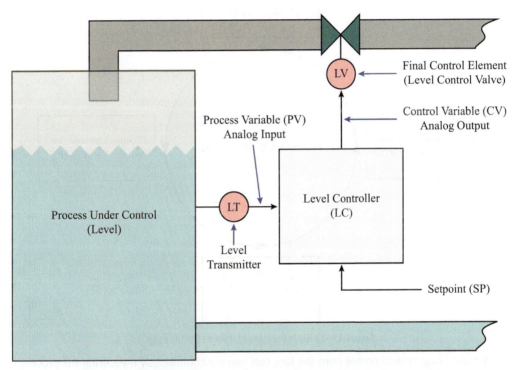

Figure 17–21 Elements of a Process Control Loop (System)

In the system shown in Figure 17–21, a level transmitter (LT), a level controller (LC), and a control valve (LV) are all used to control the level of the liquid in the storage tank. The control system is designed to maintain the level in the storage tank at some predetermined level from the bottom of the tank. This predetermined level is called the process setpoint or **setpoint (SP)**. It will be assumed that the rate at which the liquid leaves the tank varies. The level transmitter is the measurement device that converts the physical level in the tank into a corresponding electrical analog signal, also referred to as the process variable (PV). The process controller reads the level measurement and compares it against the SP or desired level. If the level in the tank is not at the desired level, then the process controller produces a series of corrective actions that are sent to the control valve in an attempt to bring the process under control. The control valve is referred to as the control variable (CV) or final control element. The CV or final control element is a device that when operated, exerts a direct influence on the process under control. Devices such as valves, variable speed drives, actuators, and heating elements are all examples of final control elements.

As you can see from this example, a process control system consists of four key elements: the *process*, the measurement or PV, the *process controller* containing the SP, and the final control element or CV. When all four elements are present, it is considered a closed process control loop. Figure 17–22 shows a simple block diagram of a closed process control loop.

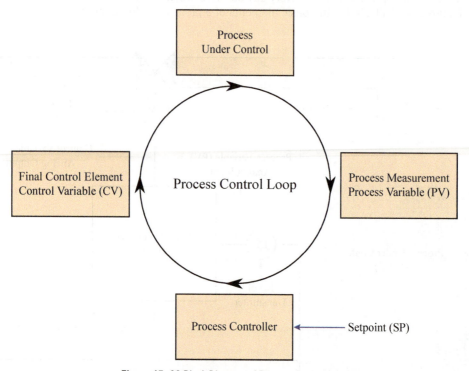

Figure 17–22 Block Diagram of Process Control Loop

The term *closed loop control* comes from the fact that you are continuously measuring the process under control to determine if corrective action is required to maintain the process within the desired limits (setpoint), whereas with open loop control there is no continuous measurement and corrective action taking place to maintain the process within limits. For example, in Figure 17–21, if you were to manually adjust the valve position based on current conditions, there would be no guarantee that the level would remain the same in the tank because of changing conditions. It is like operating blind.

The process controller can be a stand-alone device or a software instruction that is part of a larger control system. A process controller has analog inputs and outputs that are used to monitor and control the process. In addition to the process connections, the process controller will also have a means to enter the desired SP, as well as the ability to make adjustments to various parameters that are designed to tune the process controller for a given process. Some process controllers that are used for temperature control may have the analog output converted to a time proportioning on/off output for driving a heater or cooling unit.

The remainder of this section is devoted to describing the Allen-Bradley Logix 5000 Proportional Integral Derivative (PID) instruction. The PID instruction is a software type of closed loop controller, like that described above.

The Allen-Bradley Logix 5000 PID instruction is an output instruction that is placed in either a periodic or continues task. Because the PID instruction uses a time base calculation, its execution needs to be synchronized with the sample rate of the PV signal. The PID instruction cannot be updated any faster than the PV is being updated. Preferably, the PV is updated at least two times faster than the PID instruction is allowed to calculate for best results. The update time of the PID instruction is dependent on the physical properties of the process under control. For very slow process conditions like temperature loops, an update time of several seconds or minutes may be all that is required to obtain good control. For faster loops, like flow and pressure, the update time may need to be one second or faster. Because the PID instruction will perform a new calculation every time the instruction is scanned and is enabled (true), the easiest way to provide the desired PID update time is to place the PID instruction into a periodic task set to the desired update time. When placed in a periodic task, place the PID instruction on an unconditional rung, as shown in Figure 17–23a. If the PID instruction is placed in a continues task, you must use a timer to trigger the execution of the PID instruction at the desired update time, as shown in Figure 17–23b.

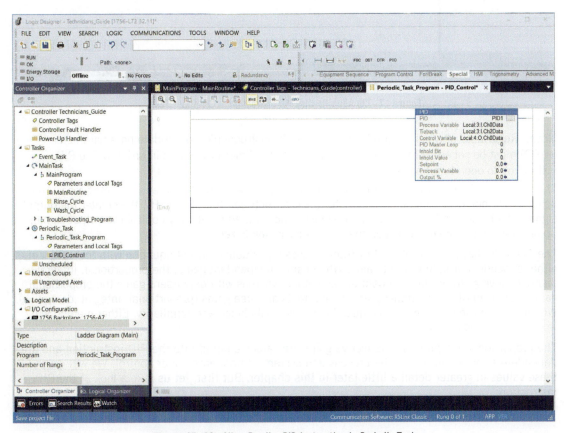

Figure 17–23a Allen-Bradley PID Instruction in Periodic Task

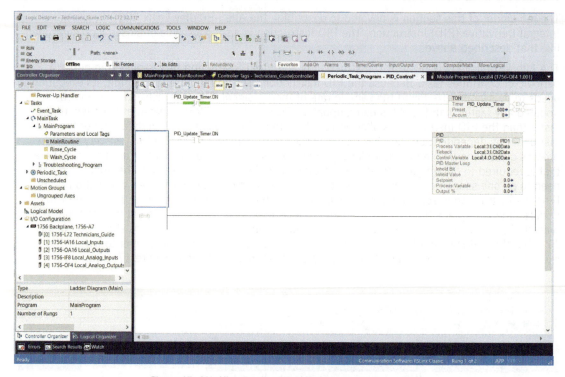

Figure 17–23b Allen-Bradley PID Instruction in Continues Task

Note: Regardless of which method you use, the PID instruction's loop update time parameter (.UPD) must be set equal to the periodic task rate or the timer preset rate in order for the PID instruction to operate properly.

The PID instruction controls its output based on the error between the SP and PV being measured using a mathematical equation. The greater the error between the SP and PV, the greater the output signal will change, and vice versa. The PID instruction will automatically control the output signal, up or down, until the error between the SP and PV is nearly zero.

The Allen-Bradley Logix 5000 PID instruction uses a positional form of equation with the option of either dependent or independent gains. When using independent gains, the proportional, integral, and derivative gains are independent of each other, whereas with dependent gains the proportional gain is replaced with a controller gain which affects all three gains (proportional, integral, derivative). Which equation you use becomes a question of personal choice and familiarity. Either method will work to control your process.

The proportional, integral, and derivative gains are values entered into the PID instruction and are dependent upon the nature of the process and system components under control. We will cover these values in greater detail a little later in this chapter. But first, let us take a look at the PID instruction.

As you can see in Figures 17–23a and 17–23b, there are ten operands (parameters) displayed in the PID instruction block. Three are required operands, four are optional depending on your application, and three are display only. Table 17–1 provides a brief description of each of the operands in the PID instruction block.

Table 17–1 Logix 5000 PID Instruction Operands (Parameters)

Operand	Type	Description
PID	PID Structure	Assign a unique PID tag for each PID instruction in your program.
Process Variable (PV)	Tag	This is the memory location of the process value you want to control (analog input). Most often this is the input data tag for the analog input channel that the PV is wired to. There is no need to perform any scaling of the analog input value if desired. In the PID configuration parameters, you will enter the unscaled and engineering units for the PV being received by the PID instruction and the instruction will perform the conversion internally.
Tieback (optional)	Tag/Immediate	This is the memory location of the tieback value when using a hand/auto station which is bypassing the output of the PID instruction. Enter 0 if not using tieback.
Control Variable (CV)	Tag	This is the memory location of the analog output that is controlling the process (final control element). Most often this is the output data tag for the analog output channel that the CV device is wired to. There is no need to perform any scaling of the analog value. In the PID configuration parameters, you will enter the maximum and minimum values of the CV that corresponds to 0 and 100% of the PID output. If you do not enter the analog output data tag that corresponds to the analog channel directly, make sure that the tag you assign to this parameter is of the REAL data type.
PID Master Loop (optional)	PID Structure	The PID tag of the master PID loop when performing cascade control. Enter 0 if not using cascade control.
Inhold Bit (optional)	Tag	This is the current status of the inhold bit from the 1756 analog output channel that the CV signal is wired to, and is used by the PID instruction to support bumpless restart. Enter 0 if you do not want to use this feature.
Inhold Value (optional)	Tag	The data read back value from the 1756 analog output channel that the CV is wired to, and is used by the PID instruction to support bumpless restart. Enter 0 if you do not want to use this feature.
Setpoint (SP)	Display Only	Displays the current SP value for the PID instruction. You can use the PID instruction's control structure mnemonic (.SP) to change this value as desired within your logic or through an operator interface device.
Process Variable (PV)	Display Only	Displays the current scaled process value in the PID instruction.
Output %	Display Only	Displays the current output percentage value for the PID instruction (0–100%).

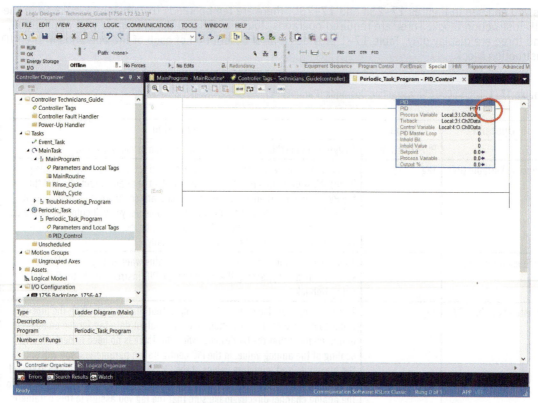

Figure 17–24 PID Instruction Setup Dialog Box

After entering the PID instruction operands, you can then open the configuration screens by clicking on the small box in the upper right corner of the instruction block, as shown in Figure 17–24. You should now see the main PID setup dialog box with five tabs (Tuning, Configuration, Alarms, Scaling, and Tag) that allow you to enter additional information to complete the configuration of the instruction, as shown in Figure 17–25.

The Tuning window is the first window to appear when you open the main PID setup dialog box. The Tuning window allows you to configure the PID instruction for the mode of operation and how the PID instruction reacts and regulates your process. Table 17–2 provides details on each parameter in the Tuning window.

Note: To avoid receiving error messages during configuration of the PID instruction, configure the parameters in the Tuning window last.

The Configuration window is selected by clicking on the Configuration tab on the PID setup dialog box. Figure 17–26 shows the configuration window. The configuration window allows you to configure how the PID algorithm is to function when in auto mode. Table 17–3 provides details on each parameter in the configuration window.

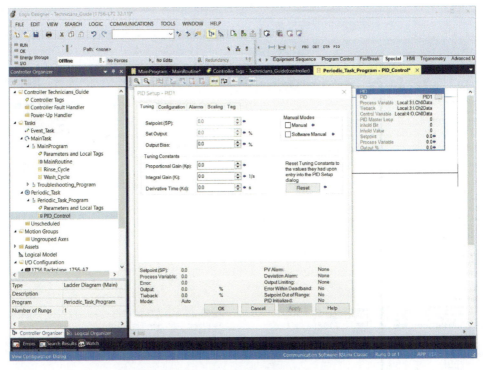

Figure 17–25 PID Tuning Configuration Window

Table 17–2 PID Tuning Window Parameters

Parameter/Field	Description
Setpoint (SP)	Enter the SP value, in engineering units, for the PID instruction to control to. For example, if the PV is measuring pressure (psi) and you want the PID instruction to maintain the pressure at 45.5 psi, enter that value in this parameter. You can also enter the SP value within your PLC logic or externally by addressing the .SP mnemonic in the PID control tag.
Set Output %	When operating in software manual mode, this value is used for the output of the PID instruction. You enter the output percentage (.SO) desired, otherwise this parameter displays the current output of the controller when in auto mode.
Output Bias	Enter an output bias percentage in this parameter if required for your application. In most cases, an output bias is not required.
Proportional Gain	Enter the desired proportional gain value in this parameter.
Integral Gain	Enter the desired integral gain value in this parameter.
Derivative Gain	Enter the desired derivative gain value in this parameter.
Manual Mode	When manual mode is desired, select either manual (.MO) or software manual (.SWM) mode. For example, if you want to set the PID output to 50% using software manual mode, select software manual mode and enter the value 50 into the Set Output% parameter (above). You can also set the manual mode in your PLC logic by addressing the desired mnemonic in the PID control tag.

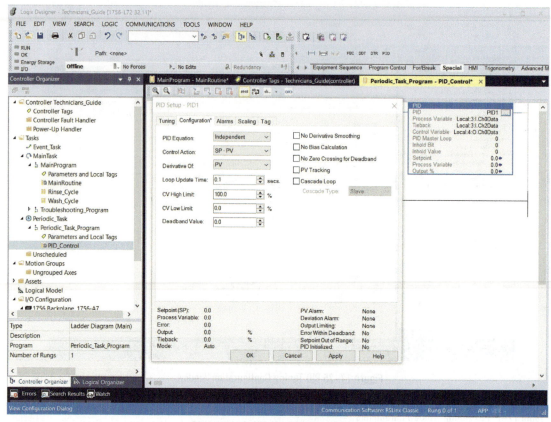

Figure 17–26 PID Configuration Window

Table 17–3 PID Configuration Window Parameters

Parameter/Field	Description
PID Equation	Select the PID equation to use, *dependent or independent* gains. As mentioned previously, select independent gains when you want all three gains to operate independently. Select dependent gains when you want a controller gain that will affect all three terms. Which PID equation you select is a matter of preference and experience.
Control Action	Select either *SP-PV* (reverse acting) or *PV-SP* (forward acting) for the PID control action. Reverse acting will cause the CV output of the PID instruction to decrease when the process variable (PV) is greater than the setpoint (SP), for example in a heating application. Forward acting will cause the CV output of the PID instruction to increase when the PV is greater than the SP, for example in a cooling application. Set this parameter to match your type of process loop application.

Derivative of	Select either *PV* or *Error*. When PV is selected, the derivative rate calculation will be based on the PV. Use PV to eliminate output spikes during SP changes. Use derivative of error for fast response to SP changes if your process can tolerate overshoot.
Loop Update Time	Enter the loop update time for the PID instruction. The loop update time should be set to the same interval time as the PID instruction is being executed. For example, if the PID instruction is placed in a periodic task, set the loop update time to the same periodic task time. If using a continuous task and timer, set the loop update time to the timer's preset. The loop update time is entered in 10-ms (0.01 s) increments. As a general rule, the loop update time should be 5–10 times faster than the natural period of the control loop. The faster the process responds to changes, the faster the loop update time needs to be.
CV High Limit	Enter the maximum PID output limit (%) allowed by the PID instruction.
CV Low Limit	Enter the minimum PID output limit (%) allowed by the PID instruction.
Deadband	Enter a deadband value. The deadband extends above and below the SP by the value you enter in this parameter. The deadband value is entered in the same engineering units as the SP. By entering a deadband value, you are telling the PID instruction not to change the output as long as the error remains within this range. Enter a value of 0 for no deadband.
Derivative Smoothing	Enable or disable this selection. When enabled, a derivative smoothing filter is used to minimize large derivative spikes caused by noise in the PV value.
Bias Calculation	Enable or disable this selection. When you select no bias calculation, you disable the back-calculation of the .BIAS term, and the PID instruction no longer provides a bumpless transfer from manual to auto when the integral control is not used. By default, the bias calculation is enabled.
Zero Crossing	Enable or disable this selection. When enabled, allows the PID instruction to continue using the error in calculating a new output value as the PV crosses into the deadband until the PV crosses the SP. Do not enable this if you are not using a deadband.
PV Tracking	Enable or disable this selection. When enabled, the SP will track the PV when in manual mode. When the PID loop is returned to auto mode, the SP value will be the same as the PV.
Cascade Loop	Enable or disable this selection. When performing cascade control with two PID loops, enable this selection.
Cascade Type	If you enable cascade loop above, select either inner (slave) or outer (master) for this PID instruction.

The Alarm window is selected by clicking on the Alarm tab on the PID setup dialog box. Figure 17–27 shows the alarm configuration window. The alarm configuration window allows you to configure alarm settings for the PID instruction. The PID alarms can be monitored by your logic, or externally, by referencing the corresponding mnemonic in the PID control structure tag. Table 17–4 provides details on each parameter in the alarm configuration window.

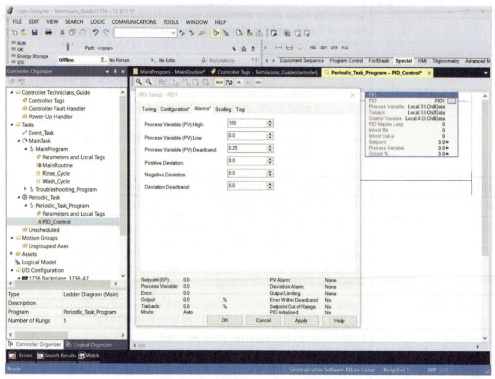

Figure 17–27 PID Alarm Configuration Window

Table 17–4 PID Alarm Configuration Window Parameters

Parameter/Field	Description
PV High	Enter a process variable (PV) high alarm value. When the PV is equal to or greater than the high alarm value, mnemonic bit .PVHA in the PID control member is set to 1.
PV Low	Enter a PV low alarm value. When the PV is equal to or less than the low alarm value, mnemonic bit .PVLA in the PID control member is set to 1.
PV Deadband	Enter a PV deadband alarm value. This is a one-sided deadband. The PV High and Low alarm bits are not set until the PV crosses the deadband and reaches the alarm value that you set in the PV High or Low alarm parameters above. Once the alarm bit is set (high or low), the alarm bit will remain on until the PV exits the deadband.
Positive Deviation	Enter a positive deviation alarm value. When the PV is above the positive deviation value, mnemonic bit .DVPA in the PID control member will be set to 1.
Negative Deviation	Enter a negative deviation alarm value. When the PV is below the negative deviation value, mnemonic bit .DVNA in the PID control member will be set to 1.
Deviation Deadband	Enter a deviation deadband alarm value. This is a one-sided deadband like the PV deadband above. The Positive and Negative Deviation alarm bits are not set until the PV crosses the deadband and reaches the alarm value that you set in the positive or negative deviation alarm parameters above. Once the alarm bit is set (positive or negative), the alarm bit will remain on until the PV exits the deadband.

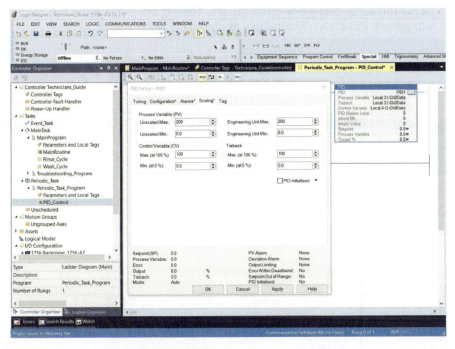

Figure 17–28 PID Scaling Configuration Window

The Scaling window is selected by clicking on the Scaling tab on the PID setup dialog box. Figure 17–28 shows the scaling configuration window. The scaling configuration window allows you to configure the minimum and maximum scaling values for the PV, CV, and tieback. Table 17–5 provides details on each parameter in the scaling configuration window.

Table 17–5 PID Scaling Configuration Window Parameters

Parameter/Field	Description
PV Unscaled Maximum	Enter a maximum PV value that equals the maximum unscaled value received from the analog input channel for the PV value.
PV Unscaled Minimum	Enter a minimum PV value that equals the minimum unscaled value received from the analog input channel for the PV value.
PV Scaled Maximum (engineering units)	Enter the maximum engineering units corresponding to the PV Unscaled Maximum value.
PV Scaled Minimum (engineering units)	Enter the minimum engineering units corresponding to the PV Unscaled Minimum value.
CV Maximum	Enter the maximum CV value corresponding to 100%.
CV Minimum	Enter the minimum CV value corresponding to 0%.
Tieback Maximum	Enter a maximum tieback value that equals the maximum unscaled value received from the analog input channel for the tieback value.
Tieback Minimum	Enter a minimum tieback value that equals the minimum unscaled value received from the analog input channel for the tieback value.

The last tab on the PID setup dialog box is the Tag tab. The Tag tab displays the PID control tag name along with general PID status information, as shown in Figure 17–29.

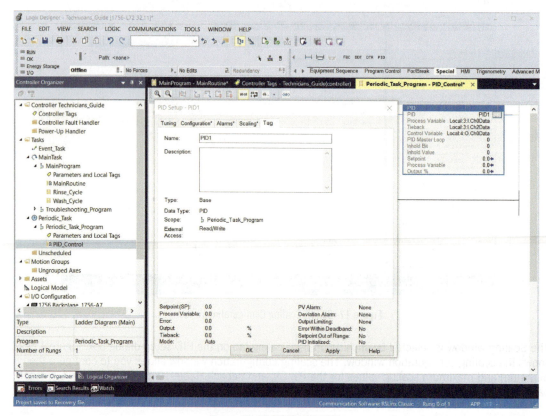

Figure 17–29 PID Tag Window

There are well over 60 mnemonics associated with a PID control structure, and for that reason they will not be covered here. Refer to the Allen-Bradley Logix 5000 Controllers General Instruction Manual for a complete list and description of each PID control structure mnemonic.

After studying the list of configuration parameters associated with the PID instruction, you should begin to have a better understanding of the instruction and its configuration requirements. Of all the PLC instructions, the PID instruction is the most complex and requires the greatest knowledge to configure and implement properly.

PID Program Example

The following example will help in understanding the application and configuration of the PID instruction. Figure 17–30 shows a basic process control loop for which we will program the Allen-Bradley Logix 5000 PID instruction.

The control strategy in this example is to maintain the level in the tank at a predetermined height (setpoint) as the outflow of the storage tank varies over time. The level is maintained (controlled)

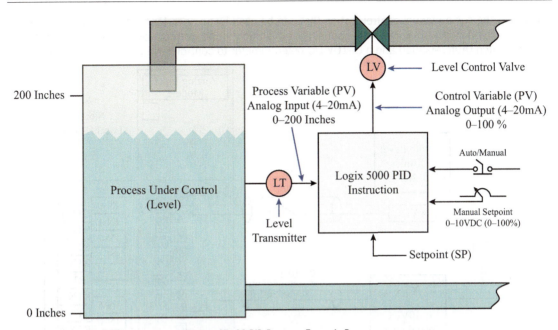

Figure 17–30 PID Program Example Process

by a flow control valve located in the inlet pipe. In simple terms, the inlet control valve is being regulated to nearly mirror the outlet flow rate and the storage tank is used to compensate for rapid changes in outlet flow. One of the first steps involved in programming any PID-type controller is to identify the system components that will be used to monitor and control the process. In our example, we will begin by first identifying the PV, then the CV, and last we will identify any auxiliary control functions (manual, etc.).

Process Variable (PV) – The process variable under control in our example is the liquid level in the storage tank, as measured in inches by the level transmitter located in the storage tank. The level transmitter is designed to continuously measure the level and output a corresponding analog signal that is representative of that level measurement. The analog signal is wired to a PLC analog input module. Notice in Figure 17–30 that the level transmitter has a range of 0–200 in. and converts that range into a 4–20 mA analog signal. The PLC analog input module is designed to convert the 4–20 mA analog signal into a digital value with a range as determined by the analog module's configuration. In our example, we have configured the analog input module's scaling feature for a corresponding value of 0.0–200.0.

Control Variable (CV) – The control variable or final control element in our example is the flow control valve located in the inlet pipe, refer back to Figure 17–30. The amount of liquid entering the tank is determined by the position or opening of the flow control valve. The position of the flow control valve in our example is determined by an analog output signal (4–20 mA) that is generated by the PLC. The digital range required to produce the 4–20 mA analog signal has been configured in the analog output module for a value of 0.0–100.0 corresponding to the percent opening of the valve.

Auxiliary Control Functions – The last item in setting up our PID instruction is to determine if there are any auxiliary control functions required for our process loop. If you refer back to Figure 17–30, you will notice that there is a manual control station that can be used by an operator to manually control the position of the flow control valve. The manual control station in our example consists of a two-position selector switch (Auto/Manual) and a 0–10V DC potentiometer to control valve position in manual.

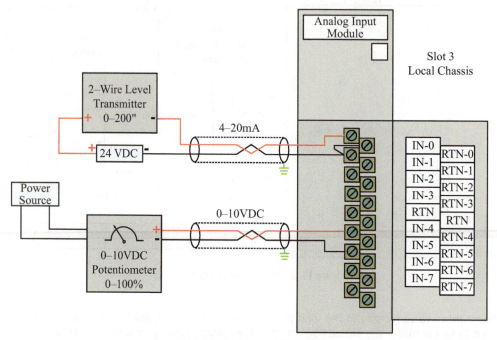

Figure 17–31a Analog Input Module Wiring

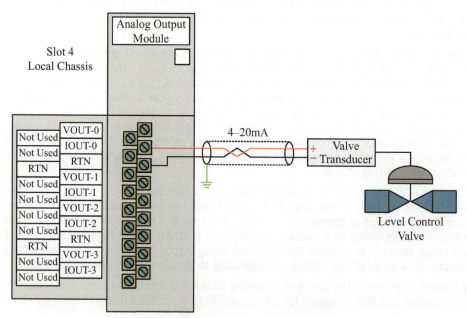

Figure 17–31b Analog Output Module Wiring

The two-position selector switch is wired to the PLC as a digital input. When the input is *true* or *ON* the selector switch is in the manual position.

The 0–10V DC potentiometer is wired to a PLC analog input module. The 0–10V DC output range of the potentiometer corresponds to a valve position of 0–100%. In order to control the PID instruction's output in manual, requires that we set the Manual Mode (.MO) bit to 1 and use the analog input value from the potentiometer as the tieback value in the PID instruction. When manual mode is selected, the PID instruction will take the tieback value (potentiometer), scale it if necessary, and place the result into the PID CV output. We have configured the analog module's scaling feature for the tieback with a value of 0.0–100.0 corresponding to the percent opening desired for the valve Figure 17–31a, 31b, and 31c shows the I/O wiring for our PID example.

In Figure 17–32, we have programmed the PID instruction for our example process. The PID logic has been placed in a periodic task that will be executed at a defined interval (time) based on the process tuning results. Notice the PID Manual Mode (.MO) is controlled by the digital input from the auto/manual selector switch. The PV memory location is the data location for channel 0 of the analog input module that is located in slot 3 of the local chassis. The

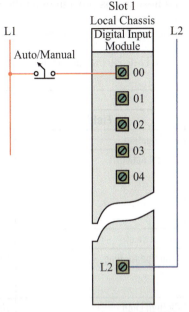

Figure 17–31c Digital Input Module Wiring

tieback memory location is the data location for channel 2 of the analog input module that is located in slot 3 of the local chassis, and the CV memory location is the data location for channel 0 of the analog output module that is located in slot 4 of the local chassis. The PID Master Loop, Inhold Bit, and Inhold Value operands are not being used so they have been set to zero (0) or off.

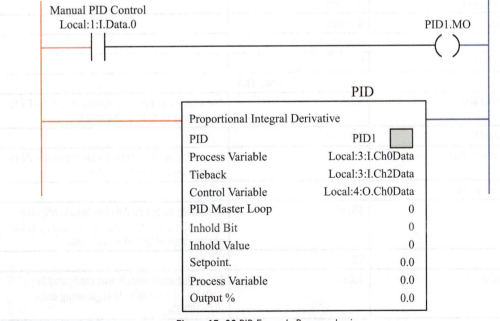

Figure 17–32 PID Example Program Logic

The next step is programming the PID instruction's parameters. If you recall from our study of the PID instruction, there are four configuration screens: tuning, configuration, alarms, and scaling. Table 17–6 shows how we have configured the PID instruction's parameters for our example process.

Table 17–6 PID Example Configuration

Configuration Tab		
Field	**Setting**	**Notes**
PID Equation	Independent	
Control Action	SP-PV	E = SP − PV (*reverse acting*). Whenever the level PV in the storage tank drops below the SP, we want the output of the PID instruction to increase, causing the control valve to open in an attempt to increase the flow into the storage tank (*reverse acting*).
Derivative of	PV	
Loop Update Time	1.0	The value (1.0) entered is only arbitrary (temporary) and should be set based on the tuning results. This value needs to match the PID instructions update time.
CV High Limit	100.0	No limit, set to maximum output %.
CV Low Limit	0.0	
Deadband Value	0.0	Disabled
No Derivative Smoothing	Disabled	
No Bias Calculation	Disabled	
No Zero Crossing	Disabled	
PV Tracking	Disabled	
Cascade Loop	Disabled	
Cascade Type	0	
Scaling Tab		
PV Unscaled Max	200.0	The analog input module already scaled the PV to engineering units (0.0–200.0 in.)
PV Unscaled Min	0.0	
PV Engineering Max	200.0	The analog input module already scaled the PV to engineering units (0.0–200.0 in.)
PV Engineering Min	0.0	
CV Max	100.0	The analog output module has been configured to accept a 0.0–100.0 value corresponding to the analog range of the valve (4–20 mA)
CV Min	0.0	
Tieback Max	100.0	The analog input module was configured to scale the tieback value to engineering units (0.0–100.0%)
Tieback Min	0.0	

Alarms Tab		
PV High	200.0	Set to the maximum PV value. (configure as desired for your process application)
PV Low	0.0	Set to the minimum PV value. (configure as desired for your process application)
PV Deadband	0.0	Not used
Positive Deviation	0.0	Not used
Negative Deviation	0.0	Not used
Deviation Deadband	0.0	Not used
Tuning Tab		
Setpoint (SP)	0.0 (desired SP)	Set this parameter to your desired process SP in engineering units. For example, if you wanted 150.0 in. you would enter 150.0. The SP can be entered through the programming software, from an operator interface device, or in programmed logic if desired
Set Output %	0.0	
Output Bias	0.0	
Proportional Gain (Kp)	0.0	This parameter is set during the tuning process (see tuning PID loops)
Integral Gain (Ki)	0.0	This parameter is set during the tuning process (see tuning PID loops)
Derivative Time (Kd)	0.0	This parameter is set during the tuning process (see tuning PID loops)
Manual Mode	-	We are using a hardwired manual station to set the manual mode (.MO) through logic

Now that we have programmed and configured our PID instruction for our example application, the last and final step is to set (adjust) the tuning parameters in the PID instruction based on the characteristics that are unique to the process under control. The next section will cover in greater detail the loop tuning parameters and several procedures for determining their values.

PID Loop Tuning

Tuning a PID loop is the process of selecting values for the PID tuning parameters (Proportional Gain (Kp), Integral Gain (Ki), Derivative Time (Kd), and Loop Update Time) so that the PID instruction is able to quickly eliminate an error between the SP and PV without causing excessive fluctuations in the process loop under control. PID loop tuning is probably one of the most difficult procedures to master and requires a knowledge of general process control, process controllers, and the process under control. Many people would say that tuning a PID loop is an art. Although not completely true, I would agree that to properly tune a PID loop does require the knowledge and skill that can only come from experience. If you thoroughly understand what each tuning parameter does, you are more likely to be able to tune a PID loop with confidence and success. Before we discuss the techniques that can be used to tune PID loops, let us first review each of the PID instruction tuning parameters.

Proportional Gain (Kp), or controller gain, refers to the amount by which the error signal will influence controller output. In other words, the controller gain changes the output of the controller by an amount proportional to the error between the SP and the PV. The higher the gain, the greater the output will change for a given error. Too much gain may cause the control loop to become unstable or oscillate. You should always try to start out with a small controller gain value and then gradually increase it while observing how the process loop responds to a change or upset.

The *Integral Gain* (Ki), or reset parameter, is used to change the output of the controller by a rate proportional to the error over time. When dependent gains are selected, the Integral Gain (Ki) parameter is expressed in minutes per repeat. This means that as long as there is an error between the SP and the PV, the integral action will add to the controller's proportional output by repeating the previous proportional action over time, and at the frequency specified in the Integral (Ki) parameter. As long as there is an error, the integral action will continue to add to the output of the PID instruction until the PV equals or nearly equals (with deadband) the SP. Some integral gain is required to eliminate the effects of offset, which can occur with proportional-only control. Keep in mind that when dependent gains are selected, the integral action is expressed in minutes per repeat, so the larger the value, the longer the time between repeats. See Figure 17–33 for an illustration of both proportional and integral response to a step change. For independent gains, it is the integral gain (1/s).

The *Derivative Time* (Kd), or rate action, is used as a method of changing the PID output in proportion to the rate of change of the PV. The faster the PV is changing, the greater the influence that the derivative action will have on the output. The derivative action acts as an anticipator, or "brake," in the control loop, helping to minimize the amount of overshoot and undershoot in the control loop's response. You typically see derivative action used when a process has considerable time constant lags, such as temperature loops. You should set the Derivative Time (Kd) parameter to 0, derivative term *off*, when controlling process loops that have a fast response time.

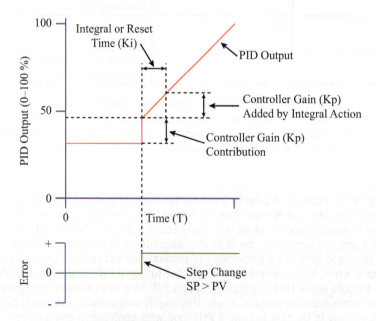

Figure 17–33 Proportional and Integral Response to Step Change

The *Loop Update Time* parameter sets the time between PID calculations and should be set to a time that is five to ten times faster than the natural period of the process under control. The faster that the process responds to a change in the PID output, the faster the loop update time needs to be. Determining the natural period of the process will be covered next under "Loop Tuning Techniques." If you are unsure, start with a loop update time of one second and adjust from there. Keep in mind that the PID Loop Update Time parameter must match the PID calculation update time (PID block execution frequency).

Loop Tuning Techniques

Manual tuning of a PID controller is really a guesswork method and should be left to those with extensive experience in tuning PID loops. The process is quite simple in that you set the integral and derivative terms to zero, adjust the proportional term first, then the integral term followed by the derivative term if used. Sounds simple, but each step requires you to make small changes, see what happens, and adjust again. For someone inexperienced this could become frustrating and take considerable time, whereas for someone with extensive experience it may be preferred and the quickest, especially with fast loops.

Today, there is software available that can be used to assist in tuning PID loops. In fact, many PID controllers have features like auto tune. Take full advantage of these features when available. Use tuning software to assist in establishing the initial tuning values for your process.

The tuning techniques that follow can be used by the technician in establishing the initial tuning values in the absence of tuning software or experience. Do keep in mind that these calculated values are only initial values and additional adjustments are likely required to fine tune the control loop. Process loop characteristics vary greatly, and these tuning techniques may not achieve reasonably good control.

In 1942, J.G. Ziegler and N.B. Nichols published two loop tuning techniques that are still used today by many control engineers and technicians. The two techniques are called the *ultimate gain method (closed loop)* and the *reaction curve method (open loop)*.

 Caution: During loop tuning it is possible that the PID CV output may become unstable and oscillate. For this reason, be sure that the process can safely tolerate such instability. Personnel safety and equipment reaction should also be carefully considered before the tuning process begins.

Ziegler and Nichols Ultimate Gain Tuning Method (Closed Loop Technique)

The ultimate gain method requires that you determine the ultimate gain and ultimate period of the process control loop. The ultimate gain is a controller gain value that will cause a sustained but stable oscillation in the PV from a slight error or upset. The period of these sustained oscillations is called the ultimate period or the natural period of the control loop. This method is conducted with the PID instruction in automatic mode with the integral and derivative terms set to zero (0) or turned off. Once you have determined what the ultimate gain (Gu) and ultimate period (Pu) of the control loop is, you can then use these two values to calculate the initial Proportional Gain (Kp), Integral Gain (Ki) or Reset Time (Ti), Derivative Time (Kd) or Rate Time (Td), and Loop Update Time parameters for the PID instruction. This method should only be used on fast response systems like pressure and flow that can tolerate oscillations.

To determine the ultimate gain and ultimate period of your control loop, perform the following steps.

Step 1.   Enter the following values into the tuning parameters of your PID instruction through the PID Setup Screen:

Proportional Gain (Kp) = 0.5 or 1.0

Integral Gain (Ki) = 0

Derivative Time (Kd) = 0

Loop Update Time = 1.0 second

Step 2.   Enter an initial SP value that you desire into the SP parameter through the PID Setup Screen.

Step 3.   You will need to observe the PV as it varies with time and with respect to the SP value you entered in Step 2. A strip chart recorder or trending chart works well for making this observation, particularly when working with slow-responding process loops.

Step 4.   Place the PID instruction in manual mode and adjust the output of the PID instruction until the PV equals or nearly equals the SP. If you are using a manual control station, then use the manual control station for this adjustment; otherwise select software manual (.SWM) and enter a value in the Set Output (.SO) parameter through the PID Setup Screen. Remember that the value you enter is a percent (0.0–100.0).

Step 5.   After the process loop is stable in Step 4, place the PID instruction in automatic (closed loop) mode.

Step 6.   While observing the relationship of the PV to the SP over time, impose an upset (create an error) on the control loop and observe the response. The simplest and easiest way to impose an upset is to change the SP by a small amount.

Step 7.   If the response curve of the PV produced by Step 6 becomes unstable (see Figure 17–34a), the proportional gain is too high. Reduce the proportional gain (Kp) and repeat Step 6 until you have obtained a sustained but stable oscillation of the PV, as shown in Figure 17–34b.

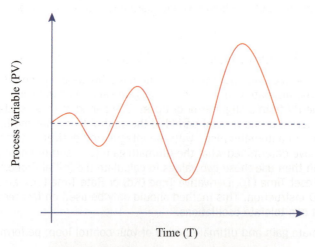

Figure 17–34a Unstable Process Response Curve

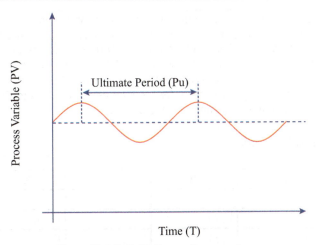

Figure 17–34b Stable Process Response Curve

Step 8. If the response curve of the PV produced by Step 6 dampens out over time, as shown in Figure 17–34c, then the proportional gain is too low. Increase the Proportional Gain (Kp) and repeat Step 6 until oscillations begin to occur, as was shown in Figure 17–34b.

Step 9. When you obtain a stable response in which the PV is oscillating above and below the SP in an even manner, record the values of the ultimate gain (Gu) and ultimate period (Pu). The ultimate gain is the current value in the Proportional Gain (Kp) parameter. The ultimate period is the time of one cycle, measured between successive peaks (see Figure 17–34b). The ultimate period is also referred to as the natural period of the process.

Step 10. Return the PID instruction to manual mode and gain control of the process. At this time you may stop controlling the process if desired.

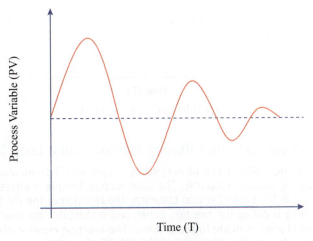

Figure 17–34c Damped Process Response Curve

Step 11. Set the Loop Update Time parameter to a value that is from 5–10 times faster than the ultimate (natural) period as observed in Step 9. If, for example, the ultimate period is 30 seconds and we desire a loop update 10 times faster than the ultimate period, then the Loop Update Time parameter would be set to 3 seconds. Do not forget that the PID instruction update time must also be the same.

Step 12. Determine the tuning parameters for the loop using the Gu and Pu recorded in Step 9, and one of the equations shown in Table 17–7.

Table 17–7 Closed Loop PID Equations

Control	Kp	Ti	Td	Ki	Kd
P Only	0.5 * Gu				
PI	0.45 * Gu	Pu / 1.2		Kp / Ti	
PID (series)	0.6 * Gu	Pu / 2	Pu / 8	Kp / Ti	Kp * Td

Step 13. Place the PID instruction in auto mode. If you have an ideal process, the PID controller should achieve a decay ratio of one-quarter wave when an error occurs, which Ziegler and Nichols defined as good control as shown in Figure 17–35. In reality, the tuning parameters just calculated and entered into the PID instruction are more likely just a good starting point, and additional adjustments of the tuning parameters will be required.

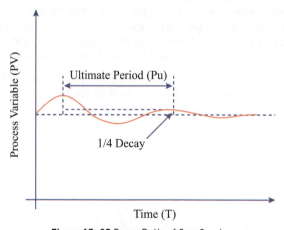

Figure 17–35 Decay Ratio of One-Quarter

Ziegler and Nichols Reaction Curve Tuning Method (Open Loop Technique)

The second method of tuning control loops developed by Ziegler and Nichols was based on data taken from a reaction curve during an open loop test. The open loop technique involves making a manual step change in the output of the controller and observing the reaction of the PV over time. This observed reaction over time is called the reaction curve, and it indicates the reaction of the control system loop when a step change is made to the process. The reaction curve is obtained by using a strip chart recorder or trending chart and recording the PV signal over time after a step change has been made to the controller output. An example reaction curve is shown in Figure 17–36.

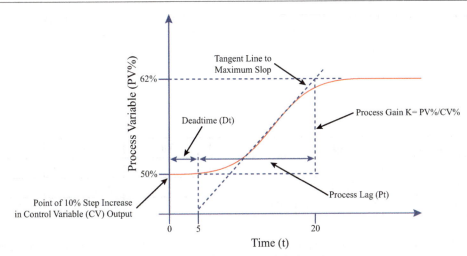

Figure 17–36 Process Reaction Curve

In Figure 17–36, the sloped line drawn tangent to the curve at its point of maximum slope is the process reaction rate and denotes how fast the process reacted to the step change made in the controller output. The inverse of this line's slope is a measure of the severity of the process lag as measured in time (Pt). The process dead-time (Dt) can be determined by measuring the time from when the step change was made to the intersection of the tangent line with the baseline. The process gain (K) is a measure of how much the PV increased relative to the size of the step change.

Perform the following steps to calculate the Proportional Gain (Kp), Integral Gain (Ki) or Reset Time (Ti), and Derivative Time (Kd) or Rate Time (Td) parameters for the PID instruction using the reaction curve method.

Step 1. Place the PID instruction in manual mode and adjust the output of the PID instruction until the PV is at about 50% of its range. If you are using a manual control station then use the manual control station for this adjustment; otherwise, select software manual (.SWM) and enter a value in the Set Output (.SO) parameter in the PID Setup Screen.

Step 2. Turn on your strip chart recorder or trending chart and allow the system to stabilize. You should be trending the PV and CV at this time. Make sure that your strip chart recorder or trending chart is set fast enough to capture the reaction curve.

Step 3. Manually introduce a 10% step change (50–60%) in the output of the PID instruction. Record on the chart the time when you made the step change. Record the reaction curve of the PV to the step change.

Step 4. Once you have recorded the PV reaction curve, you may stop controlling the process at this time if desired.

Step 5. Using the process reaction curve recorded in Step 3, draw a line tangent to the reaction curve at its point of maximum slope. Next, draw a line that is the inverse of the tangent line just drawn and measure the severity of the process lag in time (Pt). Determine the process dead-time (Dt) by measuring the time from when the step change was made to the intersection of the tangent line with the baseline. Finally, measure the process gain (K) by measuring how much the PV increased relative to the size of the step change in the CV.

Step 6. Compute the values for the Proportional Gain (Kp), Integral Gain (Ki), Reset Time (Ti), Derivative Time (Kd), and Rate Time (Td) parameters from one of the equations in Table 17–8.

$$K = \Delta PV\% / \Delta CV\%$$

Table 17–8 Open-Loop PID Equations

Control	Kp	Ti	Td	Ki	Kd
P	Pt / (Dt*K)				
PI	0.9*(Pt / (Dt*K))	3.3*Dt		Kp / Ti	
PID	1.2*(Pt / (Dt*K))	2*Dt	0.5*Dt	Kp / Ti	Kp*Td

Step 7. After calculating the values for the Proportional Gain (Kp), Integral Gain (Ki), Reset Time (Ti), Derivative Time (Kd), and Rate Time (Td) parameters, enter them into the PID instruction using the Setup Screen.

Note: Integral Gain (Ki) and Derivative Time (Kd) are used with independent gains. Reset Time (Ti) and Rate Time (Td) are used with dependent gains.

Step 8. The PID instruction is now ready to be placed in the auto mode. As in the previous method, if you have an ideal process, the PID controller should achieve a decay ratio of one-quarter wave when an error occurs. See Figure 17–35 for an example of a decay ratio of one-quarter wave. As stated before, the tuning parameters just calculated and entered into the PID instruction are more likely just a good starting point, and additional adjustments of the tuning parameters may be required.

One of the major advantages of the reaction curve method over the ultimate gain method is that the tuning of the PID parameters can be done much quicker when tuning very slow process loops. Another advantage of the reaction curve method is that there is less process disturbance, which can be important when dealing with sensitive or critical processes.

PID Tuning Considerations

There is no single method to tune PID loops and tuning is really a compromise between speed and robustness, and performance and stability. Take the time to understand the process to be controlled and check for things like hysteresis, linearity, noise, interactions, process gain, etc. Processes typically fall into three general types: self-regulating, integrating, and runaway. Other considerations include: tuning for load changes or SP changes; is the process fast or slow responding; are there interacting PID loops; does the process require cascade or feedforward control; etc. **KNOW** your process before attempting to program and tune PID instructions.

This is only intended to be an introduction to PID control in order to provide the technician with a basic understanding. There are entire books devoted to the subject of process control as well as those that specialize in the field of process control and instrumentation.

Chapter Summary

Process control signals fall into two types: those that measure the process, and those that control the process, referred to as *process variables* (*PVs*) and *control variables* (*CVs*). These process control signals are typically transmitted as low-level electrical analog signals to and from devices like PLCs. In order for analog input signals to be used in the PLC program, they must first be converted into a corresponding digital value by an *Analog-to-Digital Converter* (A/D converter), or in the case of an analog output signal a *Digital-to-Analog Converter* (D/A converter). When working with digital values in the PLC that represent analog signals, you may be required to convert, or scale, the digital values from one range to another. In order to make this linear conversion from one range to another, mathematical formulas can be employed, or in the case of the Allen-Bradley Micro800 and some older PLC controllers scaling instructions can be used.

Process control can be defined as a means by which you regulate a process. All process control systems consist of four key elements: the process, the measurement or PV, the process controller with SP, and the final control element or CV. When all four elements are present, it is considered a *closed loop control* system. An *open loop control* system on the other hand does not have the PV, process controller, and CV providing automatic regulation of the process. A process controller is a device or software instruction that regulates a process by monitoring an analog input from a PV and outputting an analog output to a final control element based on a SP. The process controller is often referred to as a PID controller, as in the case of the Allen-Bradley Logix 5000 PID instruction. PID stands for *proportional, integral,* and *derivative* control that uses a mathematical formula to regulate a process based on the error between the SP and PV. Loop tuning is the process of finding the right values for the proportional, integral, and derivative terms in the PID equation so the PID controller can keep the process under control.

Key Terms

process variable

offset

setpoint

integral gain

slope

control variable

proportional gain

derivative time

Review Questions

1. If an analog input has a digital range of 0–32,767 and you wish to scale that range between 200 and 1500°F, what would be the *slope* (*rate*) and *offset* values required for the scaling formula?
2. If the current digital input value in Question 1 was 12,537, what would be the scaled value in °F?
3. A 4–20 mA analog input signal is converted to a corresponding range of 3,277–16,384 by the A/D converter. What would be the digital value that corresponds to 9.34 mA?

4. The transmitter producing the 4–20 mA signal in Question 3 is a pressure transmitter with a range of 0–250 psig. What is the pressure reading when the transmitter is putting out 17.42 mA?

5. List the six parameters that must be entered into the Micro800 Scaler (SCP) instruction.

6. What are the four key elements that make up a closed loop control system?

7. PID stands for what?

8. J.G. Ziegler and N.B. Nichols in 1942 published two loop tuning methods that are still used today. What are those two methods called?

Chapter 18

Function Block Diagram and Structured Text Programming

Learning Objectives

After completing this chapter, you should have the knowledge to:

- Describe function block diagrams.
- Use function block elements.
- Describe structured text programming.
- Apply structured text components.

Function Block Diagram Programming

Function Block Diagram (FBD) programming is a method of programming that uses function blocks to make decisions or perform calculations. FBD programming is typically found in process control applications where there is much more data handling and calculations, as compared to discrete machine control applications. There are many types of function blocks available. In fact, Allen-Bradley's Logix 5000 PLCs have over 80 different function blocks available to perform various tasks.

A function block takes one or more inputs, makes a decision or calculation, and then generates one or more outputs. An output of one function block can also be the input to other function blocks. In the case of the Allen-Bradley Logix 5000 PLCs, the user must create a function block routine to program and use function block instructions. A function block routine is typically configured in a periodic task or executed using a Jump-to-Subroutine (JSR) instruction from the main routine or another routine.

The following examples will illustrate the FBD programming features of the Allen-Bradley Logix 5000 PLCs. The function block diagram is made up of function block elements (Figure 18–1). The elements consist of the function blocks themselves and the elements used to get information into and out of the function blocks.

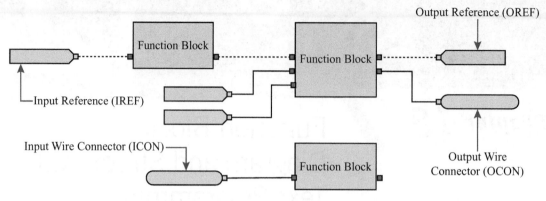

Figure 18-1 Function Block Elements

The function block performs an operation on an input value or values and produces an output value or values. As you can see in Figure 18-1, there are four basic input and output elements used with function blocks: Input Reference (IREF), Output Reference (OREF), Input Wire Connector (ICON), and Output Wire Connector (OCON).

The Input Reference (IREF) supplies a value from an input device or tag to a function block input. The OREF sends a value from the function block to an output device or tag. The ICON and OCON transfer data between function blocks when they are far apart or on different sheets, or to disperse data to several points in the routine.

Each function block uses a tag to store configuration and status information. The RSLogix 5000 programming software will automatically create a tag for the function block when it is created. This default tag can be used as is, or you can assign a different tag. For the IREF and OREF elements, you have to create a tag or assign an existing tag.

The order of execution (flow of data) is done by wiring elements together, as shown in Figure 18-2. The actual location of the function blocks does not affect the order of execution. As you can see in

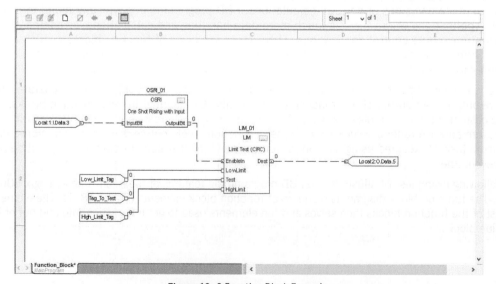

Figure 18-2 Function Block Example

Figure 18–2, there are two types of wire symbols: a solid line indicates a SINT, INT, DINT, or REAL value data path, and a dashed line indicates a BOOL value (0 or 1) data path.

This gives the reader a very basic overview of the FBD programming language. To further understand and use this programming language, refer to the Logix 5000 Controllers Function Block Diagram Manual.

Structured Text Programming

Structured text is a textual programming language similar to C++, Java, Python, or Lua that uses statements to define what to execute. It is best used for complex mathematical operations or specialized array/table loop processing.

Because structured text programming language is difficult to monitor and troubleshoot, it should be left to special applications that, once debugged, rarely require troubleshooting by the technician. Many computer programmers will feel right at home programming in structured text.

Structured text can contain the following components: assignments, expressions, instructions, constructs, and comments. Each will be discussed to give the reader a basic understanding of their function. Before getting started, it is worth mentioning that structured text is not case sensitive. You also use tabs and carriage returns to make the structured text easier to read.

Assignments

You use an **assignment** statement to assign values to tags. The operator (symbol) that is used to indicate an assignment statement is ": = ". You terminate the assignment statement with a semicolon ";". Figure 18–3 shows the syntax for an assignment statement along with several examples.

The tag must be a BOOL, SINT, INT, DINT, or REAL data type. If you use a BOOL tag, then the expression must be of the BOOL type; otherwise, use a numeric expression.

You can also create a non-retentive assignment by using the operator (symbol) for a non-retentive assignment "[:=]". For example, *tag* [:=] *expression* is a non-retentive assignment and the tag value is reset to zero (0) each time the PLC controller is placed into run mode.

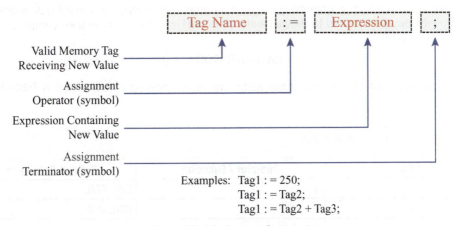

Examples: Tag1 : = 250;
           Tag1 : = Tag2;
           Tag1 : = Tag2 + Tag3;

Figure 18–3 Assignment Statements

Expressions

An **expression** can be a tag name, equation, or comparison. When writing an expression, you can use any of the following elements in the expression:

- a memory tag that stores a value
- an immediate value or number
- function elements such as: LOG, TAN, etc.
- operator elements such as: + , - , <, >, And, etc.

When writing expressions, you can use any combination of uppercase and lowercase letters. For example, the AND operator can be entered as AND, And, or and, all three are acceptable. When writing a complex expression, use parentheses to group expressions within expressions. This will make the whole expression easier to read and ensure that the expression executes in the desired order. The example below shows the use of parentheses.

$$tag1: = (tag2 + tag3) * (tag4 - tag5);$$

Expressions can be either BOOL or numeric expressions. A BOOL expression produces a value of 1 (true) or 0 (false), whereas a numeric expression calculates an integer or floating-point value.

A BOOL expression uses BOOL tags and operators (relational and logical) to compare values or check if conditions are true or false. For example, the expression "tag1 > 150" uses a relational operator to check if the value stored in tag1 is greater than 150. If tag1 is greater than 150, the expression is true or 1; otherwise, the expression is false or 0. You typically use BOOL expressions to condition the execution of other logic. This will become more apparent later in this chapter.

A numeric expression uses arithmetic operators and functions to calculate a value. For example, the numeric expression "tag1: = tag2 + 10;" adds 10 to the value stored in tag2 and places the result into tag1.

A numeric expression can also use bitwise operators to return a value based on two values. In the example below, DINT tag1 and tag2 are logically ANDed together and the result is placed into tag3 (DINT).

$$tag3: = tag1 \text{ AND } tag2;$$

Many times in structured text programming you nest a numeric expression within a BOOL expression. In the example below, 10 is added to the value stored in tag1 and the result is then compared to see if it is greater than 100.

$$(tag1+10)>100$$

To calculate arithmetic values, use one of the arithmetic or function operators shown in Table 18–1.

Table 18–1 Arithmetic and Function Operators

To / For	Operator / Function	Optimal Data Type
Add	+	DINT, REAL
Subtract	-	DINT, REAL
Multiply	*	DINT, REAL

Exponent (x to the power of y)	**	DINT, REAL
Divide	/	DINT, REAL
Module-divide	MOD	DINT, REAL
Absolute Value	ABS	DINT, REAL
Arc Cosine	ACOS	REAL
Arc Sine	ASIN	REAL
Arc Tangent	ATAN	REAL
Cosine	COS	REAL
Radians to Degrees	DEG	DINT, REAL
Natural Log	LN	REAL
Log Base 10	LOG	REAL
Degrees to Radians	RAD	DINT, REAL
Sine	SIN	REAL
Square Root	SQRT	DINT, REAL
Tangent	TAIN	REAL
Truncate	TRUNC	DINT, REAL

To compare two values, use one of the Relational Operators shown in Table 18–2.

Table 18–2 Relational Operators

Comparison	Operator	Optimal Data Type
Equal	=	DINT, REAL
Less Than	<	DINT, REAL
Less Than or Equal	<=	DINT, REAL
Greater Than	>	DINT, REAL
Greater Than or Equal	>=	DINT, REAL
Not Equal	<>	DINT, REAL

To compare the bits within values, use one of the Bitwise Operators in Table 18–3.

Table 18–3 Bitwise Operators

For	Operator	Data Type
Bitwise AND	&, AND	DINT
Bitwise OR	OR	DINT
Bitwise Exclusive OR	XOR	DINT
Bitwise Complement	NOT	DINT

To check if conditions are true (1) or false (0), use one of the Logical Operators in Table 18–4.

Table 18–4 Logical Operators

For	Operator	Data Type
Logical AND	&, AND	BOOL
Logical OR	OR	BOOL
Logical Exclusive OR	XOR	BOOL
Logical Complement	NOT	BOOL

Table 18–5 Order of Execution

Order	Operation
1	( )
2	Function (…)
3	**
4	-(negate)
5	NOT
6	*, /, MOD
7	+, - (subtract)
8	<, <=, >, >=
9	=, < >
10	&, AND
11	XOR
12	OR

The operators you write into an expression are not necessarily performed from left to right, but rather in a prescribed order. Operations of equal order are performed from left to right. When expressions contain more than one operator or function, group them in parentheses "( )" as this will ensure the correct order of execution and make it easier to read the expression. Table 18–5 shows the order of execution for the various operations.

Instructions

Structured text statements can also be instructions. An **instruction** is a stand-alone statement that performs a given task such as copying the contents of one array into another array or jumping to another routine. A structured text instruction executes each time it is scanned unless it is used within a construct. Constructs will be covered next.

A structured text instruction uses parentheses to contain its operands, and depending on the instruction, there can be zero, one, or multiple operands. Always terminate an instruction with a semicolon ";". The example below shows the structured text format for a Copy File instruction. Where "tag1_array[0]" is the source array, "tag2_array[0]" is the destination array, and 10 is the number of elements to copy.

COP(tag1_array[0], tag2_array[0], 10);

Another example of a structured text instruction is the Jump-to-Subroutine instruction shown below.

JSR(*routine_name, input_count, input_parameter, return_parameter*);

Notice how the operands of the instruction are contained within the parentheses and separated by commas. Also notice that the instruction is terminated with a semicolon.

Even though they are similar, instructions differ from functions in that instructions cannot be used in expressions; only functions can be used in expressions. Refer to the Logix 5000 Controllers Structured Text Manual for a complete listing of the available structured text instructions.

Constructs

A **construct** is a conditional statement used to trigger structured text code (statements). Constructs can be programmed singly or nested within other constructs. Always terminate a construct with a semicolon ";". Constructs are very powerful statements and one of the key elements in structured text programming. Table 18–6 lists the available constructs.

Table 18–6 Constructs

IF ... THEN
CASE ... OF
FOR ... DO
WHILE ... DO
REPEAT ... UNTIL

IF ... THEN

Use the IF ... THEN construct to do something if or when specific conditions occur. For example, *IF* the motor faults, *THEN* turn on the alarm light. *IF* the water is greater than 100°, *THEN* start the cooling pump. The syntax for the IF...THEN construct is shown below:

IF *bool_expression* THEN

<statement>;

END_IF;

The operand can either be a BOOL tag or an expression that evaluates to a BOOL value. The example below shows a BOOL tag operand.

IF motor_fault THEN

alarm_light : = 1;

END_IF;

Where "motor_fault" is a BOOL tag that is either 1 (true) or 0 (false). If the BOOL value is 1 or true, then the statement is executed. If the "motor_fault" tag contained a value of 1, then the tag "alarm_light" would be set to a value of 1. The example below shows an expression that evaluates to a BOOL value.

IF tank1_temperature > 100 THEN

cooling_pump : = 1;

END_IF;

where tag "tank1_temperature" contains a numeric value that is compared to an immediate value. If the value stored in "tank1_temperature" is greater than 100, then the statement is true and the tag "cooling_pump" would be set to a value of 1 or *ON*.

The IF...THEN construct also has two optional statements that can be used with it, ELSIF and ELSE. The ELSIF statement allows you to select from several possible groups of statements, where each ELSIF represents an alternative path. The ELSE statement does something when all of the IF or ELSIF conditions are false. The following examples will help to illustrate the use of the ELSIF and ELSE statements.

Example 1 (IF...THEN...ELSE)

IF motor_fault THEN

alarm_light : = 1;

ELSE

alarm_light : = 0;

END_IF;

As you can see in the above example, the ELSE statement turns the alarm light off (0) if the condition of the IF statement is false.

Example 2 (IF...THEN...ELSIF)

IF tank1_temperature > 100 THEN

cooling_pump : = 1;

ELSIF tank1_temperature < 75 THEN

cooling_pump : = 0°

END_IF;

In this example, the ELSIF provides an alternative path if the temperature is not greater than 100°. In this case, if the temperature is below 75°, then turn the cooling pump off. As you can clearly see, we have created an IF, THEN, and ELSIF statement that turns the cooling pump on anytime the temperature in the tank is above 100° and leaves the cooling pump on until the temperature drops below 75°.

The table in Table 18–7 summarizes the various combinations of IF, THEN, ELSIF, and ELSE.

Table 18–7 IF, THEN, ELSIF, and ELSE

If you want to ...	And ...	Then Use This Construct
Do something if or when conditions are true.	Do nothing if conditions are false.	IF ... THEN
	Do something else if conditions are false.	IF ... THEN ... ELSE
Choose from alternative statements based on input conditions.	Do nothing if conditions are false.	IF ... THEN ... ELSIF
	Assign default statements if all conditions are false.	IF ... THEN ... ELSIF ... ELSE

CASE ... OF

Use the CASE ... OF construct to select what to do based on a numerical value. For example, a recipe number that determines which ingredients to add in a batching process. The syntax for the CASE ... OF construct is:

CASE *numeric_expression* OF

selector1:   <*statement*>;

selector2:   <*statement*>;

      "   "

        "   "

selectorN: <*statement*>;

ELSE

<*statement*>;

END_CASE;

The *numeric_expression* operand can either be a tag or an expression that evaluates to a number. The *selector* operand is an immediate number of the same type as the *numeric_expression* operand. The "N" represents the last *selector* number. The *statement* is the statement to execute when the *selector* is true. The ELSE is optional and will only execute the statements if the *numeric_expression* does not equal any of the *selectors*.

The *selector* values can be a single value, multiple distinct values, a range of values, or a combination of distinct and range of values. The syntax for the *selector* values is shown in the example below:

value: <statement>;

value1, value2, value3, valueN: <statement>;

value1..valueN: <statement>;

value1, value2, value3..valueN: <statement>;

Use a comma (,) to separate each value and two periods (..) to identify a range.

The following example will help to illustrate the CASE...OF construct.

CASE batch_recipe_number OF

                    1:                ingredient_valve1 : = 1;

                              ingredient_valve2 : = 1;

                              ingredient_valve3 : = 0;

                              ingredient_valve4 : = 0;

                2,3,4:   ingredient_valve1 : = 0;

                              ingredient_valve2 : = 1;

                              ingredient_valve3 : = 1;

                              ingredient_valve4 : = 1;

                5..7:   ingredient_valve1 : = 1;

                                   ingredient_valve2 : = 1;

                                   ingredient_valve3 : = 1;

                                   ingredient_valve4 : = 1;

            8,9..12: ingredient_valve1 : = 1;

                                   ingredient_valve2 : = 1;

                                   ingredient_valve3 : = 0;

                                   ingredient_valve4 : = 0;

    ELSE

    ingredient_valve1 : = 0;

    ingredient_valve2 : = 0;

    ingredient_valve3 : = 0;

    ingredient_valve4 : = 0;

    END_CASE;

In the above example, the value stored in the tag "batch_recipe_number" is compared to the selector values for a match. If a match is found, then the ingredient valves are either opened or closed based on the selector statements. If no match is found then the ingredient valves are closed.

FOR...DO

Use the FOR...DO construct to do something a specific number of times before doing anything else. For example, clear an array of bits or check an array of part numbers for a match. The syntax for the FOR...DO construct is:

    FOR *count* : = *Initial_value* TO *final_value* BY *increment*

    DO

    <statement>;

    END_FOR;

The *count* operand is a tag that stores the count position as the construct executes, assign a unique tag to this operand. The *initial_value* operand specifies the initial value for the count. This can be a tag, expression, or immediate number. The *final_value* operand specifies the final value for the count, which determines when to exit the loop. This can be a tag, expression, or immediate number. The *increment* operand is optional and is the amount to increment the count by each time through the loop. If you don't specify an *increment*, the count increments by 1.

 Caution: The controller does not execute any other statements in the routine until it completes the FOR...DO loop. This could lead to a processor fault if the loop time is greater than the watchdog timer for the task.

The following examples illustrate the FOR...DO construct.

Example 1

```
FOR count_tag : = 0 TO 20 DO
array1[count_tag] : = 0;
END_FOR;
```

Example 2

```
FOR count_tag : = 0 TO 100 DO
        IF part_number = part_number_array[count_tag] THEN
        found_tag : = 1;
        EXIT;
        ELSE
        found_tag : = 0;
        END_IF;
    END_FOR;
```

In the first example, array elements 0–20 are cleared in the array called "array1". Notice how the *count_tag* is used as the subscript in the array tag. Since the *increment* operand was omitted, the count increment defaulted to 1. In the second example, the value in the tag called "part_number" was compared against an array of numbers (part_number_array) for a match. If a match was found, the BOOL tag "found_tag" was set to 1 and the loop was stopped by using the EXIT statement. If a match was not found then the "found_tag" was set to 0.

WHILE...DO

Use the WHILE...DO construct to keep doing something as long as certain conditions are true. The syntax for the WHILE...DO construct is:

```
WHILE bool_expression DO
        <statement>;
END_WHILE;
```

The *bool_expression* operand is a BOOL tag or expression that returns a value of 1 (true) or 0 (false). The *statement* is what is executed as long as the *bool_expression* returns a 1 (true).

 Caution: The controller does not execute any other statements in the routine until it completes the WHILE...DO loop. This could lead to a processor fault if the loop time is greater than the watchdog timer for the task.

REPEAT...UNTIL

Use the REPEAT...UNTIL construct to keep doing something until certain conditions are true. The REPEAT...UNTIL construct executes the statements in the construct first before checking if conditions are true. If conditions are not true, then the controller executes the statements within the loop again. The syntax for the REPEAT...UNTIL construct is:

    REPEAT

        <statement>;

    UNTIL *bool_expression*

    END_REPEAT;

The *bool_expression* operand is a BOOL tag or expression that returns a value of 1 (true) or 0 (false). The *statement* is what is executed as long as the *bool_expression* returns a 0 (false).

 Caution: The controller does not execute any other statements in the routine until it completes the REPEAT...UNTIL. This could lead to a processor fault if the loop time is greater than the watchdog timer for the task.

Comments

You can add comments to your structured text program to help describe how your program works. This will make your program much easier to understand by someone else or if you must work with it at a later date. Comments will not affect the execution of the program and are downloaded to the controller memory.

You can add comments on a single line, at the end or within a line of structured text. The following syntax is used for entering comments.

    //comment - single line only

    (**comment**)

    /**comment**/

Example Comments:

    //Check for high temperature conditions

    IF water_temp > 100 THEN....

    IF water_temp (*sludge tank 1 water temperature*) > 100 THEN....

    pressure_valve1 : = 1; /*open the pressure valve on tank 1*/

Structured Text Programming Example

The following structured text program example loads the correct temperature setpoints into a working temperature array based on which recipe the operator has selected. It also monitors the temperatures for abnormally high conditions.

    //Check for operator input and then select the correct temperature setpoints to load

    IF operator_select THEN

    CASE recipe_number OF

        1:      tank_temp[0] : = 150;

              tank_temp[1] : = 175;

              tank_temp[2] : = 200;

```
                    tank_temp[3] : = 100;
                    batch_start : = 1;
        2:          tank_temp[0] : = 125;
                    tank_temp[1] : = 150;
                    tank_temp[2] : = 175;
                    tank_temp[3] : = 75;
                    batch_start : = 1;
        3:          tank_temp[0] : = 200;
                    tank_temp[1] : = 250;
                    tank_temp[2] : = 300;
                    tank_temp[3] : = 150;
                    batch_start : = 1;
        4..5:       tank_temp[0] : = 225;
                    tank_temp[1] : = 300;
                    tank_temp[2] : = 325;
                    tank_temp[3] : = 180;
                    batch_start : = 1;
ELSE    tank_temp[0] : = 100;
        tank_temp[1] : = 100;
        tank_temp[2] : = 100;
        tank_temp[3] : = 75;
        batch_start : = 0;
END_CASE;
END_IF
//Monitor batch temperatures during batching operation
IF batch_start THEN
FOR count : = 0 TO 3 DO
IF temp[count] => tank_temp[count] + 20 THEN
batch_start : = 0;
OVER_TEMP_ALARM : = 1;
END_FOR;
END_IF
(*End of Program*)
```

Chapter Summary

Function Block Diagram (FBD) programming uses function blocks to make decisions or perform calculations and is typically found in process control applications. A function block takes one or more inputs, makes a decision or calculation, and then generates one or more outputs.

Structured text programming language uses statements to define what to execute and is similar to BASIC programming. It is best used for complex mathematical operations or specialized array/ table loop processing. Structured text programming contains assignments, expressions, instructions, constructs, and comments.

Key Terms

assignment   expression

instruction   construct

Review Questions

1. A function block diagram is made up of function block _____.
   a. references
   b. assignments
   c. elements
   d. operators

2. What is the function of the Output Wire Connector (OCON)?

3. "OREF" stands for what?

4. Structured Text programming is similar to what type of programming language?
   a. Ladder Logic
   b. SFC
   c. Word
   d. BASIC

5. What are constructs?

6. The operator (symbol) that is used to indicate an assignment statement is _____ .
   a. ";"
   b. "*"
   c. "^"
   d. ": = "

7. List any two constructs discussed in this chapter.

8. A dashed line indicates what type of data path in a function block diagram?

9. What type of expression is "tag1>150"?

10. A structured text instruction executes once unless it is used within a construct.

    a. True

    b. False

Chapter 19

Sequential Function Chart Programming

Learning Objectives

After completing this chapter, you should have the knowledge to:

- Describe the building blocks used to create a sequential function chart program.
- Describe steps and transitions.
- Develop a basic sequential function chart program.

Sequential Function Charts

Sequential function chart (SFC) programming is a method of programming that is similar to designing a flowchart of your process. You program steps and transitions that are arranged like a flowchart and, when executed, control your process in a prescribed order. Some advantages of using SFC programming to identify your process steps are:

- easier to organize and read (graphical representation)
- faster execution of your logic
- faster and easier troubleshooting
- faster to design and debug

Sequential function chart programming is most often used with applications that are sequential in nature and have definable steps. For example, a burner control application would have definable steps that would typically be executed in sequential order. Figure 19–1 shows an example of what a burner control application might look like in an SFC format.

An SFC program is made up of steps, actions, and transitions. The steps are the major building blocks of any SFC program. They perform timing and counting functions as well as executing any actions associated with them. Actions are used for turning tags ON or OFF and performing other functions. Transitions are conditions that must be met before moving on to the next step.

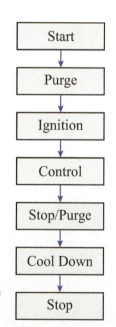

Figure 19–1 SFC Burner Control Process

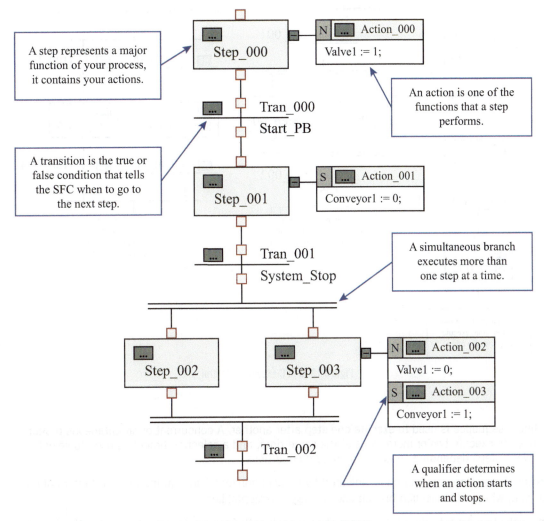

A step represents a major function of your process, it contains your actions.

An action is one of the functions that a step performs.

A transition is the true or false condition that tells the SFC when to go to the next step.

A simultaneous branch executes more than one step at a time.

A qualifier determines when an action starts and stops.

Figure 19–2 SFC Program Key Elements

The key elements of an SFC program are shown in Figure 19–2. Notice how an SFC program can have simultaneous branches that execute more than one step at a time.

In Figure 19–2, Step_000 is the first step to be executed and its actions performed until transition Tran_000 is true, at which time execution is moved to Step_001. The default names for each step are shown, but other names can be given to a step to help define its purpose such as Purge, Ignition, etc. The first two steps are linear steps, meaning they execute one right after the other, whereas the next two are concurrent. Concurrent steps are steps that are executed simultaneously.

When developing an SFC program, you must organize the execution of your steps to match the process. You can use linear, concurrent, and branch sequences to control the execution of your steps. Figure 19–3 shows an example of each type.

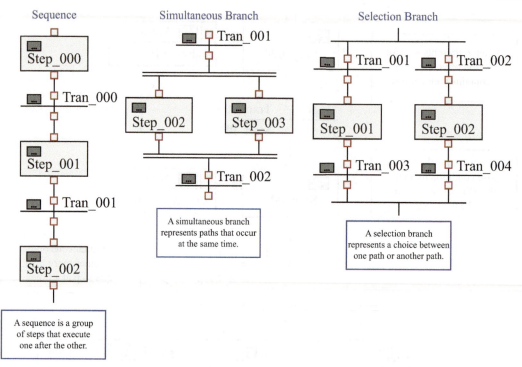

Figure 19–3 SFC Execution Control

A linear sequence is used to execute one step after another. A concurrent or simultaneous branch is used to execute two or more steps at the same time, and a selection branch is used to select between steps depending on logic (transition) conditions.

Note: A simultaneous branch is indicated by parallel horizontal lines at the top and bottom of the branch, whereas a selection branch uses a single horizontal line.

In a selection branch, the SFC program checks each path from left to right; the first path with a true transition is executed and all others are ignored. If all path transitions are false, the program starts over with the first path. A path can have just a transition and no step if desired. This is useful if you want to bypass the entire branch group. See Figure 19–4.

Creating Steps

Steps should be created that represent the major functions of your process. A **step** contains the actions to be performed during that part of the sequential process. When creating a step, a tag is created that provides information about the step. The tag name, by default, follows the step number created, such as Step_000, Step_001, Step_002, etc. You can change the tag name to something that is more meaningful, such as Initialize, Purge, Ignite, etc. The information contained in the tag structure for an SFC step (step properties) is shown in Figure 19–5. The members of the tag provide a substantial amount of information about the step such as, how long has the step been active, is this the first scan or the last scan, etc. All of this information can be used as needed when programming and monitoring your SFC program. Refer to the detailed SFC programming manual for a complete explanation of each tag member.

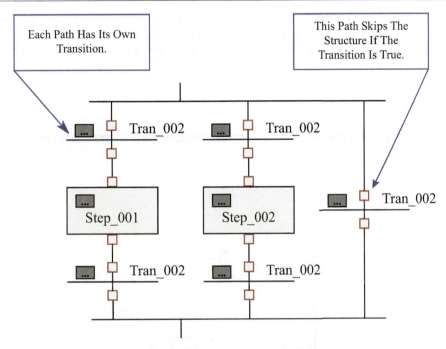

Figure 19–4 SFC Selection Branch with Bypass

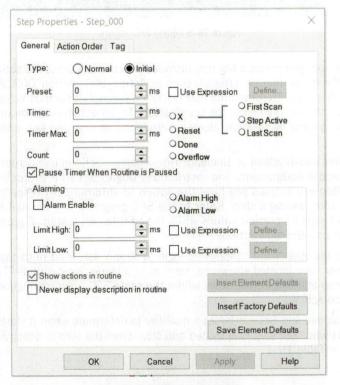

Figure 19–5 SFC Step Properties

You add **actions** to steps to perform functions such as execute a program subroutine, opening a valve, start a motor, or initializing a mode. To add an action to a step, right-click on the step and then choose *Add Action*; see Figure 19–6. A step can have multiple actions associated with it.

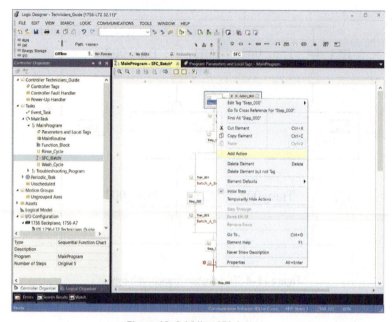

Figure 19–6 Adding SFC Actions

When creating an action, you create a tag that provides information about the action. The tag name, by default, follows the action number created, such as Action_001, Action_002, Action_003, etc. You can change the tag name to something that is more meaningful, such as Valve Open, Motor1 Start, etc. The information contained in the tag structure for an action is shown in Figure 19–7. This should look familiar as some of the same elements are found as in a step tag, which was discussed earlier.

Actions can be either non-Boolean or Boolean. Non-Boolean actions use Structured Text programming to execute assignments and instructions (logic) or call a subroutine. When programming non-Boolean actions you have the option to automatically reset the assignments and instructions before leaving a step. Refer to the SFC programming manual for details. Otherwise, all data keeps its current values when the SFC leaves a step. Figure 19–8 shows several examples of non-Boolean actions.

A Boolean action on the other hand contains no logic. It simply sets a bit in its tag structure that can be monitored by other logic located elsewhere, such as in a ladder logic routine. This method allows you to reuse a Boolean action multiple times within the same SFC program. Figure 19–9 shows an example of a Boolean action.

Each action (non-Boolean and Boolean) uses a qualifier to determine when it starts and stops. By default, actions start when the step is activated and stop when the step is deactivated. A list of the qualifiers is shown in Table 19–1.

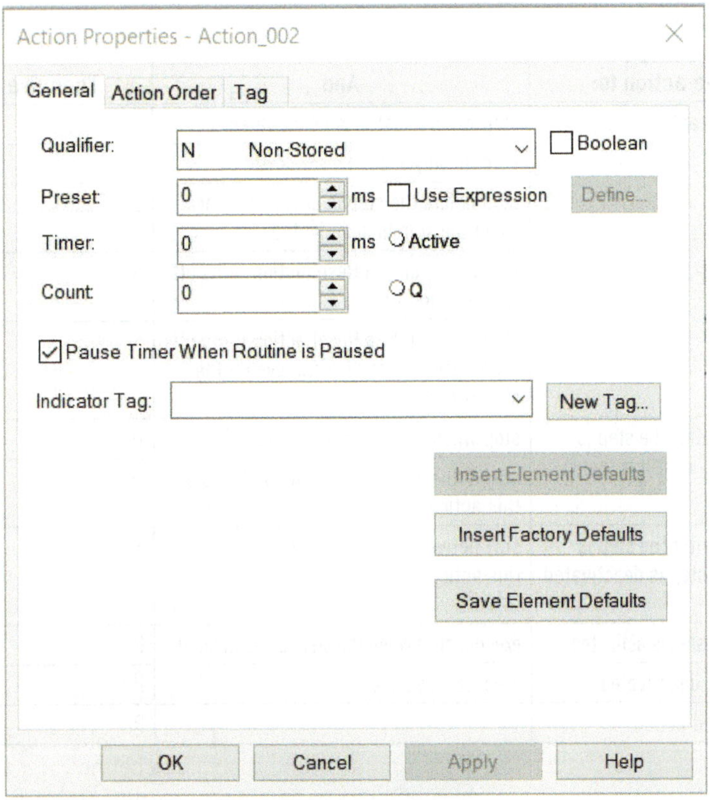

Action Properties - Action_002 ✕

General | Action Order | Tag

Qualifier: N   Non-Stored   ∨   ☐ Boolean

Preset: 0   ⏶⏷ ms   ☐ Use Expression   Define...

Timer: 0   ⏶⏷ ms   ○ Active

Count: 0   ⏶⏷   ○ Q

☑ Pause Timer When Routine is Paused

Indicator Tag: [∨]   New Tag...

Insert Element Defaults

Insert Factory Defaults

Save Element Defaults

OK   Cancel   Apply   Help

Figure 19–7 Action Properties

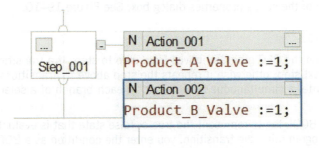

Figure 19–8 Non-Boolean Actions

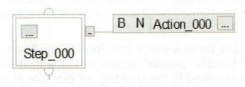

Figure 19–9 Boolean Action

Table 19–1 Action Qualifiers

If you want the action to:	And	Then use this qualifier:
Start when the step is activated...	stop when the step is deactivated.	N
	execute only once.	PI
	stop before the step is deactivated or when the step is deactivated.	L
	stay active until a Reset action turns off this section.	S
	stay active until a Reset action turns off OR a specific time expires, even if the step is deactivated.	SL
Start a specific time after the step is activated and the step is still active...	stop when the step is deactivated.	D
	stay active until a Reset action turns off this action.	DS
Start a specific time after the step is activated, even if the step is deactivated before this time...	stay active until a Reset action turns off this action.	SD
Execute once when the step is activated...	execute once when the step is deactivated.	P
Start when the step is deactivated...	execute only once.	PO
Turn off (reset) a stored action.		R

When a step contains multiple actions, the order of execution is from top to bottom as shown in the "Action Order" window of the step's properties dialog box. See Figure 19–10.

Creating Transitions

Transitions control when the SFC program moves from step to step. If a transition is true, the SFC program goes to the next step; otherwise it repeats the step above it. Transitions occur between steps in a sequence, after a simultaneous branch, and in each branch of a selection branch. See Figure 19–11.

Each transition uses a BOOL tag to represent the true or false state that is evaluated by the SFC program. When programming the transition, you enter the condition as a BOOL expression that uses BOOL tags, relational operators, and logical operators to compare values or check if conditions are true or false. The BOOL expression is entered in structured text format. You can also call a subroutine that contains an End Of Transition (EOT) instruction that returns the state of the conditions to the transition. Figure 19–12 shows examples of both a BOOL expression and subroutine transitions.

You can use the step millisecond timer within a transition to signal when the step has run the required time or too long and the SFC program should go to the next step. This is a very good use of the millisecond timer contained in the step tag. An example of this is shown in Figure 19–13.

Figure 19–10 Action Order Window

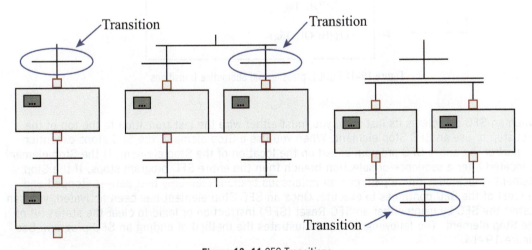

Figure 19–11 SFC Transitions

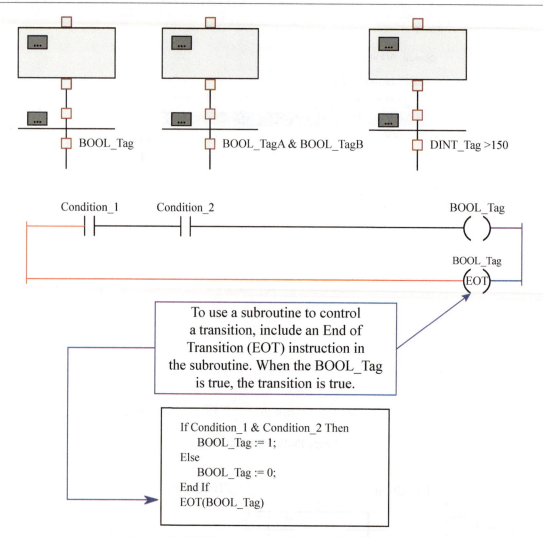

Figure 19–12 BOOL Expression and Subroutine Transitions

When an SFC completes its last step, you must either wire the last transition to the top of the first step or use an SFC Stop element. When you use a Stop element, the SFC stops execution for part or the entire SFC program based on the location of the Stop element. If the Stop element is located after a sequence or selection branch then the entire SFC program stops. If the Stop element is located within a path of a simultaneous branch, then only that path is stopped and the rest of the SFC continues to execute. Once an SFC Stop element has been activated, you can restart the SFC by using either an SFC Reset (SFR) instruction or logic to clear the status bit of the Stop element. The following example illustrates the method of ending an SFC program. See Figure 19–14.

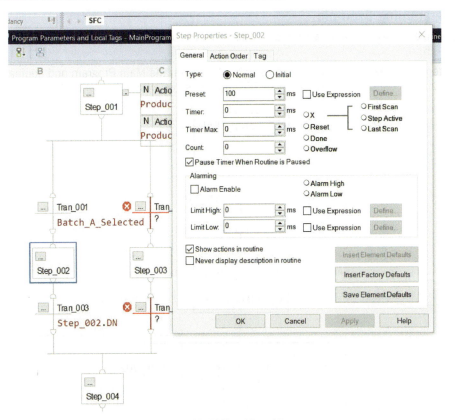

Figure 19–13 Timed Transition

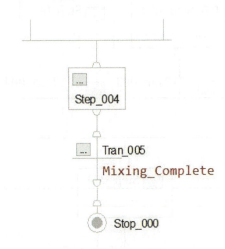

Figure 19–14 Ending SFC Program

SFC Programming Example

In this example, we will program a simple SFC program to control a batch operation as shown in Figure 19–15.

The first step is to create the SFC routine by right-clicking on the Main Program and adding a new routine as shown in Figure 19–16. The new SFC routine is given the name SFC_Batch.

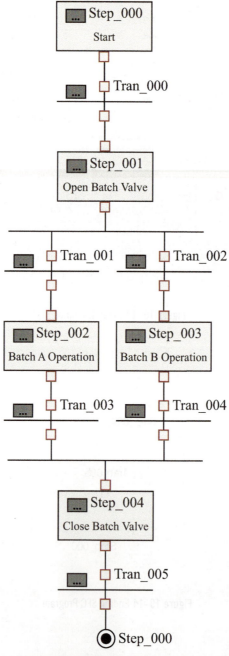

Figure 19–15 SFC Batch Operation Example

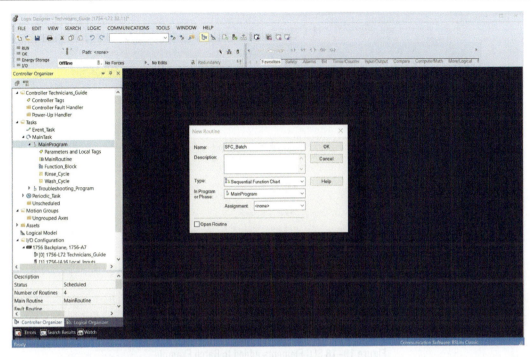

Figure 19–16 Creating a New SFC Routine

Double-click on the new routine (SFC_Batch) and the SFC program window will appear in the right pane; see Figure 19–17. Note the SFC toolbar at the top of the window. The toolbar is used to select the SFC elements to be added to the program.

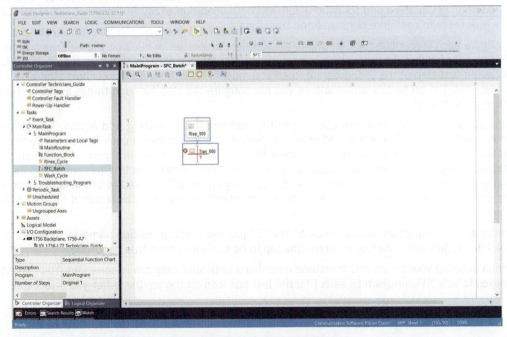

Figure 19–17 SFC Programming Window

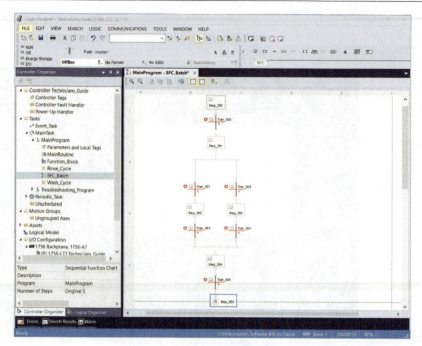

Figure 19–18 SFC Elements Added to Batch Example

Programming a SFC consists of dragging and dropping program elements from the toolbar onto the SFC program window replicating your SFC steps and transitions. After your steps and transitions have been placed, the last step is to configure your elements and add your actions. In Figure 19–18, we have added all of our steps and transitions for the batch operation example shown in Figure 19–15. We are now ready to configure our steps and transitions.

To configure each step, right-click on the step and select properties or click on the ellipses button located on the step box. The properties screen will appear for that step as shown in Figure 19–19. If this is the first step of the SFC program, then you must choose the Initial checkbox; otherwise select Normal. The properties screen allows you to configure the step for such things as preset time, alarming, action order, etc.

Actions are added to a step by right-clicking on the step and then selecting Add Action. To configure an action, right-click on the action and select properties or click on the ellipses button located on the action box. The properties screen will appear for that action as shown in Figure 19–20. Select the appropriate properties for that action and then select "OK". After selecting the properties for the action, you enter the action statement by double-clicking on the action window in the action box if a Non_Boolean action was selected. Continue to add actions to your steps as required for your SFC program.

To configure each transition, double-click on the "?" just below the transition name as shown in Figure 19–21, and enter the tag or subroutine tag to be evaluated for a true condition.

Continue entering your action and transition conditions until your program is complete. You can add text boxes to your SFC program by selecting the text box icon on the toolbar. After you select the

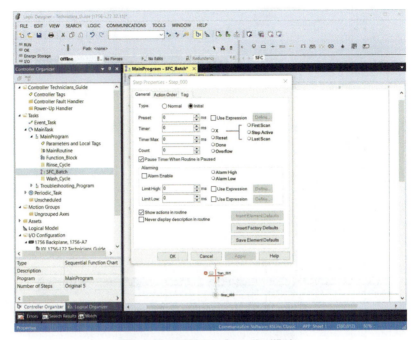

Figure 19–19 SFC Step Properties Window

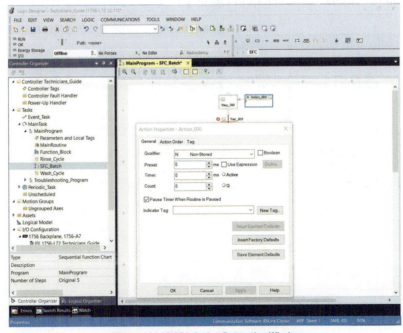

Figure 19–20 SFC Action Properties Window

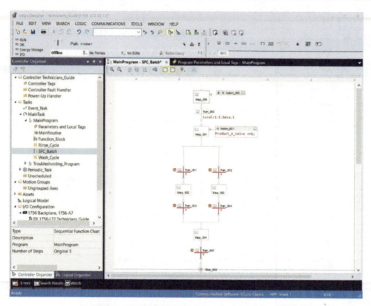

Figure 19–21 SFC Transition Configuration

text box on the toolbar, a text box will appear on your SFC programming window that you can move around. The pushpin symbol in the text box allows you to attach the text box to an SFC element by dragging the wire to the element, as shown in Figure 19–22.

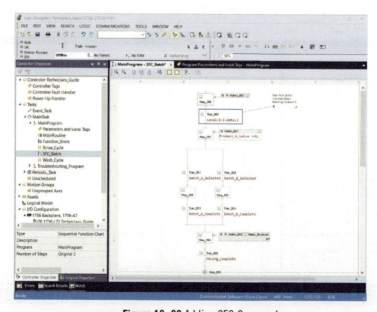

Figure 19–22 Adding SFC Comments

Only a general overview of SFC programming has been presented in this chapter. For a complete description on how to program the Logix 5000 PLCs using SFC programming language, refer to the Logic 5000 Controllers Sequential Function Charts Programming Manual.

Chapter Summary

Sequential function chart (SFC) programming is like designing a flowchart of your process using steps and transitions. Steps contain actions that are used for turning tags ON or OFF and performing other functions. Transitions are the conditions that must be met before moving on to the next step(s). Steps can be configured as linear steps (one step after another), concurrent steps (executed simultaneously), or selected branch (selected between steps).

SFC programming is easy to organize, read, design, and troubleshoot. It is most often used with applications that are sequential in nature and have definable steps.

Key Terms

step

action

transition

Review Questions

1. List the key elements of an SFC program.
2. You can use _____, _____, and _____ to control the execution of your SFC steps.
3. Each step and transition is assigned a tag, which provides information about the step or transition.
   a. True
   b. False
4. Actions can be one of two types. List the two types and their differences.
5. A transition can use a subroutine to determine the state of the transition. What instruction is used in the subroutine for this?
   a. END
   b. OTE
   c. SFC
   d. EOT

Chapter 20

Understanding Communication Networks

Learning Objectives

After completing this chapter, you should have the knowledge to:

- Understand networking principles.
- Describe the different network categories.
- Explain the different network topologies.
- Understand the media used to construct a network.
- Explain the different methods used to access a network.
- Understand what a network protocol is and how packets work.
- Understand the different industrial communication protocols.

With PLCs being used in greater numbers and distributive control becoming the norm, the control systems of today are more complex, requiring various communication schemes to tie these system components together. Passing, or exchanging, information is not only a desire but a requirement in many control systems today. A communication scheme can be as simple as having two PLCs 10 ft apart passing information between them, to as complex as a plant-wide control and information network in which PLCs, **human machine interfaces** (HMIs), intelligent I/O devices, and information technology (IT) systems are able to exchange information. This chapter on networking is not intended to provide the reader with a comprehensive knowledge of communication networks or schemes, but rather to give a general overview of the control and information networks that are being used today in many industrial plants and factories.

History

Data communication has become an integral part of modern control systems and continues to evolve as technology advances at an ever-increasing rate. One of the first methods used to communicate between PLCs was a pair of wires used to connect a digital output of one PLC to the digital input of a second PLC. As simple as this method may seem, it did satisfy some of the most important criteria that we look for in many modern industrial control and information networks today, such as:

- real-time or nearly real-time control
- high data integrity
- high noise immunity
- reliability in harsh industrial environments

As industrial plants began to implement PLCs in a distributive or modular approach to controlling equipment, it was necessary to develop fast, secure, and reliable communications schemes to tie the various PLCs and their systems together. PLC manufacturers began developing their own control networks to meet this ever-growing demand. These were considered proprietary networks since they were only compatible with the manufacturer's own equipment. These proprietary networks were often called highways, data highways, or control networks. Many of these proprietary networks continue to function today in many plants and factories across the world.

With the advancements in digital communications technology and the increased number of PLC manufacturers, it became clear that the proprietary networks of the past were limiting the ability for modern control systems to work together. This also became apparent to plant engineers who were under pressure to find ways to integrate the various PLCs, intelligent I/O devices, information, and HMI systems. Today, most manufacturers of PLC and control equipment support open communication networks. Open networks are based on international standards developed through industry associations. This has opened the door for control equipment manufactured by different vendors to communicate across a common network.

Networking Principles

A communication network exists when two or more devices are connected together by some type of media for the sole purpose of exchanging information.

When devices are connected to a network media, they are often called **nodes**. Nodes can be PLCs, personal computers, HMIs, intelligent I/O devices, routers, and switches, to name just a few. Some of these devices may also be called by other names like "stations," "network devices," or just plain "devices." Nodes are typically divided into two classes of devices, those that produce and consume data (i.e., computers, PLCs, HMIs) and devices that only receive and forward the data (i.e., repeaters, switches, routers, bridges). See Figure 20–1.

Modern data communications uses a digital method of sending and receiving binary information by way of low-level DC digital pulses. These digital pulses represent binary digits or bits that are sent as a bit stream called a **packet** over a common media format (i.e., wires, cables, optical fibers) to the multiple devices connected to the network media. A packet is a unit of data represented by a series of digital signals (being either *ON* or *OFF*) that represents the message to be transmitted. Think of a packet as a letter that you would mail. The length of the bit stream or packet size is dependent on many factors such as data length, format, and protocol. These digital signals that

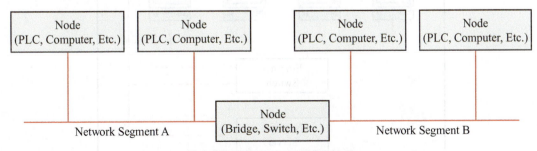

Figure 20–1 Nodes and Network Media

make up the messages are what is used to communicate information between devices. The remote I/O communications network covered earlier is an example of a digital communication network that allows PLCs to communicate with remote I/O devices over a common media. Messages, packets, and protocols will be covered in more detail later in this chapter.

You will often hear the term **bandwidth** when working with communication networks. Bandwidth is the speed at which information can be transferred. Network bandwidth can be thought of as the ability of a network to pump data through the communication media. In other words, it's like the pipe that brings water to your house: the bigger the pipe, the greater the volume of water that can be delivered to your house in a given amount of time. Network bandwidth is usually expressed in *bits per second* transmitted, abbreviated bps. Most networks operate in the millions of bits per second range, so the abbreviation Mbps is typically used. Theoretically, a 10 Mbps network can be transmitting data at the rate of 10 million bits per second, but this is not always the case. Some networks, such as the Ethernet, may really only be operating at 40%–50% of the rated bandwidth, depending on the mode of operation and hardware devices used.

Network Categories

The geographic area that networks encompass typically categorizes communication networks. Two of these categories, Local Area Networks (LANs) and Radio Area Networks (RANs), are quite common in the control and information networks of many factories and cities. The following is a description of the four main network categories:

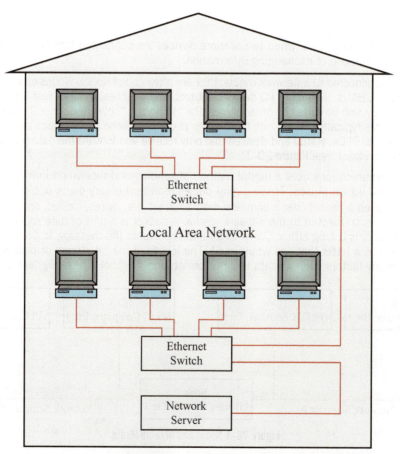

Figure 20–2 Local Area Network (LAN)

- **Local Area Networks** (LANs) are typically high-speed, low-error data networks covering a relatively small geographic area (up to several thousand meters). A LAN connects such devices as PLCs, computers, HMIs, printers, servers, and other devices in a single building or geographically limited area (see Figure 20–2). Most industrial control and information networks fall in this category. Not all LANs are the same. LANs have standards or protocols that specify the cabling and signaling methods used. Some of these standards or protocols are proprietary and others are not.

- **Metropolitan Area Networks** (MANs) are data communications networks that generally span a metropolitan area. A MAN can be found, for example, connecting the departments of a city government if those departments are scattered throughout the city. A MAN network will typically use transmission devices provided by a local common carrier such as the telephone or cable TV provider. A MAN is larger than a LAN but smaller than a WAN. This category is used primarily for information networks like those found in large cities.

- **Wide Area Networks** (WANs) are data communications networks that encompass a broad geographic area, such as between cities or counties. A WAN will typically use transmission devices provided by common carriers such as telephone and cable TV providers. These transmission devices can include copper cable, optical fiber, microwave, and satellite communications devices. A WAN is a network that spans the largest geographical area. When you connect to the Internet you are connecting to a WAN.

- **Radio Area Networks** (RANs) are data communications networks that use radio or microwave signals as the network media to communicate between devices. This category of network is typically found in supervisory control and data acquisition (SCADA) systems. You may find this type of network used by a city municipal water department, for example, to monitor and/or control the city's water reservoirs and pumping stations that may be located throughout the city (see Figure 20–3). This

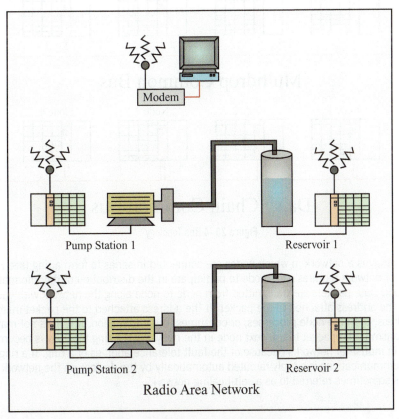

Radio Area Network

Figure 20–3 Radio Area Network (RAN)

type of network can sometimes provide a cost savings over traditional methods like leasing phone lines from a local common carrier. In some cases, where no physical media exists or would be impractical to install, this may be the only option.

Network Configurations

Communication networks can be physically constructed in several different configurations. These configurations are called network **topologies**. A network topology can be in the form of a *bus*, *star*, or *ring* configuration. Each topology serves different network requirements, and in many cases more than one topology might be used to construct a network. The following is a description of the three primary network topologies.

- **Bus Topology** is a network in which nodes are connected to a common bus in a daisy chain or multidrop fashion, as shown in Figure 20–4. In the common bus topology, each device is capable of receiving all information packets on the bus, and network communication can occur between any two devices without having to pass the information through a central network controller (as is the case with a star topology). Common bus topologies are well suited to industrial control applications since each device has equal priority and can exchange information at any time. Many industrial I/O networks are of the bus topology. When adding or removing devices from this type of topology, very little, if any, reconfiguration is required. The main disadvantage with this topology is that all devices share a common bus and a break in the bus can affect many devices.

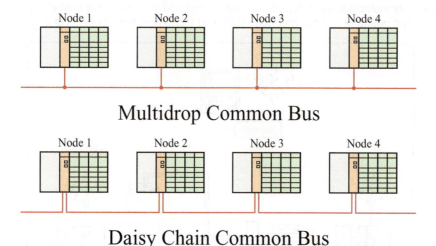

Multidrop Common Bus

Daisy Chain Common Bus

Figure 20–4 Bus Topology

- **Ring Topology** is a network in which nodes are connected in series to form a ring (see Figure 20–5). This type of network requires each node to participate in the distribution of the information along the network. Data packets are transmitted from node to node along the network with each node checking the address attached to the packet. If the address attached to the packet matches the node address, then the node processes, or consumes, the information. If it does not match, then the node retransmits the packet to the next node in the network. The ring topology is becoming more common in industrial networks because of the fault tolerance abilities of a ring. If a ring segment breaks, communications is simply rerouted automatically by the devices on the network. A ring network is sometimes referred to as a self-healing network.

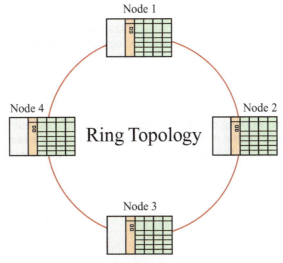

Figure 20–5 Ring Topology

- **Star Topology** is a network in which each node is connected to a network controller (see Figure 20–6). The network controller has the task of passing or retransmitting the information received from one node to the other nodes connected to it. The network controller is responsible for all communication routing on the network. In this type of topology, a failure of a single device does not cause the entire network to fail, but a failure of the network controller could cause an entire network to fail. The network controller is typically an intelligent device such as a network switch or repeater (hub). This type of topology traditionally has not been used in industrial control networks because of the dependence on the network controller, the installation cost (more cable required than in the bus topology), and the once unreliable data transfer time. This is no longer the case with more recent technology advancements and availability of industrial hardened equipment. The star topology is often used in networking computers in business offices and becoming more common in industrial applications.

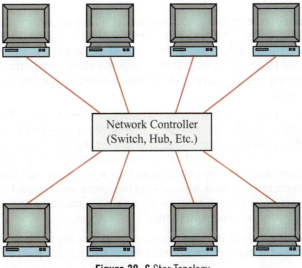

Figure 20–6 Star Topology

Network Media

The network media is the wires, cables, and optical fibers used to physically connect the various communication devices together within a network topology. The most commonly used media includes *twisted-pair*, *coaxial*, and *optical fiber* cables. To implement a communications network you must have a good understanding of the media and hardware used to construct a network.

Twisted-Pair Cable

Twisted-pair is the basic networking cable used today in most offices, buildings, and many industrial plants. This type of cable is relatively inexpensive and has a fair degree of noise immunity. The name comes from the way that each pair of wires are twisted around one another, as shown in Figure 20–7.

Figure 20–7 Twisted-Pair Cable

Twisted-pair cables help reduce crosstalk and noise susceptibility by reducing magnetic coupling between pairs. Some networks using twisted-pair wiring, such as Ethernet, use balanced transmission. In balanced transmission, two wires are used to carry the electrical signal, and each carries a signal (and noise) that is of equal potential with the other conductor, but with opposite polarity. The receiving device only measures the difference between the two conductors. This has the effect of canceling the noise.

Twisted-pair cable can be unshielded or shielded. Unshielded Twisted-Pair cable (UTP) is the most common cable used in offices, buildings, and areas that have low levels of electrical noise. Shielded Twisted-Pair cable (STP), on the other hand, greatly improves the electrical noise immunity of the cable and is normally the type used in industrial network applications. It must be noted that when using STP cable, maximum shielding can be achieved only if the shield is properly terminated. Remember that the purpose of the shield is to pick up the noise and conduct it to ground. The cable, connectors, and equipment must provide a continuous path to ground for the noise, or shielded cable is no better than unshielded cable and in most cases worse.

When twisted-pair cables are used in industrial control networks, the manufacturers of the equipment will typically specify the type and quality of the cable to be used for optimum performance. A shielded twisted-pair cable with an outer jacket that can withstand high temperatures, dirt, oil, solvents, and abrasions is typically specified. Recommendations given by the manufacturers should be followed to minimize network problems.

The twisted-pair cables used in most offices, buildings, and factories for Ethernet communication networks are rated by a category (CAT) numbering system representing the quality and data rate of the cable. The following are the most common of the CAT cable ratings being used.

- CAT 5—Category 5 cable is a multi-pair (usually 4) cable used primarily for data transmission rates up to 100 Mbps.

- CAT 5e—Category 5e cable is the same as CAT 5 cable, except that it is manufactured to a higher standard with transmission rates up to 1000 Mbps. CAT 5e cable is recommended for all new installations.

- CAT 6—Category 6 cable has data transmission rates up to 400 MHz (10 gigabit). Category 6 cable is the fastest communications cable available that is unshielded. This is a high-end cable and the emerging favorite. It should be noted that Category 6 cable is not recommended for industrial applications because it can carry higher frequencies better than Category 5e cable, which makes it more susceptible to the higher-frequency noise from arc welders, motors, VFDs, and other sources of noise.

- CAT 7—Category 7 cable is a shielded multi-pair high-performance cable with transmission rates up to 700 MHz. This high-performance cable requires special connectors and connecting devices to accommodate the shielded pairs.

- CAT 5i—Category 5i cable is an industrial grade cable that has a higher temperature range and an outer jacket that is more resistant to abrasion, oils, chemicals, and solvents.

All communication cables must be connected at each end. In some cases the means of connection is as simple as stripping the cable back and connecting the wires on the proper terminals, and at other times a connector is used. Most twisted-pair cables used for Ethernet networks use an RJ-45 type connector, as shown in Figure 20–8.

Figure 20–8 RJ-45 Connector

There are industrial RJ-45 type connectors available that are equipped with threaded boots that provide protection from moisture, liquids, and vibration. The equipment that the cable is to be connected to determines the type of connection. Cable terminations are a common source of network problems, due to such factors as improper terminations, improper wiring, and mechanical or environmental damage. Care should be taken when terminating communication cables to minimize network problems.

Coaxial Cable

Coaxial cable was, and still is, widely used in industrial control networks because of its excellent immunity to electrical noise generated by factory equipment and offers the best performance of all copper cables. On the other hand, coaxial cable is hard to work with and costs more than twisted-pair cable. The basic structure of a coaxial cable is a center conductor surrounded by a dielectric substance, shield, and outer jacket (see Figure 20–9).

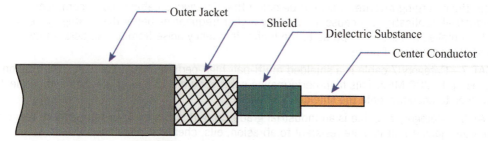

Figure 20–9 Basic Structure of a Coaxial Cable

Coaxial cables have characteristic impedances (50 ohms, 75 ohms, etc.), and must be terminated at their ends for proper operation. For example, if coaxial cable is used in a bus topology in which the ends are not connected to a device that terminates the cable, a special connector called a terminator must be placed on the ends, as shown in Figure 20–10.

Figure 20–10 Coaxial Cable Terminator

The terminator absorbs all energy reaching it and this prevents reflections that can cause network problems. Terminators must have the proper resistance value (50 ohms for 50-ohm coaxial cable, etc.) and be installed for proper network operation. The connector type typically used with coaxial cable is the BNC connector, as shown in Figures 20–11a and 20–11b.

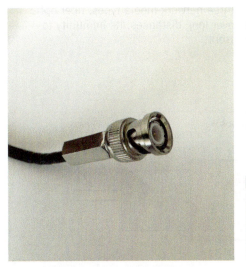

Figure 20–11a BNC Connector

Figure 20–11b T-Tap and BNC Connectors

Fiber Optic Cable

Fiber optic cables use thin glass strands called optical fibers to carry data signals by means of light. The process is simple, digital electrical signals are converted to light signals and transmitted down a glass fiber. At the other end, the light signal is converted back to a digital electrical signal. A single optical fiber strand consists of two parts, the inner core made of glass and the outer cover or cladding. The cladding helps to confine the light inside the core by acting as a mirror to reflect the light down the core (see Figure 20–12).

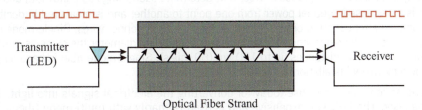

Optical Fiber Strand

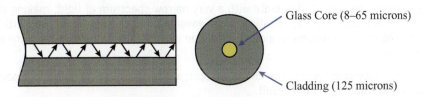

Glass Core (8–65 microns)

Cladding (125 microns)

Figure 20–12 Optical Fiber Strand and Signal Conversion

The optical fibers themselves are enclosed within a protective structure. This structure typically consists of a fiber coating, buffer, strength member, and outer jacket, as shown in Figure 20–13.

Most fiber cables consist of multiple optical fibers. Of the three network media types, fiber optic is considered to be the best because of its high data rates over long distances, its immunity to electrical noise, and the fact that it carries no electrical current.

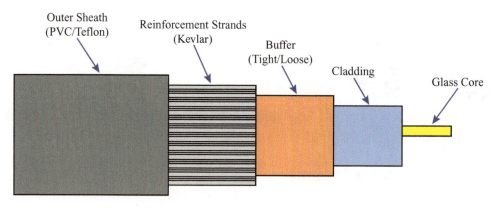

Fiber Optic Cable (Single Fiber)

Figure 20–13 Optical Fiber Cable

Optical fibers come in two types: *multi-mode* and *single-mode*. Multi-mode fiber has a core diameter of around 50–62.5 microns (1 micron = 1 millionth of a meter) with a cladding diameter of about 125 microns, and can have transmission distances up to 12 km without the use of repeaters. Multi-mode fiber has traditionally been the optical fiber of choice for most industrial communication networks because of its compatibility, ease of coupling, cost, and sufficient bandwidth. Not all multi-mode fibers are the same however, there are several grades of fiber that determine such things as bandwidth and *attenuation*. **Attenuation** is a decrease in optical power from one point to another and is measured in decibels (dB). You could compare attenuation in optical fibers to the I^2R loss in copper wires. Single-mode fiber on the other hand, has a core diameter around 8 microns with a cladding diameter the same as multi-mode, and can have transmission distances up to 100 km (60 miles). Single-mode fiber is becoming the optical cable of choice for many installations.

There are two optical devices in use today for converting the electrical signals into light. They are the LED and the laser. The LED is inexpensive and is used primarily with multi-mode fibers. The LED emits a wide band of light that lowers the bandwidth of the signal that the fiber can carry, making it impractical for single-mode fibers. Lasers, on the other hand, have a very narrow and intense beam, are high-speed capable devices, and operate with a very narrow spectrum of light, making them ideally suited for single-mode fiber applications. The wavelength of the light employed with most optical fiber systems is in the infrared region. The human eye cannot see light in the infrared region of the light spectrum.

Caution: Never look directly into the end of an optical fiber that is connected to a power source or permanent eye damage could result.

Fiber optic connectors provide an easy means to connect the optical fiber to the equipment with as little loss of power as possible. With optical fiber connectors, the challenge becomes one of alignment of the light-carrying optical cores to the optical cores on the equipment connectors.

Any misalignment of the fiber cores means a loss of optical power at the connection. Remember that the optical core diameter of a multi-mode fiber is only about 50 microns—about the width of a human hair, if not smaller. Fiber connectors use what are called ferrules to hold the fiber cores. A ferrule has a precision hole that the fiber core is inserted into that provides for accurate positioning of the fiber. Ferrules are typically made of ceramic, plastic, or stainless steel. Epoxy has traditionally been used to secure the fiber in the ferrule, but recent advancements in manufacturing have developed alternate methods that are fast, efficient, and do not require the equipment and skill level of the traditional epoxy methods. Fiber optic connectors come in different types. Two of the most common used in industrial applications are the ST and SC connectors, as shown in Figures 20–14a and 20–14b. Some newer fiber optic connectors to hit the market are the LC and MT-RJ connectors, which are about half the size of the ST and SC connectors.

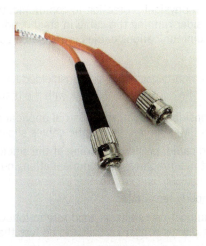

Figure 20–14a "ST" Fiber Connector

Figure 20–14b "SC" Connector

Reducing Electromagnetic Interference

Industrial environments are notorious for producing electromagnetic interference (EMI) in communication and other low-voltage cables. In order to minimize EMI problems, proper selection and installation of communication cables cannot be overemphasized. The following recommendations can help minimize EMI problems.

- Survey the environment the cable will be located in. What kinds of sources of EMI may be present, such as power cables, high-voltage sources, motors, transformers, welders, variable-frequency drives, generators, etc. that must be considered. This would be a good time to also consider other environmental concerns such as temperature extremes, moisture, oils, chemicals, solvents, abrasion, vibration, shock, etc.

- To the extent practical, plan communication cable runs to avoid potential EMI sources.

- Do not route copper communication cables near high-voltage power sources and/or power cables.

- To provide maximum EMI and physical protection to communications cables that are located within industrial plants, consider running the cable in steel conduit. This provides excellent EMI protection. Never run communications cables in the same conduit or raceway with power wires or cables.

Note: Some manufacturers recommend against running unshielded cable in metal conduits, as it may affect the electrical performance of the cable. Always check with the manufacturer.

- Shielded cable should be considered for most industrial applications. For maximum noise immunity or in areas of high EMI sources, fiber optic cables should be considered.

- When you are using shielded cable, proper grounding of the shield is important. You should always follow the equipment manufacturer's recommendations on grounding shielded communication cables.

- Make sure all cables are properly terminated.

- Avoid sharp bends in communication cables. A good rule to follow is that the radius of the bend should be four (4) times the diameter of the cable and no less than 1 in.

- Connectors should be properly selected for the environment and network type.

- Good cable termination practices cannot be overemphasized. Failure in this regard is the source of most communication problems.

- Install communication equipment in enclosures to protect against moisture, dirt, and other contaminants. When steel enclosures are used they also offer a degree of EMI protection.

Network Addressing

Devices connected to a network must have some means of identification in order for messages to be received only by the intended device or devices. A network address, also called a device or node address, is like your home address. In order for the mail carrier to deliver a letter intended for you, they must identify your house by its address. In communication networks the same basic principle applies. Each node connected to the network is assigned an address. When a data packet (message) is transmitted across a network media, each node connected to the network checks the destination address attached to the packet. If the address matches the node's address, then the packet is accepted, if not, the packet is discarded. The format and assignment of an address for a node on a network depends on the type of network and the protocols in effect for that network.

On networks, the address assigned to a device is a unique number such as 1, 10, 77, 115, etc. On some, it can be a series of numbers such as 256.256.168.23 separated by a character(s). The address can be assigned to a device by setting switches physically located on the device or through software. On many industrial control devices, the address is assigned using DIP or rotary switches, as shown in Figure 20–15.

On some networks, the assignment of node addresses is done automatically by a network controller across the network media by first scanning the network and then assigning each node a unique address. On others, the network address is programmed into the device using software and stored in memory on the device.

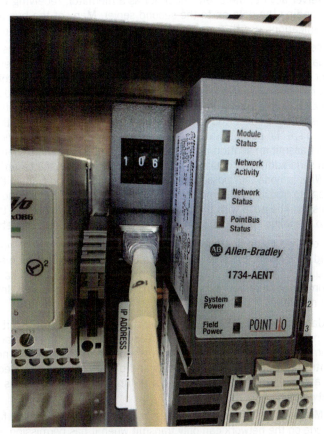

Figure 20–15 Device Address Switches

Network Access Methods (Access Protocols)

PLCs, computers, and other devices must employ a common network access-control method. An *access-control* method defines how and when devices can access and communicate information across the network. There are many methods used but the most common ones are *polling* or *client/server*, *token passing*, and *carrier sensing multiple access with collision detection* (CSMA/CD). The following is a description of how each of these network access-control methods work.

- **Polling or Client/Server Access Method**—In a polling or client/server access-control method, a single PLC or computer is designated as the client and all other devices or nodes connected to the network are designated as servers. In polling, the client is programmed to interrogate, or poll, each server device in sequence to see if it has data to transmit. The client will send an inquiry to a server and then wait a predetermined time for the server to respond. If the server does not respond within the allotted time, the client will assume that the server is dead, or inactive, and continue to poll the next server in sequence. When the client has polled the last device in the sequence, the client then repeats the process. In a client/server access-control method, the servers can only respond to the client, not to each other. If a server wishes to send data to another server device, the client must act as a mediator, receiving the data from one server and then sending the data on to the second server. Many of the bus networks used in industrial communications use this method of access-control or provide the capability.

- **Token Passing Access Method**—In a token passing network, all devices or nodes connected to the network have equal access to the network. There is no client and each is a peer to the other. Another common name for this type of network access method is *peer-to-peer*. In the token passing method, each node on the network is allowed to send data packets directly to other nodes and each node has a scheduled turn and allotted time to send data. In order for a node on the network to send data, it must first possess what is called the token, and only one node can possess the token at any one time. The token is a special packet that, when received by a node, gives that node the exclusive but temporary right to transmit an allotted amount of data. If a node has more than the allotted data to transmit, it will send additional data packets each time it receives the token until it has sent the entire message. When a node that possesses the token has completed transmitting a data packet or if it has no message to transmit, it immediately passes the token to the next node in ascending order of the node addresses. The node with the highest address passes the token to the node with the lowest address, so the token continues to circulate in a loop throughout the network.

Each node has a defined token holding time. If for any reason a node does not pass the token within the allotted time given, the originating node then begins polling node addresses in ascending order until it finds a node that will accept the token. In a token passing network there is a maximum number of node addresses available, limiting the number of devices that can be connected to the network.

The size of the data packet is defined as well as the token hold time, so it is possible to calculate the network access time. For this reason, the token passing access method was the most common network access method used in industrial control networks for passing time-critical control information between PLCs and other controllers. A second advantage is that a failure of any one node does not cause a failure of communications between the other nodes on the network.

- **CSMA/CD Access Method**—The previous two access methods (client/server and token passing) stipulate that a node on the network can only transmit when it has permission, either from the client or by possessing the token. In the CSMA/CD access method, each node has equal right to attempt to transmit without waiting for permission. Whenever a device is ready to transmit data, it checks the network for the presence of traffic on the network. If the network is clear, the device then transmits its data. If the network is busy, the device waits until the network is clear. A "collision" occurs when two or more devices attempt to transmit at the same time. When a collision occurs, a collision detection signal is sent to all devices on the network. Each of the colliding devices must then back off and wait a brief but random time before beginning the process again.

A collision is not an event to be avoided, but simply a method used to arbitrate access to the network. The resolution of a collision occurs very quickly. The two devices transmitting almost immediately abort their transmission and wait a random amount of time before reattempting the transmission. The number of attempts with collision and a random number determine the *back off time*. The back off time is typically in the microsecond range but can be in the millisecond range if there are significant collisions occurring.

Because it is impossible to predict the amount of time required for all colliding devices to successfully complete their transmission, the CSMA/CD access method may not be the best choice for time-critical control networks. It is worth noting that data transmission updates are typically processed in a fast (millisecond) time frame. Advances in network bandwidth and hardware (active switches) have made this access method more widely accepted in the industrial control industry.

Network Protocols

Network communication protocols are sets of formal rules describing how to transmit and share data across a network. Without these formal rules, devices on a network would not be able to understand each other (they must speak the same language). Low-level protocols define the electrical and physical standards to be observed, bit and byte ordering, transmission execution, error detection, and correction of the bit stream. High-level protocols deal with the data formatting, including the syntax of messages, the computer-to-computer dialogue, character sets, sequencing of messages, etc. The Open Systems Interconnect (OSI) reference model defines many of the protocols used. The OSI is a reference model developed by the International Organization for Standardization (ISO) in 1978 as a framework for international standards in network architecture. The OSI model is split into seven layers, from lowest to highest, as shown in Figure 20–16.

Each layer uses the layer immediately below it and provides a service to the layer above. In some implementations a layer may itself be composed of sublayers. It must be noted that a network requires only layers 1, 2, and 7 of the OSI model to operate. Individually, each layer of the OSI model is responsible for a specific task.

Most communication networks today contain all or most of the OSI layers to allow other networks and devices to share information. When a device is sending data over the network, the data starts at the application layer and works its way down to the physical layer, where it's placed onto the network media. When a device receives a message, it is received by the physical layer and works its way up to the application layer. The final result of this effort is to ensure that the data sent from one device is the same exact data received by another device on the network (refer to Figure 20–16).

Network Messages

When it comes to network communications, the terms *message*, *packet*, and *frame* can be confusing for even the most experienced technician. It seems that many of these terms are used interchangeably. We will attempt to define and describe some of these terms.

- **Message**—Message is a common term used to describe information transmitted via a network. A message is the complete information to be conveyed regardless of size or protocols.

- **Packet or Data Packet**—A Packet is a generic term used to describe a unit of data at any layer of the OSI model, but it is most correctly used to describe the application layer data units. An application layer data unit is a packet of data exchanged between two application programs across a network. This is the highest-level view of communication in the OSI model. A single packet exchanged at this level may actually be transmitted as several smaller packets at a lower layer, as well as having extra information (headers) added for routing, etc.

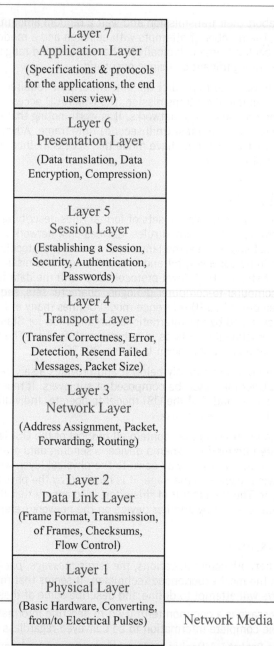

Figure 20–16 Seven Layer OSI Model

- **Frame**—Frame is a data link layer packet that contains the header, data, and trailer information required by the physical media. That is, network layer packets are encapsulated to become frames. A typical frame consists of a header section, a data section or payload, and a trailer section. A frame refers to the structural container of a packet. In most cases, a frame as outlined above is the equivalent of a data packet (see Figure 20–17).

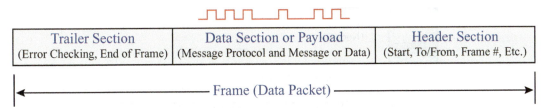

Trailer Section	Data Section or Payload	Header Section
(Error Checking, End of Frame)	(Message Protocol and Message or Data)	(Start, To/From, Frame #, Etc.)

← ——————————— Frame (Data Packet) ——————————— →

Figure 20–17 Basic Structure of a Data Frame

The following is a brief outline of the three sections of a data packet frame.

- **Header**—The header section contains instructions about the data carried by the data packet. These instructions may include: *length of data* (some networks have fixed-length data packets, while others rely on the header to contain this information); *synchronization* (a few bits that help the packet match up to the network); *packet number* (which packet this is in a sequence of packets); *protocol* (on networks that carry multiple types of information, the protocol defines what type of packet is being transmitted); *destination address* (where the packet is going); and *originating address* (where the packet came from).

- **Data**—The data section, also called the *payload*, is the actual data that the packet is delivering to the destination. If a packet is fixed-length, then the payload may be padded with blank information to make it the right size.

- **Trailer**—The trailer section, sometimes called the *footer*, typically contains a couple of bits that tell the receiving device that it has reached the end of the packet. It may also have some type of error checking. The most common error checking used in packets is Cyclic Redundancy Check (CRC). CRC is pretty simple. It works by taking the sum of all the 1s in the payload and adds them together. The result is stored as a hexadecimal value in the trailer. The receiving device adds up the 1s in the payload and compares the result to the value stored in the trailer. If the values match, the packet is good. But if the values do not match, the receiving device discards the packet and sends a request to the originating device to resend the packet.

A message transmitted over a network may be composed of many data packets arranged in a numbered sequence. As mentioned above, most network protocols have fixed-length data packets. If a message is too large for one data packet, the message is broken up into several data packets, each given a sequence number before being sent on to the network media. When the data packets arrive at the destination node, they are reassembled according to their sequence number.

The subject of protocols, packets, frames, and messages may seem overwhelming to the reader at this point, so let's use the following example to illustrate the transfer of data across a network.

Let us assume that we have the message "Feed Boiler Number 2 High Temperature Warning" that needs to be sent from the PLC controller operating the boiler to the HMI (human machine interface) computer located in the control room.

When the high temperature condition is detected, the PLC controller prepares the message for transfer across the network by breaking the message into data packets (payloads) of the proper length. Each data packet is given a header section containing such things as a packet sequence number, destination address (node address of the HMI), source address (its node address), etc. Each data packet is also given a trailer section containing the error-checking value and end of data packet flag. Once the data packets are assembled, they are called frames or data packet frames. Each data packet frame is transmitted onto the network media as a bit stream message according to the network protocol, as seen in Figure 20–18.

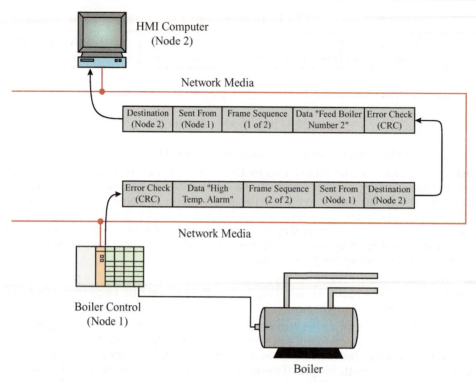

Figure 20–18 Data Packet Frames from Controller to HMI

The HMI computer checks the destination address on each packet being sent across the network. If the destination address matches, then the complete data packet frame is received. Once received, it is checked for errors. If an error is detected, the frame is discarded and a message is sent to the sending node to resend the data packet frame. If no errors are detected, the data packet frame is held until all the data packet frames have been received for that message. After all data packet frames have been received, the header and trailer sections are stripped from the frame and each payload data section is reassembled according to the packet sequence number. The message "Feed Boiler Number 2 High Temperature Warning" would then appear on the HMI computer screen alerting the operators to the problem.

The process just described is how most information is shared across a network media. It must be made clear that the protocol for the type of network you are using will determine how the messages, packets, frames, and electrical or optical signals are constructed and used.

Network Communication Instructions

Many PLC manufacturers provide network communication instructions that allow the transfer of information like the status of inputs, outputs, and tags or registers between PLCs. These programmed instructions typically can be used to read and/or write multiple data words between PLCs and other devices. Some PLC controllers provide for the transfer of information using a *producer/consumer* approach. In the producer/consumer approach, data memory locations are assigned as global producers or consumers during setup. This allows data to be shared between PLCs over a network without programming special instructions into the controllers.

When transferring critical control information between PLC controllers, you should make provisions to ensure that the data being used from other controllers is valid. Invalid data can occur when there is a loss of communications between PLCs that goes undetected. Programming heartbeat logic in one or both PLC controllers may be required. Some PLC manufacturers provide status information within the PLC on all active nodes connected to the network. This status information is updated continuously and can be used to check for valid communications.

Most of the transfer of information between PLC controllers and non-PLC devices (HMIs, computers, information systems, etc.) does not require the programming of PLC instructions to carry out the transfer. The HMI and computer information systems are typically configured during setup to poll information from the various PLCs at a predetermined interval.

Industrial Communication Networks

Industrial communication networks can be divided into three types: *I/O and device*, *control*, and *information*, as shown in Figure 20–19.

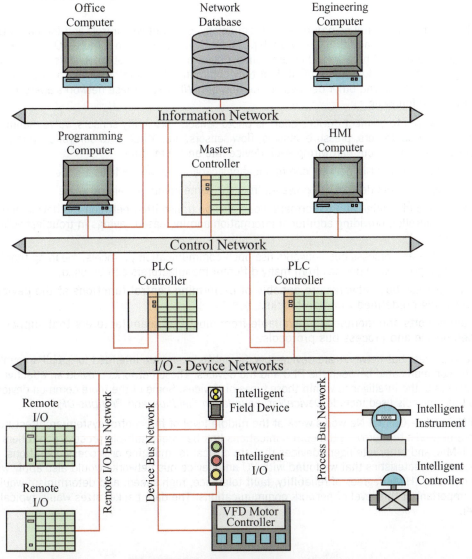

Figure 20–19 Industrial Communication Networks

Depending on the needs and requirements of the control systems on the factory floor and those of operators, supervisors, engineers, and managers, some or all of these network types may be used. Let's take a closer look at the three types of industrial networks.

I/O and Device Networks—I/O and Device Networks work at the lowest level of the control system architecture. These are networks that provide the communications link between the real-world devices and the machine or process controllers (PLCs). In previous chapters you learned about remote I/O, which is an example of a communications network that falls in this category. Other network types include device and process bus networks. Some of the characteristics that would typically be found in this type of network would be relatively small data packet sizes, client/server protocols, bus type topologies, high degree of reliability, fault tolerance, real-time operation, and deterministic operation. Remote I/O has long dominated this category of networks, providing the communication link between remote I/O devices and the PLC. Many of the older remote I/O networks were proprietary, meaning that the hardware and communications protocols used were developed by the PLC manufacturer to support only their hardware.

With advancements in digital communications technology, distributive control, and increased desire for more information, two additional network types were developed to join remote I/O networks. In this category of industrial networks, they are called *Device* and *Process* bus networks. These two network types are not unlike remote I/O in that they communicate I/O information to PLCs, and process control systems and other devices on the network. What sets these networks apart from remote I/O is the following.

- They allow you to connect devices such as photo sensors, proximity sensors, valve manifolds, smart motor controllers, pressure sensors, flow sensors, etc., directly to controllers and other control systems without hardwiring each device into an I/O module.
- Each device has built-in network communications capabilities and acts as a node on a network.
- Most often the field devices are powered from the same communications cable.
- Diagnostic and configuration information can also be transmitted between the device and master controller, providing additional information and increased savings in troubleshooting and setup cost.
- Most device and process bus networks use open communication protocols, meaning that they are not proprietary, so devices from many different manufacturers can be used.
- Some process bus networks are capable of performing control functions at the device level using predefined function blocks.
- Scanner ports and modules are available from most PLC manufacturers that support the open device and process bus protocols.

Device and process bus networks are device-specific networks and are intended for small amounts of data to be communicated between the PLC and the devices. These networks provide an effective way to gain access to the intelligence within those types of devices. Some of the more common device and process bus protocols used include *DeviceNet*, *Foundation Fieldbus*, and *Profibus-DP*.

Control Networks—Control Networks work at the middle level of the control system architecture. These are networks that provide the communications link between multiple process controllers (PLCs), HMIs, and other intelligent devices that are critical for machine or process operations. Some of the same characteristics that we found with I/O and device bus networks would also apply with control networks. High degrees of reliability, fault tolerance, high speed, and determinism would be equally important at this level of network communications. The data packet sizes would typically also be larger.

Control networks provide for the communications of critical machine or process information between PLCs, programming devices, intelligent controllers, and HMI stations. As control and plant engineers have worked toward a more *distributive* control architecture, the need for reliable communications between these various devices has also evolved. A distributive control architecture is one in which PLCs are distributed throughout the factory floor, controlling single machines or processes, rather than a single or central PLC controlling all the various machines or processes. The advantages of a distributive control architecture are increased processing speed with smaller controllers, no single source point of failure for complete shutdown of operations, decreased troubleshooting time, reduced expansion and reconfiguration limitations, etc. Most of the early control networks installed were proprietary networks. As more and more PLCs of different manufacturers found their way onto the factory floor, many from equipment suppliers, there became an ever-increasing need to easily integrate these PLCs. To the benefit of many plant and control system engineers, most PLC and control system manufacturers are building their equipment today with one or more of the open system network protocols. Some of the more common control networks found today include *Modbus Plus*, *ControlNet*, *Profibus-DP*, *Ethernet/IP*, *Modbus TCP/IP*, and others.

Information Networks—Information Networks work at the highest level of the control system architecture. These are networks that provide the communications link between controllers (PLCs) on the factory floor and the business information systems used by supervisors, process engineers, managers, and inventory and production management systems. These networks also provide the backbone for plant-wide integration of control systems.

As PLCs multiplied on the factory floor, so did the need to retrieve information that was in them to help increase production, correct process errors sooner, track products, and monitor production. With factory floor information in the hands of supervisors, managers, and process engineers, decisions can be made more quickly, problems averted, and money saved. The days of the daily paper production reports are nearly gone, thanks to information networks. Some of the characteristics typically found in information networks include large data amounts, high network compatibility with existing business systems networks, and reliability. Some common information network protocols include *Ethernet TCP/IP*, *Ethernet/IP*, *Modbus/IP*, *ProfiNet*, and others.

Industrial Protocols

The following section provides a brief description of some of the more common industrial control and information protocols being used by PLC manufacturers and engineers to implement their industrial automation systems. The protocols mentioned in this section are but a few of the many being used today. Because the descriptions are brief, they are intended only to give the reader a general overview of the protocol.

DeviceNet—DeviceNet, originally developed by Allen-Bradley (Rockwell Automation), is an open system protocol based on the Controller Area Network (CAN) technology. DeviceNet is a digital network that uses a trunk-line/drop-line topology (bus topology) in which node devices (i.e., sensors, motor controllers, small I/O blocks) can be connected by either daisy chain or on short drop lines from the main trunk. Power and signal are provided on the same network cable using separate twisted-pair busses. DeviceNet supports both isolated and non-isolated physical layer design of devices. An opto-isolated option allows externally powered devices (i.e., smart motor controllers, solenoid valve manifolds) to share the same bus cable. The end-to-end network distance varies with data rate and cable thickness. A maximum end-to-end distance of 1,640 feet can be achieved using thick cable and a communication rate of 125 kbs. Each DeviceNet network supports up to 64 nodes. DeviceNet systems can be configured to operate with a client/server or peer-to-peer

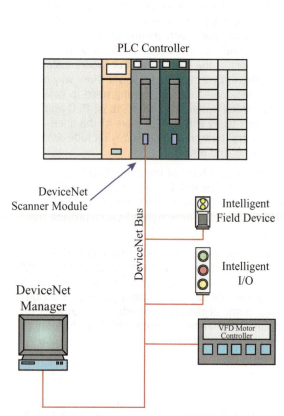

Figure 20–20a DeviceNet Network

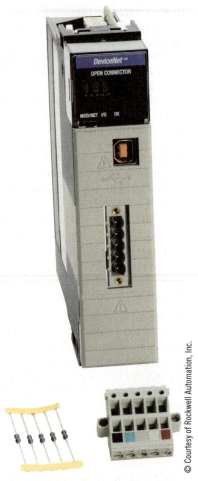

Figure 20–20b ControlLogix DeviceNet
Scanner Module

© Courtesy of Rockwell Automation, Inc.

communication method. Figure 20–20a shows an example of a DeviceNet network and
Figure 20–20b shows a picture of an Allen-Bradley ControlLogix DeviceNet Scanner Module.

Foundation Fieldbus—Foundation Fieldbus was established in 1994 to create a single international
fieldbus standard for hazardous environments and is an open bus standard that enables devices
of different manufacturers to be integrated into one system. Fieldbus technology was intended to
replace the conventional 4–20 mA analog wiring methods found in process control applications with
a digital communications network. Fieldbus technology supports bus-powered field devices such as
process sensors, actuators, and I/O. Foundation Fieldbus protocol allows for the capability to distribute the control functions to the devices using predefined function blocks. This allows field devices
the capability of assuming process control functions, thus reducing the risk of system failure and the
amount of I/O and control equipment needed. The Foundation Fieldbus protocol is based on the OSI

reference model and operates only at layers 1, 2, and 7 of the model, as is the case with most open fieldbus systems. Foundation Fieldbus uses two communication bus systems: the slow, intrinsically safe H1 bus and a higher-level, high-speed Ethernet HSE bus. The H1 bus is used at the device level using shielded twisted-pair cable in a bus topology. Other topology combinations are possible when equipped with junction boxes. Field devices are connected to the bus using tee connectors and short drops called *spurs*. The maximum length of an H1 segment without repeaters is 6,000 feet including spurs, with a maximum number of devices limited to 32 per segment. Data transmission rates for the H1 bus are 31.25 Kbps. The H1 bus of the Foundation Fieldbus uses a central communication control system called a Link Active Scheduler (LAS) that controls and schedules the communication on the bus. This device is called the Link Master. During network setup of the LAS, a transmission schedule is constructed. The schedule determines when devices process their function blocks and when it is time to transmit data. The LAS also allows for unscheduled transmissions for things like device setup and diagnostic data when needed. The LAS handles unscheduled transmissions using a token passing access method. The higher-level HSE bus is based on standard Ethernet technology and runs at 100 Mbps. (Ethernet will be covered later in this section.) The HSE bus allows for the high-speed integration of controllers (PLCs), workstations, HMIs, and H1 bus subsystems using a bridge device. Figure 20–21 shows a Foundation Fieldbus architecture.

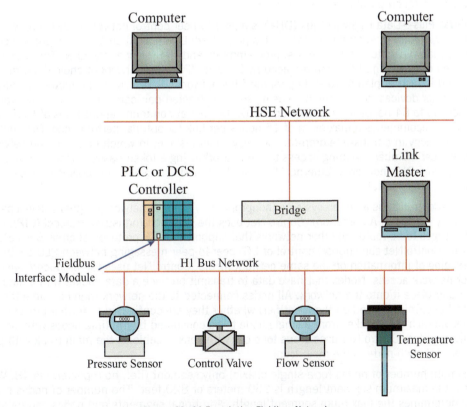

Figure 20–21 Foundation Fieldbus Network

Profibus—Profibus is a group of protocols originally designed by Siemens and adopted into the European standards in 1996. Profibus is an open bus system with protocols (Profibus-FMS, Profibus-DP, and Profibus-PA) that are compatible with each other. Profibus-PA is equivalent to the Foundation Fieldbus just discussed. In fact, the physical bus design is the same, and it exhibits many common features such as function block control at the device level, client/server, and peer-to-peer communications. Profibus-DP is optimized especially for communication between automation systems and decentralized field devices. It is designed to handle time-critical communications. Profibus-FMS is optimized for communication between automation systems (i.e., PLCs, HMIs) as well as data exchange with field devices.

Modbus Plus—Modbus Plus is a high-speed, peer-to-peer LAN network developed by Modicon (currently Groupe Schneider). The original Modbus network, one of the first, was introduced by Modicon in 1979 and was a client/server network that permitted a host computer to communicate to one of several PLCs to perform programming, data transfer, upload/download, and other host operations. Modbus Plus is a local area network that allows host computers, PLCs, and other devices to communicate as peers using twisted-pair cable media. Modbus Plus networks can support up to 32 nodes at distances up to 1,500 feet at communications rates of 1 Mbps. Additional nodes and distances can be achieved by using repeaters. Modbus Plus networks use a bus topology in which nodes on the network function as peer members, gaining access to the network using a token passing access protocol.

Data Highway Plus—Data Highway Plus (DH+) is a peer-to-peer control network developed by Allen-Bradley (Rockwell Automation, Inc.). This proprietary LAN was Allen-Bradley's primary control network for many years to link their PLCs, programming, and HMI devices together. The DH+ network is considered obsolete and has been replaced by Ethernet/IP as the network of choice, but many factories and industrial plants are still operating DH+ networks. A DH+ network allows computers, PLCs, and other devices to communicate using shielded twisted-pair cable media. DH+ networks can support up to 64 nodes at distances up to 10,000 feet at communications rates of 57.6 Kbps. Allen-Bradley recommends a maximum of 15 nodes per link for optimal performance. DH+ networks use a bus topology in a trunk-line/drop-line or daisy chain fashion, in which nodes on the network function as peer members, gaining access to the network using a token passing access protocol. Routing between DH+ networks, Ethernet/IP, ControlNet, and DeviceNet networks is possible using ControlLogix bridge modules.

ControlNet—ControlNet is an open systems (nonproprietary) control network originally developed by Allen-Bradley (Rockwell Automation). ControlNet uses the Common Industrial Protocol (CIP), which allows messages to be routed to other networks that support the CIP protocol, such as DeviceNet and Ethernet/IP. ControlNet can support control of I/O, peer-to-peer messaging between PLC controllers, and messaging of information on the same network media. ControlNet uses the producer/consumer model for network access. Nodes that have data to transmit produce a data packet with a data identifier and place it onto the network. All nodes connected to the network then consume the data packet and determine (based on the identifier) whether they are configured to further process the data. This allows a data packet from a single node to be consumed by multiple nodes without having to be retransmitted. ControlNet uses a bus topology with coaxial cable as the main trunk with passive taps to connect each node to the main trunk.

The maximum number of nodes for a single coaxial-only segment (i.e., no repeaters) is 48. With 48 nodes, the maximum segment length is 250 meters or 820 feet. The number of nodes per segment determines the maximum segment length. Additional segments and nodes can be added by using repeaters. The maximum number of nodes that can be connected to a single ControlNet network is 99. The data transmission rate for ControlNet is 5 Mbps.

Ethernet—The original Ethernet was developed as an experimental LAN in the 1970s by the Xerox Corporation and operated on coaxial cable at a data rate of 3 Mbps using CSMA/CD protocol. The success of the original Ethernet led to the development of the 10 Mbps Ethernet through a joint effort by Digital Equipment Corporation, Intel Corporation, and Xerox Corporation in 1980. The American National Standards Institute (ANSI) and Institute of Electrical and Electronics Engineers (IEEE) published Ethernet as an official standard in 1985 as ANSI/IEEE std.802.3-1985. It has become known as the IEEE 802.3 standard.

Ethernet is the single most common network technology used in the world today to connect personal computers (PCs) and workstations within offices, buildings, and across the Internet. Some of the reasons that the Ethernet protocol has monopolized the computer networking industry is that it is easy to understand, implement, manage, and maintain, and is relatively inexpensive. Initially, Ethernet was not a consideration for the factory floor because it was slow and response time varied greatly based on network traffic. With the development of high bandwidths, inexpensive Ethernet switching technologies, improved error checking, and deterministic advancements, Ethernet/IP has emerged as the standard in control systems today. Ethernet/IP will be covered next.

The term *Ethernet* refers to LAN products covered by the IEEE 802.3 standard that defines what is commonly known as the CSMA/CD protocol. Three data rates are currently defined for operation over optical fiber and twisted-pair cables, 10 Mbps (10Base-T Ethernet), 100 Mbps (Fast Ethernet), and 1000 Mbps (Gigabit Ethernet). Ethernet networks consist of network nodes or devices that connect to interconnecting media. The nodes can fall into two classes of devices. The first class of

devices are called *Data Terminal Equipment* (DTE), which produce and consume the data packet frames on the network. Typical devices are computers, servers, printers, PLCs, etc. The second class of devices are called *Data Communication Equipment* (DCE), which receive and forward data packet frames across the network. Typical devices are repeaters (hubs), routers, and switches. Switches are the most common DCE device being used and can be either managed or unmanaged. Figure 20–22 shows an 8-port unmanaged industrial Ethernet switch designed for today's modern control systems. The media options available to connect the different devices include unshielded twisted-pair cable (UTP), shielded twisted pair-cable (STP), and optical fiber cable.

Ethernet networks can have different network topology configurations including point-to-point, bus, and star. Since the early 1990s, the star-connected topology has been used as the standard topology in many applications. Data terminal equipment (DTE) is connected to a central repeater (known as a hub) or to a network switch that has

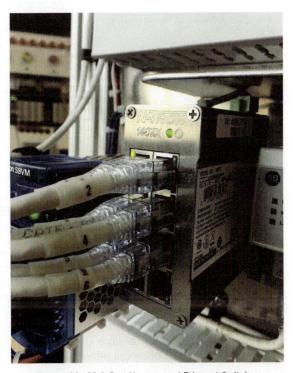

Figure 20–22 8-Port Unmanaged Ethernet Switch

multiple connection ports. Each DTE device is connected in a point-to-point configuration with the central repeater or switch. The maximum distance that a DTE device can be located from the central repeater device is typically 265 feet but can depend on the network data rate and media used. Most Ethernet networks are made up of many small star topologies interconnected to form the network, as shown in Figure 20–23.

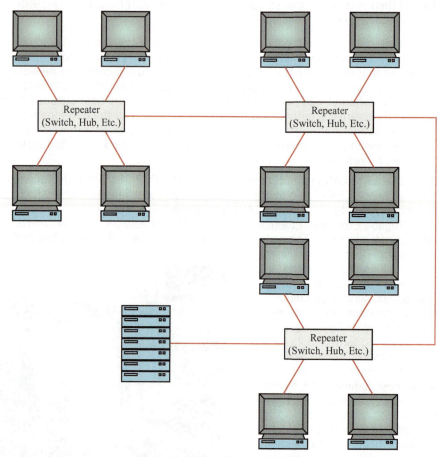

Figure 20–23 Typical Ethernet Topology

The Ethernet protocol itself operates at only layers 1 and 2 of the OSI reference model. The IEEE 802 standard divides the data link layer (layer 2) of the OSI model into two sublayers called the *Media Access Control* (MAC) sublayer and the MAC-client sublayer. The MAC-client sublayer provides the interface between the MAC sublayer and the upper layers of the OSI model. The MAC sublayer is responsible for the data frame assembly before transmission, frame disassembly and error-checking during and after receipt of data frames, media access control including initiation of frame transmission, and recovery from transmission failure. The format of an Ethernet data frame is shown in Figure 20–24. The physical layer (layer 1) specifies the transmission data rate (Mbps), signal encoding, and the type of media interconnecting the devices.

The IEEE 802.3 standard currently requires that all Ethernet MACs support half-duplex operation and the CSMA/CD access method. There are physical network size limits for half-duplex

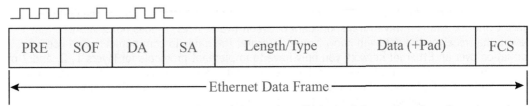

| PRE | SOF | DA | SA | Length/Type | Data (+Pad) | FCS |

◄──────────────── Ethernet Data Frame ────────────────►

Preamble (PRE) - Alternating pattern of ones and zeros that tells the receiving stations that a frame is coming.
Start-of-Frame (SOF) Delimiter - Pattern of ones and zeros indicating start of frame information.
Destination Address (DA) - Indicates which station(s) should receive the data frame.
Source Address (SA) - Identifies the sending station.
Length/Type - Indicates the length of the data field or optional data frame format.
Data (+Pad) - Data field containing the message, padded if required for proper length.
Frame Check Sequence (FCS) - Cyclic redundancy check used to check frame for errors.

Figure 20–24 Structure of an Ethernet Data Frame

operation based on the network data rate. Full-duplex operation is an optional capability that allows simultaneous two-way transmission between two devices (point-to-point link). Full-duplex transmission is much simpler than half-duplex transmission, because full-duplex uses two pairs of network wires or links for sending and receiving data frames. Media contention and data collisions are all but eliminated with full-duplex transmission, leaving more time available for transmissions and in effect doubling the link bandwidth. In order for full-duplex transmission to take place, the physical layers (cables and network equipment) must be capable of supporting full-duplex operation.

Ethernet devices implement only the bottom two layers of the OSI protocol model, and for this reason they are typically implemented as *Network Interface Cards* (NICs), or as built-in Ethernet network interface ports. The NICs and interface ports are identified by a three-part naming convention based on the physical layer standards. The three parts of the naming convention are data transmission rate, transmission method, and the media type. For example: 10BaseT stands for 10 Mbps data transmission rate, baseband transmission method, and two twisted-pair cable as the media type. Most Ethernet networks today are constructed using category 5e (CAT 5e) and terminated at each end by 8-pin RJ-45 connectors.

As mentioned, a DTE device is connected to a central repeater (known as a hub) or network switch having multiple connection ports. Each DTE device is connected in a point-to-point configuration with the central repeater or switch. When central repeaters such as hubs are used, they typically accept only one data packet frame at a time and then resend it to all active ports on the repeater. Switches, on the other hand, have ports with I/O frame buffers that isolate the port from traffic being sent at the same time on other ports. Multiple internal data paths allow data packet frames to be received on one port and then rapidly switched to the appropriate output port. Switches typically build and maintain internal tables that map Ethernet addresses to ports. The switch uses the addressing information within each Ethernet data packet frame to forward the frame only to the port connected to the destination device. This may not seem like that much of a difference between a switch and hub but it has a major effect on network operation. Because switches provide access to a high-speed network bridge (switch) that interconnects only two ports, the collision domain in the network is reduced to a series of small domains in which there are only two devices. This markedly increases network bandwidth and begins to increase the determinism of an Ethernet network, making Ethernet attractive for industrial control systems. For this reason network repeaters or hubs are all but obsolete for industrial and large networks.

In summary, Ethernet itself is simply a protocol that works at the bottom layers of the OSI stack (layers 1 and 2), and is a way to transport data between two devices; it does not guarantee that a device that receives the data will know how to interpret the data. To communicate and use information over an Ethernet network you only have to implement protocols for layers 1, 2, and 7 of the OSI model as a minimum. Remember that the applications that utilize this data operate at the highest level of the OSI model (layer 7), and if the applications are not using the same protocols at this level they will not be able to interpret the data received.

EtherNet/IP—EtherNet/IP is traditional Ethernet combined with an industrial application layer protocol targeted to industrial automation. This application layer protocol is the same Control and Information Protocol (CIP) that is used in both DeviceNet and ControlNet networks discussed earlier. EtherNet/IP was developed by Rockwell Automation, Inc. and released in early 2000 as a replacement to ControlNet, and to fulfill the desire to provide control, configuration, and collection of data on one network. Now, Ethernet/IP is owned and maintained by ODVA (Open DeviceNet Vendors Association) and is considered the network protocol of choice for modern control and information networks in manufacturing today.

EtherNet/IP is based on standard Ethernet and takes advantage of many of the features originally developed for use in the IT field, such as DHCP and switching technologies. The media used for EtherNet/IP is the same as that used for traditional Ethernet networks such as fiber optic, CAT 5 and CAT 6 Ethernet cables, along with active communication devices like Ethernet switches. In fact, industrial Ethernet switches are at the center of any Ethernet/IP star and switch-level ring topology. The primary difference between traditional Ethernet and Ethernet/IP is the quality (robustness) of the media and active network components used. Industrial environments introduce their own challenges and the network equipment installed must be specifically designed to live in those kinds of environments.

EtherNet/IP is built on the widely accepted CIP protocol standard. This standard organizes networked devices (nodes) as a collection of objects. Each object is simply a grouping of related data values in a device. Every CIP device on the network is required to have an Identity, Message Router, and Network object. The Identity object contains device identity data values called attributes. Attributes for an Identity object include vendor ID, date of manufacturer, device serial number, and other identity data. The Message Router object is an object that routes explicit request messages from object to object in a device. The Network object contains the physical connection data for the object which includes the IP address and other data describing the interface to the Ethernet port on the device. There are Application objects that have data for your specific device such as motor or analog data objects. The same objects implemented in two or more devices behaves identically from device to device. A grouping of objects in a device is referred to as that device's "Object Model". The Object Model is based on the producer–consumer model which allows the exchange of information between a sending device (producer) and many receiving devices (consumers) without requiring data to be transmitted multiple times by a single source to multiple destinations. The use of the CIP protocol along with an advanced Ethernet switch-based infrastructure is what allows EtherNet/IP to provide real-time or nearly real-time efficient communications on the factory floor making EtherNet/IP acceptable for modern industrial control networks.

Keep in mind, Ethernet/IP is an application layer protocol that is transferred inside an Ethernet TCP/IP packet over traditional Ethernet media and active communication devices. Because Ethernet technology is so widely used commercially, being able to bring Ethernet to the factory floor using EtherNet/IP has leveraged many of the existing resources and training already in place.

There are many devices available that support Ethernet/IP from PLCs, switches, and I/O devices to integrated motion and safety devices. Many of these products simplify the IP addressing with rotary switches on the product as was shown in Figure 20–15. These rotary switches provide for 254 Ethernet/IP addresses (devices) per network. Figure 20–25 shows an Allen-Bradley 1756-ENT2 Ethernet/IP module that can be installed into an I/O chassis to provide high-speed data transfer between Logix 5000 PLCs and remote I/O devices or connect to multiple EtherNet/IP network topologies. These EtherNet/IP modules can be purchased in both copper and fiber optic media, as well as single or dual ports with switching capabilities.

© Courtesy of Rockwell Automation, Inc.

Figure 20–25 Allen-Bradley EtherNet/IP Module

It should be mentioned that many EtherNet/IP devices support CIP services like CIP Security which provides for secure data transport across an EtherNet/IP network. It lets CIP-connected devices to authenticate each other before transmitting and receiving data. CIP Security and standard Ethernet security features can be used as part of a multi-layer approach to network security. Other available CIP services include CIP Safety, CIP Motion, CIP Sync, and CIP Energy.

This chapter provides the technician with a very broad overview of the types of communication networks and protocols that can be found in industrial plants today. There are many additional resources available to expand the technician's knowledge of industrial communication networks including books, whitepapers, and Internet resources from which to draw.

Chapter Summary

Communication networks provide the means for devices like PLCs, computers, and intelligent I/O devices to share or pass information using a common media. A communication network is typically called a Local Area Network (LAN), but other names like WAN, MAN, and RAN are also used depending on physical size. How the network cables are routed and the role that network devices play in the transmission of data is called the network topology. The three basic topologies are bus, ring, and star. The most common media types used in the construction of a network include coaxial, twisted-pair, and fiber optic cables. One or all three of these types could be used in the construction of a single network. In order for devices to share a common media, a method called Network Access-Control is used. Network access-control is like a traffic cop that determines when devices can access the network to transmit their information, and it also resolves any conflicts. The three most common network access-control methods include client/server, peer-to-peer, and CSMA/CD. When devices share common media, they must speak the same language. This common language is called the network protocol. A network protocol is a set of rules that determines how a message is compiled into data packets, converted into electrical signals, transmitted onto the network media, received only by the intended device, and then converted back into the original message without any loss of information or translation.

Different types of networks are used in industrial automation and control based on the type of information to be communicated. An I/O or device network is used to transmit the information between field devices (i.e., sensors, actuators, motor controllers, etc.) and the controllers that are controlling the machines or process. Control networks link the controllers, HMIs, and other control systems together. Information networks link the factory floor to the business side of the company and can also provide the backbone to link the smaller control networks together in a large plant.

Key Terms

human machine interfaces

nodes

packet

bandwidth

Local Area Network

Metropolitan Area Network

Wide Area Network

Radio Area Network

topologies

Bus Topology

Ring Topology

Star Topology

attenuation

Review Questions

1. When devices are connected to network media they are typically called what?
2. Network bandwidth is usually expressed in
   a. meters.
   b. bytes per hour.
   c. bits per second.
   d. packets per second.
3. What category of network would you typically find monitoring and/or controlling a city's water reservoirs and pump stations, which may be located in remote areas?
4. Star topology is a network topology in which each device is connected:
   a. to the other devices in a daisy chain fashion.
   b. to a PLC.
   c. to a network controller.
   d. to a common bus.
5. When twisted-pair cable is used as the network media it can:
   a. be no longer than 100 meters.
   b. only have one pair of wires per cable.
   c. use BNC type connectors.
   d. be unshielded or shielded.
6. A single optical fiber strand consists of two parts. What are those two parts?
7. What are the three most commonly used network access-control protocols?
8. The Open Systems Interconnect (OSI) model is:
   a. used to connect two networks.
   b. a reference model of network architecture and a suite of protocols.
   c. a reference model of the physical topologies of a network.
   d. a network media connector.
9. A "frame" is a data link layer "packet" that contains information required by the physical media. What are the three basic sections that make up a frame?

10. In what type of industrial network would you typically find the client/server protocol and bus type topology used?

    a. Information Networks.

    b. Control Networks.

    c. Device Networks.

    d. All of the above.

11. The network address assigned to a device is:

    a. a unique name.

    b. a number or series of numbers.

    c. a word.

    d. a physical location.

Chapter 21

Start-Up and Troubleshooting

Learning Objectives

After completing this chapter, you should have the knowledge to:

- Describe the commissioning process for newly installed PLC equipment.
- Explain how input devices are typically tested.
- Explain the safety concerns when testing output devices.
- Describe how voltage readings are taken to check input and output modules.
- Explain how analog signals are test.

Start-Up

On new control system installations, careful commissioning procedures are necessary to protect personnel safety and prevent equipment damage. It is during the commissioning phase that the highest risk of injury or equipment damage can occur. This is due in large part to untested circuits, hardware, and software logic. Depending on the size of the system being commissioned, it is not uncommon to find errors or mistakes on such things as drawings, field wiring, programmed logic, equipment installation, etc. Because of these risks, careful commissioning plans should be developed and only experienced personnel assigned to oversee the commissioning process. On large or critical systems, it is common to find documented step-by-step commissioning procedures and commissioning engineers assigned to the commissioning team.

Prior to performing any commissioning procedure, it is important to check and verify that the system has been installed according to the manufacturer's specifications and/or engineer's design, and the installation meets all applicable local, state, and federal codes.

Before applying power to the PLCs, complete the following steps.

Step 1. Verify that the incoming power is within the PLC power supply specifications. If the power supply is dual rated, make sure that the proper voltage has been selected on the power supply. Most often, jumpers or switches are used to make this selection. Figure 21–1 shows a typical PLC power supply with voltage selection switch circled.

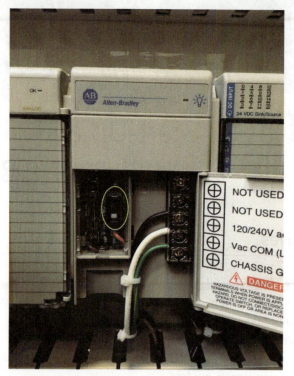

Figure 21–1 Power Supply with Voltage Selection Switch

Note: Almost all dual-rated power supplies are shipped from the factory with the voltage setting in the HIGH voltage position.

Step 2. Verify that all hardwired safety circuit(s) and emergency stop device(s) have been installed, tested, and are in the safe position.

Step 3. Check all wiring and cabling to ensure they are properly terminated and there are no loose connections.

Step 4. Connect all communication cables, making sure that connectors are fully inserted into their sockets. Secure connectors as applicable.

Step 5. Ensure that all modules are securely installed in the I/O chassis(s) and that removable terminal blocks (if applicable) are fully seated and secured.

Step 6. Clean all cabinets and equipment paying special attention to scrap wire, dust and dirt, loose or missing fasteners, and all covers are securely in place.

Note: Do not use compressed air to blowout electrical/electronic equipment or cabinets. If a portable vacuum is used, it must be one that is static free.

Step 7. Place the PLC processor key switch in the remote (REM) position, as shown in Figure 21–2.

Step 8. Check all DIP and rotary switches for proper setting(s) if applicable.

 Caution: Before proceeding make sure that all safety circuits and other emergency stop devices have been fully tested and are in the emergency *OFF* or open position. Temporarily disable power to all PLC output device circuits.

Step 9. Apply power to the PLC chassis/controller while observing the controller status light(s) for proper indication. When power is applied, the chassis power supply will provide the necessary DC voltage for the PLC and I/O modules.

Testing Inputs

Use this procedure to verify electrical continuity to input devices and the proper functioning of those devices. This procedure is an end-to-end test and must be performed while on-line with the PLC controller.

Figure 21–2 Key Switch in Remote Position

With power applied to the PLC and I/O modules, apply power to the input device circuits only, leaving power *OFF* to the output device circuits. Configure a temporary offline PLC project for the type of controller and I/O modules installed but contains no user programmed logic (code). Download this temporary project to the PLC and go on-line with the PLC. Once on-line, examine the contents of the I/O Configuration folder in the Controller Organizer window for any warning symbols. If a warning symbol is displayed, clear the warning before proceeding. From the on-line toolbar, select Test Mode and then click "Yes". The PLC should now be in the Test Mode. While in the Test Mode, input data is received and any PLC code is executed, but output data is not sent to the output modules.

Open the controller scoped tags (Controller Tags) folder in the Controller Organizer window. With the Controller Tags window open, expanded the module tag for the input module to be tested so that all input data points for that input module are visible. Ensure you are on the correct input module tag and on the Tag Monitor tab.

Where it is safe to perform, manually change the physical input devices (e.g., toggle a limit switch, push a button, etc.) one at a time while monitoring the corresponding tag in the Controller Tags window to verify that the change occurs. This process normally requires two people and a set of as-built drawings.

Additionally, with power applied to the input device circuits, any input device that is closed, or *ON*, will have its corresponding status light illuminated on the front of the input module, as shown in Figure 21–3.

Figure 21–3 Input Module Status Indicators

⚡ **Caution:** *Do not* activate input devices mounted on equipment by hand where unexpected machine motion could cause injury.

Failure of an input device to function may indicate one of the following:

- Incomplete or incorrect wiring. Check to be sure that the input device is wired to the correct input module and terminal.
- No power to the input device.
- Input module not correctly wired.
- Faulty input device.
- Input module faulted or not communicating with the PLC.

Proceed to check all input devices for proper operation before proceeding to test the output devices.

Testing Outputs

Before testing output devices, it must be determined which devices can be safely tested (turned *ON*). It is common practice to first test motor starters with the motor leads temporarily disconnected as shown in Figure 21–4. In this configuration, the motor starter itself (the output device) is tested without having to energize the motor. This prevents unwanted machine motion that might cause injury to personnel or damage to equipment.

For outputs that can be safely energized, make sure that the equipment is in a safe start-up state, all workers and equipment are in the clear, and the equipment is properly lubricated if required.

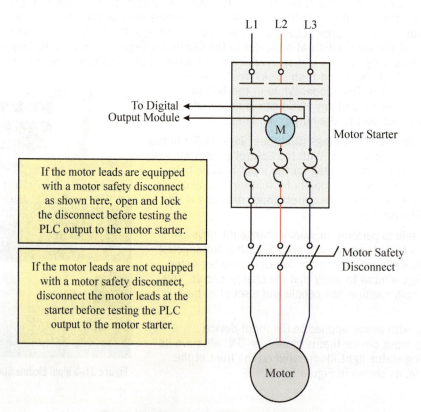

Figure 21–4 Disconnecting Motor Leads for Safety

Caution: Use extreme caution when testing output device wiring. Output devices will be activated to assess how the machinery or processes will respond. Notify all personnel of the test. Unexpected machine motion can injure personnel.

Apply power to the output circuits and reset all emergency stop circuits so outputs can be energized.

This method uses the programming software to manually turn *ON* and *OFF* each output device to be tested using the Controller Tags monitor window for the output module tag to be tested. Ensure that you have selected the correct output device address to be tested before proceeding with testing the output.

The temporary PLC project should already be downloaded to the PLC from the input tests just conducted. If not, download this temporary PLC project. As already mentioned, it is safer to have no logic programmed in this temporary PLC project as the PLC will be placed in the Run Mode and outputs will be active. From the on-line toolbar select the Run Mode and then click "Yes". On the monitor tab of the Controller Tags window, expand the data member for the output module tag to be tested. You can now force a digital output ON or *OFF*, or in the case of analog outputs, force an analog value. With the output now forced *ON*, the physical device is active and can be checked. Depending on the PLC, forcing *ON* and *OFF* outputs can be just a matter of placing a 1, 0, or value in the output memory address for the output device being tested when there is no output logic programmed in the PLC. Otherwise, you must use the force feature of the PLC controller. See forcing I/O in the controller's user manual. When all output devices have been tested, disable all forces and/or remove forces.

Final System Checkout

After all input and output device circuits have been tested and verified, the electrician or technician is ready for the final system checkout.

Reconnect any output loads (motors, etc.) that were previously disconnected for testing. In the case of motors, correct rotation needs to be established before the machine or process can be fully commissioned.

Once all I/O devices have been tested, motors reconnected, and correct rotations established, proceed as follows.

Step 1. Download the final PLC project containing all required logic to the PLC controller.

Step 2. Place the PLC in the Test Mode or disable outputs mode and check the PLC logic for any unexpected results.

Note: In the test, or disable output mode, the physical outputs will not be energized. All PLC logic will function normally but actual outputs will not be turned ON at the module level.

Step 5. Once the PLC logic and circuit operation has been verified in the test or disable output mode to the extent possible, the processor can be placed in the Run Mode for final system commissioning and testing.

Step 6. Once the PLC project has been fully commissioned and tested, it is recommended that a backup copy of the PLC project be saved for future recovery needs.

Troubleshooting

The key word to effective troubleshooting is *systematic*. To be a successful troubleshooter, the technician must use a *systematic* approach.

A systematic approach should consist of the following steps:

- Symptom recognition
- Problem isolation
- Corrective action

The technician should be aware of how the machine or process normally functions before attempting to troubleshoot a problem. When there is no prior knowledge, the best source of information, if applicable, is the operator(s). Do not hesitate to ask the operator(s) specific questions and what they think may have happened. Operators have an in-depth knowledge of the equipment they operate and can be a valuable resource to the technician.

Although the PLC can't talk, it can communicate in various ways to indicate possible problems with the hardware. There are various status lights on the PLC, power supply, and I/O modules that indicate normal and abnormal conditions. The status indicators of a typical PLC controller are as follows.

> *Run Indicator* – If this LED is *OFF*, the controller is in the program or test mode. A steady green indicates the controller is in the Run Mode.
>
> *Force Indicator* – If this LED is *OFF*, there are no tags containing I/O force values. Solid amber color indicates that I/O forces are present and enabled. Flashing amber color indicates that forces are present but not enabled.
>
> *OK Indicator* – If this LED is *OFF*, no power is applied to the controller. Flashing red indicates that the controller has detected a major fault or a firmware upgrade is required if it is a new controller. A steady red indicates that the controller is either performing a power-up, loading a project, or the controller is inoperable. A solid green indicates that the controller is operating normally.

Status lights on the I/O modules also assist with troubleshooting problems that involve input and output devices. Do not overlook these status lights as they can be a valuable troubleshooting aid to the technician.

Example Troubleshooting Scenario

If an operator reports that an electrical solenoid is not working, the first step is to find out as much information about the solenoid as possible from operators and other resources. Next, check the output module status light for the output device in question. If the output status light is *OFF*, the PLC programming software can be used to check the PLC logic (on-line) for the output address (solenoid device) in question. If the output logic is not true or *ON*, check the rung conditions to see why logically it is not being turned *ON*.

As previously discussed, output modules have status indicators that illuminate when each output is turned *ON* as well as blown fuse indicators. If the status indicator is *ON* for that output point and there are no blown fuse indicators, the problem is most likely not with the output module and controller. A voltage check at the output terminal for the device in question, as shown in Figure 21–5, will verify that the output module is working properly and the problem is either in the wiring to the solenoid, or the solenoid itself. Further voltage checks should help to locate the problem.

Note: When testing AC output modules with solid-state switches (triacs), the high internal impedance of most digital meters will show a reading of nearly full voltage when the solid-state output is in the OFF state. For accurate readings, use a voltage meter that will slightly load the output circuit under test, like a solenoid-type tester with low internal impedance. ***Do not*** use this type of tester on PLC input circuits.

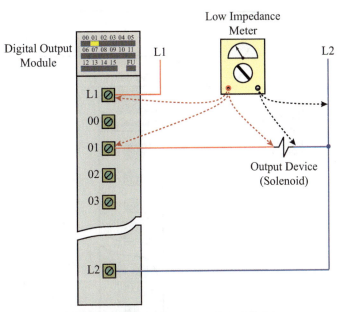

Digital Output Module

Low Impedance Meter

L1

L2

L1
00
01
02
03

L2

Output Device
(Solenoid)

Figure 21–5 Testing Voltage on an Output Module

Troubleshooting input modules follows the same basic procedure. If the input device and circuit are suspect, check the input module status light for the input device in question. An illuminated status light indicates that the input device is supplying power to the input module (contact closed) and the information is not being communicated to the controller. The input module and controller should be investigated. If, however, the status indicator is *OFF*, there could be several reasons, such as the input device is not being activated as expected, a input device is bad, faulty wiring, or input circuit power loss. Make sure the input device has been properly activated and then take voltage readings, as shown in Figure 21–6.

A voltage reading equal to the applied voltage indicates that the input device and wiring are OK and the problem may be with the input module. No voltage reading indicates a problem with either the input device, circuit wiring, or input device power source. Further voltage checks should isolate the problem.

 Caution: Use a high-impedance voltage meter when taking voltage readings on PLC input circuits. Failure to use a high-impedance voltage meter could cause unexpected machine operation by unintentionally activating inputs, such as push buttons, limit switches, etc.

Troubleshooting Analog I/O

Troubleshooting PLC analog I/O signals are often the most difficult to troubleshoot. This is due in large part to a varying signal as compared to just an *ON/OFF* signal like digital I/O. Also, analog I/O involves additional configuration settings and components to produce a reliable and accurate signal (measurement). Analog signals are also much more susceptible to external influences than digital I/O. The following guidelines can be used to help the technician isolated common analog signal problems in PLC systems.

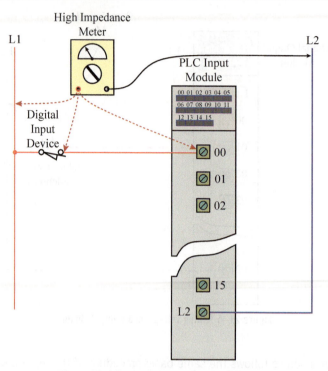

Figure 21–6 Testing Voltage on an Input Module

The first things to consider when troubleshooting any analog signal is understanding what is being reported, how long the problem has been occurring, and has there been any recent changes to the process or installation of any new equipment or wiring in the area. For example, a report of no reading (signal) is much different than one that is reported as being in error or unstable. All of this information can help the technician to mentally assess the problem and determine the next steps to be taken.

When an analog I/O device has been in service for a while without problems, and there have been no recent changes to the PLC hardware, program(s), or I/O configuration settings, the problem is most likely on the field side of the analog I/O interface, but not always. On the other hand, if this is a new installation (sensor, I/O module, PLC controller) or there have been recent changes to the PLC program logic or operator interface devices, checking the PLC first may make sense. Paying special attention to analog module configuration settings and scaling values, refer to Chapter 17. If a new analog field device has been installed, check to make sure it has been properly configured and wired. Many analog field devices have configuration settings that must be set to ensure compatibility with the analog interface equipment to which they are attached.

All PLCs have a way to monitor the analog input or output values. If you know how the analog I/O channels have been configured and the scaling ranges, you can quickly determine what the analog electrical signal should be at the terminals of the analog interface module when measured. For example, if the scaled analog input value corresponds to 38.5% of the analog channel's configured range and that range is 0–10V DC, then the voltage measured at the analog channel terminals should be close to 3.85 V. On the other hand, if an analog output value is set to 62% of its configured range and that range is 4–20 mA, then the output of the analog module for that channel should be 13.92

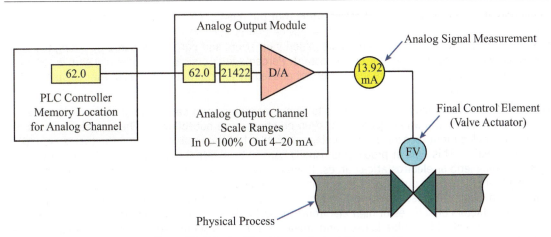

Figure 21–7 Analog Signal Conversion and Measurement

mA. As you can see, knowing where to monitor the analog values in the PLC along with the analog channel configuration settings and scaling values, will allow you to quickly confirm the expected electrical signal at the PLC analog interface terminals and field device. If these two readings do not match within the analog modules specifications, it could indicate a problem with the analog module. Figure 21–7 illustrates this means of verification.

Problems on the field side of analog inputs typically fall into one of three categories, no signal or intermittent, offset error, or an unstable reading. No signal or an intermittent loss of signal can indicate a loose or broken wire, sensor power source problem, or bad sensor. Check all wiring connections for loose or broken wires. This may seem obvious, but sometimes the most obvious can be easily overlooked.

All analog input field devices must have a power source to generate the required analog electrical signal to the PLC input. On four-wire sensors, the sensor receives its power from an external power source which can be either AC or DC. On two-wire sensors, it is the current loop (4–20 mA) itself that provides the power for the sensor and the loop typically receives its power from an external 24V DC power supply. Check and make sure that the sensor is receiving power (voltage) within the sensor's specifications. On current loops, when additional devices like panel meters, strip chart recorders, etc. are added to a current loop, the overall loop impedance may become too high and the voltage at the analog sensor may drop below the sensor's minimum voltage rating necessary for proper operation. Access to a process meter that is capable of sourcing analog signals can make quick work of performing an end-to-end test of the analog wiring and PLC interface by injecting a known signal at the sensor end (sensor removed) and confirming that signal in the PLC by monitoring the analog input value in controller memory.

Analog outputs are much easier to troubleshoot as you can force an analog value in the PLC and confirm the corresponding signal at the output terminals of the analog module, or at the field device for a complete end-to-end check. Also, analog output modules are typically sourcing modules which means they provide the necessary power and there is no external power source to check.

Analog offset errors can either be constant or a percentage. A constant offset error is found when the correct value always differs from the measured value by an additive or subtractive constant. This is also referred to as a zero offset error, where zero volts is not measured as zero. There are many potential causes for constant offset errors, but the most common are sensor calibration errors and sharing a ground circuit with some other device that is drawing significant current through the ground circuit. Percentage offset errors are called gain errors and can be corrected by taking the measured value and multiplying it by a constant. Percentage errors can be caused by gain errors in the sensor or analog input, or the loading effect caused by an interaction between the resistance of the sensor and analog input.

Unstable analog readings are also referred to as noisy readings and can be quite common in some industrial environments where good installation practices were not followed. They appear in cases where the analog input (sensor) source voltage is stable, but the measured value fluctuates around the correct value. This type of problem is usually caused by external noise entering the signal wires by electromagnetic or electrostatic sources. If analog signal wires are bundled with or near cables carrying high alternating currents or high frequency signals, it is very likely the analog signal wires will pick up electrical noise. To avoid this type of problem, do not route analog wires near high current conductors or sources of high electromagnetic fields like transformers and motors. Make sure that analog shielded cable is used and shields are properly grounded. Analog cables can also be routed inside steel conduit. The steel conduit, if properly installed and grounded, will shunt most if not all interference from external magnetic fields.

Measuring analog signals is performed with an accurate digital multimeter or process meter that is capable of reading both voltage and current. Analog voltage signals are quite straightforward to measure and do not require signal disruption. By accessing the analog terminals at the analog module and field device, the technician can easily measure the analog voltage at these points as shown in Figure 21–8.

Analog current signals, on the other hand, can be more challenging when using a multimeter as the circuit must be "broken" at some point to connect the multimeter in series with the current loop.

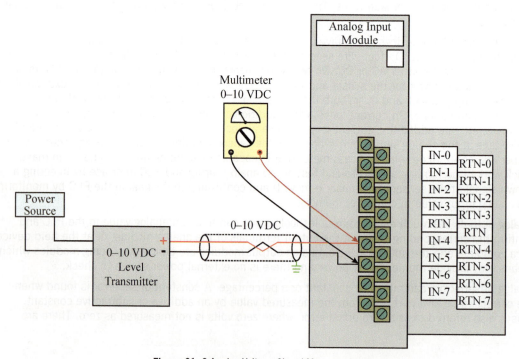

Figure 21–8 Analog Voltage Signal Measurement

Interrupting the current signal means interrupting the analog information whether that is a process measurement or signal to a final control element like an actuator. This could have adverse effects on the control system and process being controlled. You should never break an analog signal without first warning all appropriate personnel that the signal will be interrupted. Failure to do this could cause unintentional shutdown, process upset, or worse. It will also most likely cause faulty process alarms. Knowing how an analog signal is used in the control system and working with operations personnel before performing any analog testing cannot be over stated.

In many PLC control systems where analog current loops are used, special terminal blocks are installed that allow current measurements to be taken without breaking the loop. These terminal blocks incorporate a small knife switch or removable plug, and meter lead test points. By connecting the leads of a milliamp meter (multimeter) to these test points and then opening the small knife switch or removing the test plug, allows for direct measurement of the current signal without distributing the analog signal or having to un-do any wire connections in the circuit. After the measurement is taken, the knife switch is closed back in or the test plug is reinserted **BEFORE** the meter leads are removed. Figure 21–9 shows a 4–20 mA current loop being tested as described above.

A better way to measure a current signal without interrupting it, is to use a special clamp-on milliamp meter that is designed to monitor the weak magnetic fields created by the passage of the small DC currents in the analog wires. This is a completely non-intrusive method since the meter merely clamps around the wire without the need to break the circuit. A disadvantage when using clamp-on milliamp meters is the susceptibility to errors from strong external magnetic fields. Make sure external magnetic fields are not influencing your readings.

Refer to Chapter 17 for information on analog I/O configuration. Once you know the specific details (signal type, range, engineering units, configuration settings, data locations, etc.) you can properly troubleshoot a PLC analog I/O problem.

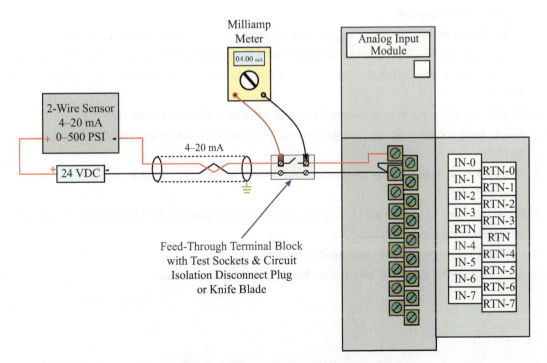

Figure 21–9 Analog Current Signal Measurement

Chapter Summary

Start-up procedures are designed to check each part of the PLC system for proper installation and operation. Safety must *always* be the overriding factor when commissioning new control systems and equipment. Care must be taken to prevent unexpected or incorrect machine motion that could cause personnel injury and/or damage to equipment.

Once the commissioning process is complete and the equipment is operational, problems or faults can occur. To successfully troubleshoot the PLC control system, a systematic approach must be used. This systematic approach includes recognizing the symptoms, isolating the problem, and taking corrective action. A variety of indicator and status lights, as well as error messages and fault codes, are available to assist the technician in troubleshooting and should not be overlooked.

The information covered in this chapter is intended to be general in nature and not specific to any particular PLC or system. For more specific information on start-up and troubleshooting procedures, refer to the manufacturer's operating manuals for additional information.

Review Questions

1. A programmable logic controller system should be installed according to the:
   a.   manufacturer's specifications
   b.   electrical inspector
   c.   "YouTube" video
   d.   equipment operator
2. At what point in the commissioning process should hardwired safety circuits be tested?
3. Describe briefly how input devices can be tested when commissioning a new PLC control system.
4. Describe briefly how output devices can be safely tested when commissioning a new PLC control system.
5. Why disconnect some output devices like motors before testing PLC outputs?
6. When taking voltage readings on digital output circuits, what instrument should be used and why?
7. When taking voltage readings on digital input devices and circuits, what instrument should be used and why?
8. Name two types of status lights on digital output modules that can assist the technician when troubleshooting digital outputs.
9. List the three steps of systematic troubleshooting.
10. What precautions should be taken when testing analog current loops?

Chapter 22

PLC Programming Examples

Learning Objectives

After completing this chapter, you should have the knowledge to:

- Better understand PLC instructions.
- Apply various PLC instructions together.
- Apply simple PLC control logic solutions.

This chapter provides the reader with several examples of PLC ladder logic programs utilizing many of the instructions covered in previous chapters. The examples presented here are intended to help the reader gain a better understanding of the various PLC instructions and how they can be combined to provide simple control logic solutions. The examples are not intended to take the place of application-specific requirements. In most cases there is more than one solution to a control problem. The right solution and subsequent logic depends on having a thorough understanding of the machine or process to be controlled.

To help in understanding the following PLC programming examples, the actual I/O memory addresses will not be shown. Instead, a description will be used followed by the type of memory address in bold type. As an example, shown in Figure 22–1 is a normally open contact (EXAMINE ON) with the description "Start PB" followed in bold by **"Digital Input"**; this memory address would be a digital input from a real-world push button labeled "Start."

The following are examples of other memory types used in this chapter.

Digital Output – A digital output address in I/O memory that a real-world device is being controlled from, such as a pilot light, motor starter, solenoid, etc.

Start PB
"Digital Input"

Figure 22–1 Digital Input Description Example

Digital Memory – A Boolean address in user memory that is used primarily for internal or dummy relays, such as "Control Relay CR-1."

Analog Input – A word address in I/O memory containing a decimal value from a real-world analog input device such as a temperature sensor, pressure transmitter, etc.

Analog Output – A word address in I/O memory containing a decimal value being sent to a real-world analog output device such as a valve, variable speed drive, panel meter, etc.

Decimal Memory – A word address in user memory containing a decimal value from calculations or simply used to store values for comparison type instructions.

Timer and counter address will only have descriptions such as "½ Second Flasher Timer," "½ Second Flasher Timer Done," etc.

Example 1—PUSH ON/PUSH OFF Circuit

In some applications it is desirable to have a single, N.O. momentary push button that is used to both turn *ON* and *OFF* a digital output(s) such as lights, solenoids, etc., or an internal digital memory address. When one begins to think about the solution and subsequent logic it becomes clear that there is certainly more than one solution possible, as is often the case. Which is the best solution is a question that challenges all PLC programmers. Only after carefully studying what the output controls and taking into account any safety or fail-safe considerations can the best solution be chosen.

In this example of the "Push ON/Push OFF" circuit, we have chosen to control a light and always have it be in the *OFF* state when the PLC begins executing the program on a restart of the processor (fail-safe position on restart). Figure 22–2 shows the logic we have selected. In this example, we are using a common logic circuit called the *flip/flop* circuit to provide the means of alternating between the two states, *ON* and *OFF*. The basic *flip/flop* circuit is shown in Rung 2 of Figure 22–2. It gets the name *flip/flop* because any time the memory address labeled "Flip/Flop Trigger" is true or *ON* and the PLC processor scans the rung, the output (light) will change state or alternate, hence the name *flip/flop*. In Rung 1, the real-world push button is used to turn on the "Flip/Flop Trigger" output that is used to change the state of the flip/flop circuit. The "ONS" instruction in Rung 1 is a one-shot instruction used to keep the "Flip/Flop Trigger" output *ON* for only one scan, which is required for proper operation of the flip/flop circuit. Refer back to Chapter 10 for a review of the "ONS" instruction.

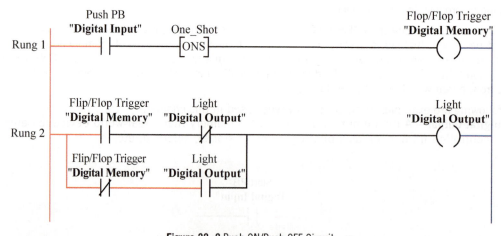

Figure 22–2 Push ON/Push OFF Circuit

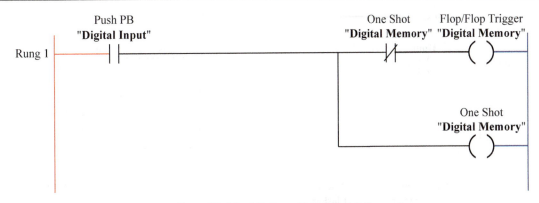

Figure 22–3 Push Button with One-Shot Coil

The basic flip/flop circuit shown here is a very universal circuit that can be used for many different applications, as you will soon see.

If the PLC you are using does not have a One-Shot instruction, one can be made by using the logic shown in Figure 22–3 in place of Rung 1 in Figure 22–2.

Example 2—½ Second Pilot Light Flasher Circuit

Often when designing control circuits it is desirable to flash pilot lights at various times to alert operators to abnormal or pending conditions. The power and flexibility of PLCs makes this a very easy addition to any control circuit design. In the example presented here, we will design a universal flasher circuit that will provide for a digital memory bit labeled "½ Second Flasher" that can be used throughout your PLC program to flash pilot lights as desired. The half (½) second flash rate used in this example is a common flash rate for pilot lights, but any flash rate can be chosen based on your application or needs. The same basic *flip/flop* circuit used in Example 1 will also be used here to provide the basis for our flasher. As you will recall from Example 1, any time the "Flip/Flop Trigger" memory bit is *ON*, the output of the *flip/flop* circuit will change state. In that example, the push button and one-shot were used to turn *ON* the flip/flop trigger. If we replace the rung containing the push button, one-shot, and flip/flop trigger output with an on-delay timer set for ½ second that automatically resets itself, we will have our ½ second flip/flop trigger. Figure 22–4 shows the flasher circuit using the on-delay timer and *flip/flop* circuit.

After reviewing the flasher circuit in Figure 22–4, did you notice that we did not use a one-shot instruction as was required in Example 1? That is because the on-delay timer in Rung 1 is self-resetting, meaning the done bit of the on-delay timer is only *ON* for one scan.

Example 3—Motor Starter Fault-Monitoring Logic

When controlling industrial motors with PLCs, it is often desirable to monitor and know when motor starters fail to operate. If a motor starter should fail to operate, then that information can be used in your PLC program to provide for the safe shutdown of other equipment, as well as alert an operator or maintenance person to the problem. Motor starters can fail to operate for many reasons, but some of the more common reasons are: motor overload trips, low or no control voltage, open starter coils, welded contacts, etc. The logic presented in this example is designed to detect a failure of a motor starter to pull in and/or drop out. In order to detect a motor starter failure, a normally open auxiliary (aux) contact

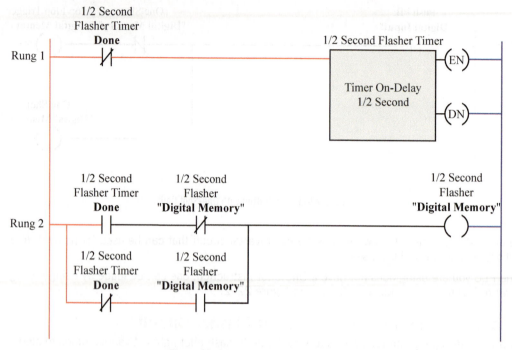

Figure 22–4 1/2 Second Flasher Circuit

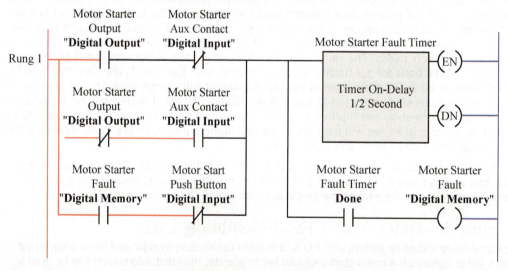

Figure 22–5 Motor Starter Fault-Monitoring Logic

mounted on the starter must be wired to an input module of the PLC (refer to Chapter 10, "Overload Contacts," for additional information on this input and the circuit presented here). The status of this input, as well as the PLC output to the motor starter, becomes the basis for the fault detection circuit shown in Figure 22–5.

In this circuit logic, a single motor starter can be monitored for a failure to both *pull in* and *drop out* by evaluating the state of the output with that of the input. If the output to the motor starter is *ON* and the input from the starter auxiliary contact is *OFF*, then this would be considered a failure to pull in, and conversely, if the output to the starter is *OFF* and the input is *ON* then this would be considered a failure to drop out. This logic is shown in the first two branches of Rung 1 in Figure 22–5. If a motor starter failure is detected, then the third branch acts as a holding circuit for the fault condition until it can be acknowledged and reset by an operator or maintenance person. In our example, we have chosen to use the "Motor Start" push button to also act as the fault reset button, but you could just as well use a separate reset button, or for that matter, any other condition that fits your application or needs.

The on-delay timer in Figure 22–5 is required to allow for the physical movement (time) of the motor starter when energizing and de-energizing the starter. The ½ second time delay chosen for the on-delay timer is typically adequate for most applications, but can be adjusted as necessary. If the on-delay timer should time out, then the "Motor Starter Fault" memory bit will turn *ON* and remain *ON* until reset. This "Motor Starter Fault" memory bit can be used in your PLC program as necessary to disable the motor starter output rung, flash pilot lights, shut down other equipment or take whatever action is desired.

Example 4—Three-Wire Motor Control Logic with Fault Monitoring, Pilot Light, and Flasher Circuit

In this example, we will show how the ½ second pilot light flasher and motor starter fault logic shown previously can be used in a basic three-wire motor control circuit. In this example, we will use both start and stop push buttons to control the motor. In addition to starting the motor, the start push button will also be used to reset a motor starter fault. Along with the start and stop push buttons, a pilot light will be used to indicate when the motor is running, as well as when a motor fault occurs. Shown in Figure 22–6 is the logic for Example 4.

Rung 1 in Figure 22–6 is your basic three-wire motor control circuit for controlling the motor output. In addition to the start and stop push buttons in Rung 1, you will also notice a normally closed contact labeled "Motor Starter Fault" that is in series with the motor output. This contact is used to turn *OFF* the output to the motor starter when a motor fault is detected. Rung 2 in Figure 22–6 is the motor starter fault monitoring logic that was shown in Example 3 and used here to monitor the motor starter. Rungs 4 and 5 set up a ½ second flasher circuit to be used in Rung 3 for flashing the motor running pilot light when a motor fault occurs. Keep in mind that this flasher circuit could be located at the beginning of your program and used in any pilot light circuit in your program as desired.

Example 5—Time-Based Events

Often, it is desirable to trigger an event at some predetermined time. If the time is in seconds or minutes, then a basic on-delay or off-delay timer may be all that is needed, but if the time is in days, weeks, etc., then additional PLC logic is required. In the examples presented here, our goal is to show two methods that can be used to trigger an event at a predetermined time that a timer instruction alone is unable to do. In the following two examples, we shall assume we are operating a ball mill that requires service by the plant maintenance personnel every 168 hours or 7 days of ball mill operation. The event to be triggered at the end of the 168 hours is a maintenance warning light that is used to alert the maintenance personnel that the ball mill requires service. In this first example, we will use a counter and timer instruction to keep track of the number of hours the ball mill has operated. In the second example, we shall accomplish the same thing, but instead of a counter and timer, we shall use math and compare instructions along with a timer.

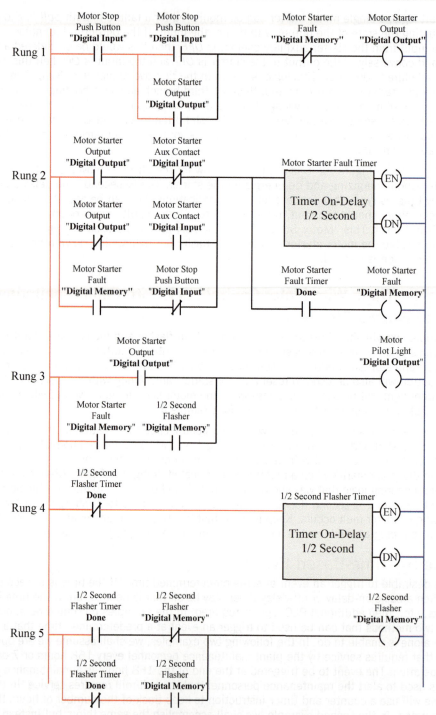

Figure 22–6 Motor Control and Monitoring Logic

The logic for the time-based event using a counter and timer is shown in Figure 22–7.

In Rung 1 of Figure 22–7, the on-delay retentive timer is enabled whenever the ball mill is running. A retentive on-delay timer was chosen so that any accumulated time would not be lost in the event the mill was stopped and then restarted. The on-delay timer is set with a preset time of 3600 seconds or one hour. When the on-delay timer has timed out at the end of one hour, the done bit

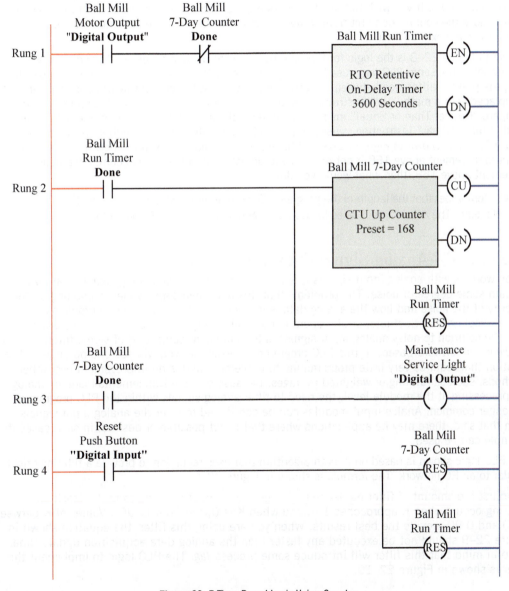

Figure 22–7 Time-Based Logic Using Counter

of the timer is used to increment the counter (CTU) in Rung 2. At the same time the counter is incremented, the retentive on-delay timer is also reset, allowing it to begin another timing cycle. The counter instruction in Rung 2 is programmed with a preset count of 168 counts, which is equal to 168 hours and the time required to trigger the maintenance warning light. When the counter accumulated value equals the preset, the done bit of the counter is used in Rung 3 to turn *ON* the maintenance warning light, alerting the maintenance personnel that the ball mill requires service. A "Reset" push button is used to reset the counter and timer to begin the operation over as shown in Rung 4. The "Reset" push button could be an integral part of the maintenance warning light. The done bit of the counter is also used to keep the timer and counter from counting above the preset time of 168 hours. If you wish to track the hours beyond 168 for display on an HMI for example, then remove the counter done bit from Rung 1. Just remember that the accumulated value of a counter has a maximum value.

Shown in Figure 22–8 is the logic for the time-based event using a timer, math, and compare instructions. The same PLC logic used in Rung 1 of Figure 22–7 is also used in this example. In fact, the primary difference between the two examples is that the counter instruction in Figure 22–7 is replaced with a math "Add" instruction, and the counter done bit in Rung 3 is replaced with a compare "Greater Than or Equal" instruction. To reset the accumulated hours in the data memory word, a math "Clear" instruction is used in place of the counter reset instruction in Rung 4 of Figure 22–7, as shown in Figure 22–8. In this example, we have chosen to allow the accumulated hours to increment above 168 if not reset. If you do not wish to have the accumulated hours increment above 168, then what would you do?

Note: Remember that the length of the program affects scan time, which in turn affects timer accuracy and total time. The actual time it takes to count to 168 hours may be 168 hours plus.

Example 6—Analog Signal Filter Algorithm

When working with analog input signals it is not uncommon to have analog signals that may contain some degree of noise. The problems that this noise can cause varies depending on the severity of the noise and how the analog data is intended to be used. Common problems include digital displays that fluctuate, erratic operation of analog output devices, process disturbances, etc. The solution to noisy analog input signals is to implement some kind of signal filtering, either hardware or software-based. In this PLC programming example, you will see one possible software solution that requires very little programming to implement and is an improvement over other methods, such as running or weighted averages. Because of the advancements made in analog I/O processing at the module level, the need to filter analog signals within the PLC program is no longer common. Analog input modules can be configured to filter the analog input signals. With that said, there may be applications where that is not possible or desired, in such cases this example can be used.

The PLC logic shown is based on a math algorithm that uses recursion to provide a filtering effect similar to an RC network. The formula is shown in Figure 22–9.

To adjust the amount of filtering needed, change the value of K in the formula. Maximum filtering occurs when K approaches 1.0, and when K is 0, the filter is *OFF*. Values of K between 0.80 and 0.90 provide the best results. When you are using this filter, the equation shown in Figure 22–9 should not be executed any faster than the analog data acquisition update time. Keep in mind that this filter will introduce some process *lag*. The PLC logic to implement this filter is shown in Figure 22–10.

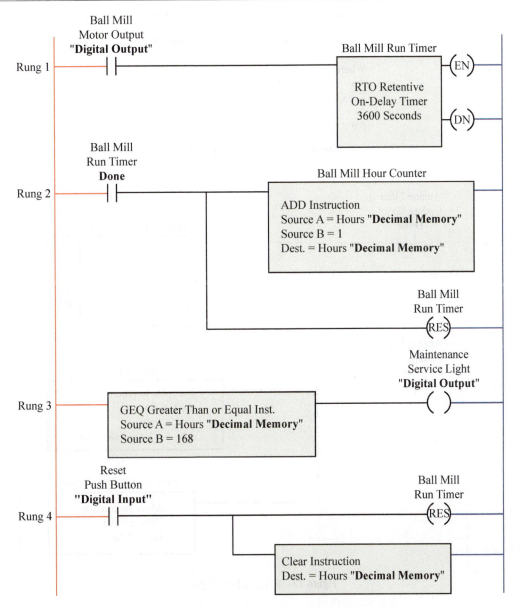

Figure 22–8 Time-Based Logic Using Math and Compare

The Allen-Bradley "Compute" instruction is being used in this example, but you could easily replace the compute instruction shown here with the required individual math instructions. In this example, we have arbitrarily set our filter update time to one-quarter of a second, as indicated in our on-delay timer in Rung 1. Make sure you adjust the on-delay timer in Rung 1 to match your analog data acquisition update time. The "Move" instruction in Rung 2 stores the current filtered analog out value for use in the next calculation as the previous value.

$$Ao = ((1 - K) * Ai) + (K * Ap)$$

Where:
Ao = Filtered Analog Out
K = Gain (0.0 – 1.0)
Ai = Analog Input
Ap = Previous Filtered Analog Out

Figure 22–9 Analog Filter Equation

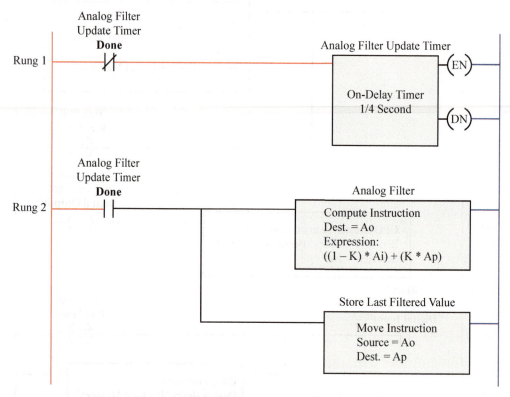

Figure 22–10 Analog Filter Logic

Example 7—Parts Conveyor Tracking Logic

In many industrial plants it is often desirable to track objects as they are transported from one end of a conveyor system to the other. There can be many reasons for tracking an object including tracking the physical location, tracking the information about an object, or just simply keeping track of the number of objects. Every tracking requirement is unique, both in the hardware and PLC logic needed.

In this example, we will track the physical location and length of an object (part) as it travels the length of a 50-foot conveyor. Located 25 feet from the beginning of the conveyor system is a spray nozzle that is designed to spray a cleaning solution onto each object as it passes. Because of physical and environmental constraints we are unable to mount a sensor at the spray nozzle to detect the presence of each object as it passes the sprayer. The only location where an object can be detected is at the very beginning of the conveyor system by means of a photoelectric sensor. For this reason, we must track the physical location along with the length of each object as it travels down the conveyor system in order to turn *ON* and *OFF* the spray nozzle at the appropriate time. In addition to the photoelectric sensor, an incremental encoder is mounted on the conveyor system and is designed to produce an *ON/OFF* pulse for every 1 inch that the conveyor travels. These two digital PLC inputs, photoelectric sensor and incremental encoder, will be used to track the position and length of each object as it travels down the conveyor system.

Based on the above information we know that the length of the conveyor is 50 feet or 600 inches, which also equals 600 encoder pulses; therefore, the spray nozzle located at 25 feet would equal 300 encoder pulses. By taking the encoder input and using it to operate a "Bit Shift Left" instruction with a length of 600 bits, we are able to have a bit location that corresponds to each inch of conveyor length. Figure 22–11 shows the conveyor, photoelectric sensor, spray nozzle, and 600-bit array laid out on the length of the conveyor system. As you study Figure 22–11, the spray nozzle is located at bit location 300, which we know equals 25 feet. For each inch that the conveyor system travels, all bits in the array are shifted one position to the left toward bit position 600.

If we use the photoelectric sensor input as the input bit address to the "Bit Shift Left" instruction, we will be setting bit position 1 of the array to *ON* or 1 any time that the photoelectric sensor is blocked by an object. Each time the conveyor moves 1 inch, the status of the photoelectric sensor, *ON* or *OFF*, is loaded into the bit array and shifted one position to the left. If an object on the conveyor system is 10 inches in length and blocks the photoelectric sensor as it passes, then our bit array will have 10 consecutive bits that will be *ON*, matching the length of the object (see Figures 22–12a and b). As the object travels down the conveyor system toward the spray nozzle, the 10 bits matching the object's length are also being shifted through the bit array, tracking the object's position on the conveyor system. When the first bit of the string of 10 bits reaches the 300th bit location in the array, the spray nozzle is energized, spraying cleaning solution on the object, and it remains *ON* until all 10 bits have passed the 300th bit location (see Figure 22–12c).

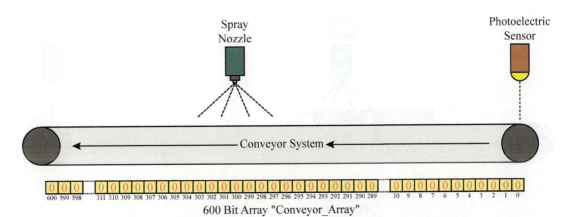

600 Bit Array "Conveyor_Array"

Figure 22–11 Conveyor Tracking System Layout

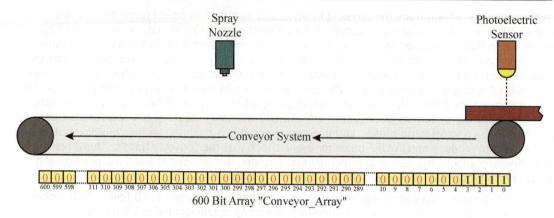

Figure 22–12a Object at Photoelectric Sensor

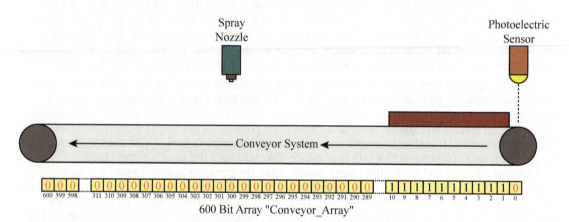

Figure 22–12b Object Past Photoelectric Sensor

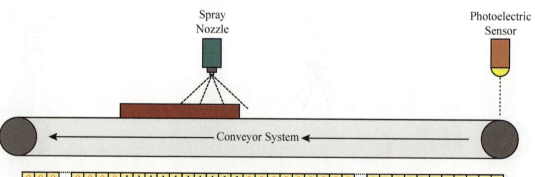

Figure 22–12c Object at Spray Nozzle

Shown in Figure 22–13 is the PLC logic for the bit array tracking example described above. The bit shift left instruction is shown with bit array "Conveyor_Array[0]". The "Conveyor Running" digital input in Rung 1 and 2 is there to ensure that the spray nozzle output operates only when the conveyor is moving.

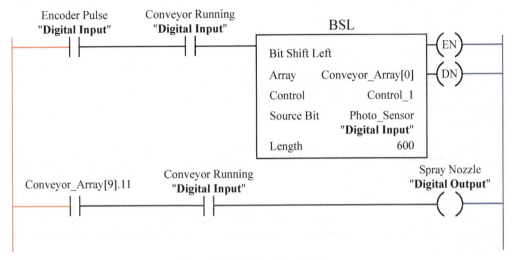

Figure 22–13 Bit Array Tracking Logic

If the conveyor system can be operated in both the forward and reverse directions, then the addition of a "Bit Shift Right" instruction with the same array address as that of the bit shift left should be used. In addition to the bit shift right instruction, each bit shift instruction must be enabled *only* when the conveyor system is operating in the appropriate direction

Note: When using an encoder with standard digital PLC input modules, make sure that the encoder pulses are not changing faster than can be read by the PLC, including scan time or missed pulses could occur, causing inaccurate or erratic operation. Many PLC manufacturers have special encoder modules or high-speed digital inputs for high-speed applications.

Chapter Summary

After working through the examples presented in this chapter, you should begin to see how the various PLC instructions can be combined to provide control logic solutions. As was stated at the beginning of this chapter, there is typically more than one solution to a problem and the right solution depends on having a thorough understanding of the machine or process to be controlled. Along with understanding what you are controlling, you must also understand how the various PLC instructions work and how the PLC executes them. Think of the various PLC instructions as tools in your tool bag, they are only as good as the person who knows how to use them.

When developing PLC logic, always think about the potential impact it may have on personnel safety and machine or process operations. Never create an unsafe condition; always look to prevent it. SAFETY FIRST!

Programming Challenge

Design a PLC program to control two pumps in duplex operation. Duplex operation involves alternating pumps/motors for equal wear and to ensure both are fully operational if needed. This is a common control strategy when one pump is adequate for normal operation and a second pump is installed in the event the primary pump fails or if one pump is not adequate for peak design.

The system to be programmed involves a large wastewater sump that is equipped with both low and high float switches to detect wastewater level. The float switches are wired to PLC digital inputs (normally open contacts). The sump is equipped with two pumps but only one is required for normal operation. When the pumps are in auto, they alternate each time the sump fills. Each pump is equipped with an Auto-Off-Manual selector switch, pilot light, Stop and Start push buttons, and motor starter with auxiliary contacts.

When the wastewater level reaches the sump high float switch, one pump is started and allowed to run until the wastewater level drops below the low level float switch. The next time the wastewater fills up the sump, the second pump will run and empty the sump. Each time the sump fills up, the two pumps will alternate taking turns to empty the sump as long as each pump is selected for auto operation. If one pump is not in auto, the other pump becomes the primary pump to empty the sump each time the sumps fills up. When a pump is in "Manual", the Stop and Start push button can be used to start and stop the pump regardless of wastewater level. A pump's pilot will illuminate when the pump is running.

Use the following PLC I/O to design the PLC ladder logic for the duplex pump operation described above:

I/O Type	Device	Description
Digital Input (NO)	Sump Low Level Float Switch	Input is true when the wastewater in the sump is above the low-level float switch.
Digital Input (NO)	Sump High Level Float Switch	Input is true when the wastewater in the sump is above the high-level float switch.
Digital Input (NO)	Pump 1 Selector Switch – Auto Mode	Input is true when Pump 1 is in Auto Mode.
Digital Input (NO)	Pump 1 Selector Switch – Manual Mode	Input is true when Pump 1 is in Manual Mode
Digital Output	Pump 1 Pilot Light	Pilot light is ON when Pump 1 is running.
Digital Input (NO)	Pump 1 Start PB	When true, starts Pump 1 in Manual Mode.
Digital Input (NC)	Pump 1 Stop PB	When true, Pump 1 is allowed to run in Manual Mode.
Digital Input (NO)	Pump 1 Motor Starter Aux Contact	Input is true when Pump 1 motor starter is energized.
Digital Output	Pump 1 Motor Starter	When true, Pump 1 motor starter is energized (pulled-in) and Pump 1 runs.
Digital Input (NO)	Pump 2 Selector Switch – Auto Mode	Input is true when Pump 2 is in Auto Mode.
Digital Input (NO)	Pump 2 Selector Switch – Manual Mode	Input is true when Pump 2 is in Manual Mode
Digital Output	Pump 2 Pilot Light	Pilot light is ON when Pump 2 is running.

I/O Type	Device	Description
Digital Input (NO)	Pump 2 Motor Starter Aux Contact	Input is true when Pump 2 motor starter is energized.
Digital Input (NO)	Pump 2 Start PB	When true, starts Pump 2 in Manual Mode.
Digital Input (NC)	Pump 2 Stop PB	When true, Pump 2 is allowed to run in Manual Mode.
Digital Output	Pump 2 Motor Starter	When true, Pump 2 motor starter is energized (pulled-in) and Pump 2 runs.

Challenge

Incorporate motor fault monitoring logic into each pump motor circuit. How will this fault condition be communicated, what impact will it have on duplex pump operations, and how will the operator acknowledge and resume pump operations after a motor fault?

Take this to the next level by adding a high/high float switch that, when activated, both pumps must then run until the sump is empty.

Glossary

ACTION: The actions to be taken when a Sequential Function Chart program step is active.

ADDRESS: A location in PLC controller memory.

ANALOG: A continuous signal that depends directly on magnitude (voltage or current) to represent some condition. For example, a voltage might represent the speed of a motor (5V corresponding to 200 rpm, 10V corresponding to 400 rpm, etc.).

ANALOG INPUT: A continuous electrical signal that depends directly on magnitude (voltage or current) to represent a process condition that a PLC is monitoring. For example, a voltage signal might represent the temperature in a boiler (5V corresponding to 750 degrees, 10V 1500 degrees, etc.).

ANALOG OUTPUT: A continuous electrical signal that depends directly on magnitude (voltage or current) to control a process condition and supplied by a PLC. For example, a current signal might represent the position of a valve (12ma corresponding to 50% opening, 16ma 75% opening, etc.).

ANALOG INPUT MODULE: A module that converts an analog input signal to a binary or decimal number for use by the processor.

ANALOG OUTPUT MODULE: A module that provides an output proportional to a binary or a decimal number provided to the module by the processor.

AND LOGIC: Logic that has a series relationship. Two devices that are in series would both have to be true to pass the logic. Devices one AND two need to be true for the logic to pass.

ARRAY: An array is similar to a data file and lets you group data of the same data type using a common name. An array tag occupies a contiguous block of memory.

ASCII: Acronym for **A**merican **S**tandard **C**ode for **I**nformation **I**nterchange. It is a seven- or eight-bit code for representing alphanumeric, punctuation marks, and certain special characters for control purposes.

ASSIGNMENT: Use an assignment statement to assign values to tags when using structured text programming.

ASYNCHRONOUS SHIFT ARRAY: An instruction that shifts data one word at a time into an array or file.

ASYNCHRONOUS SCAN: A PLC program scan in which input data can change at any time during execution of the logic.

ATTENUATION: In electronics, it is the loss of signal strength as measured in decibels (dB).

BANDWIDTH: The amount of data that can be transmitted in a fixed amount of time. For digital devices, the bandwidth is usually expressed in bits per second (bps) or bytes per second. For analog devices, the bandwidth is expressed in cycles per second, or Hertz (Hz).

BAUD: The seven or eight bits that make up a character. A character can be a letter, number, symbol, etc.

BAUD RATE: A unit of data transmission speed equal to the number of code elements (characters) per second. For example, 300 baud is 30 characters—letters, numbers, symbols—per second. 1200 baud is 120 characters per second.

BINARY: A numbering system that uses a base of 2. There are two digits (1 and 0) in the binary system.

BINARY CODED DECIMAL (BCD): One of several numbering systems used with PLCs. This unique numbering system uses four binary digits to represent each decimal digit from 0 to 9. Groups of four binary digits are grouped together to display decimal numbers. Twelve bits can represent a three-digit number. Sixteen bits are needed to represent a four-digit number.

BIT: An acronym for **B**inary dig**IT**. A bit can be only one of two possible states: *ON* or *OFF*, high or low, logic 1 or logic 0, etc.

BOOL EXPRESSION: A BOOL expression uses BOOL tags, relational operators, and logical operators to compare values or check if conditions are true or false. A BOOL expression can be found in structured text and sequential function chart programming.

BOOT: A term used in the computer world to indicate that a computer has been turned on, and the software has been loaded.

BRANCH: A parallel logic path within a user program rung.

BREAKDOWN VOLTAGE: The voltage at which a disruptive discharge takes place, either through or over the surface of insulation.

BUFFER: A temporary storage area where information is held while a device catches up with the transmission speed of the data.

BUFFERING I/O: Temporarily storing the status of PLC input and/or output data during programming execution.

BUS: A central cable that connects all devices on a local area network (LAN). It is also called the backbone.

BUS TOPOLOGY: The physical configuration of a communications network in which all devices are connected to a central cable, called the bus or backbone. Bus networks are relatively inexpensive and easy to install for small networks. Many PLC I/O systems use a bus topology.

BYTE: A sequence of binary digits usually operated upon as a unit (normally eight bits).

CASCADING: A programming technique that extends the ranges of timer and/or counter instructions beyond the maximum values that normally may be accumulated.

CENTRAL PROCESSING UNIT (CPU): Another term for **PROCESSOR**.

CHARACTER: One symbol of a set of elementary symbols, such as a letter of the alphabet, a decimal numeral, a punctuation mark, etc.

CLOCK: A device (usually a pulse generator) that generates periodic signals for synchronization or timing.

CMOS: An acronym for **C**omplementary **M**etal **O**xide **S**emiconductor. A family of very low-power, high-speed integrated circuits.

COAXIAL CABLE: A type of cable that consists of a center wire surrounded by insulation and then a grounded shield of braided wire. The shield minimizes electrical and radio frequency interference. Coaxial is the primary type of cabling used by the cable television industry and is also used for many PLC networks, such as ControlNet.

CODE: A system of symbols (bits) for representing data (characters).

COMPARE FUNCTION: A program instruction that compares numerical values for "equal," "less than," "greater than," etc.

COMPATIBILITY: The ability of various specified units to replace one another, with little or no reduction in capability.

CONSTRUCT: A conditional statement used to trigger structured text code when using structured text programming.

CONTROL VARIABLE (CV): An electrical or pneumatic signal that is used to control the position, speed, etc., of devices such as valves, motors, and pumps that have a direct influence on the process in which they are placed.

COUNTER: A device that can count up or down in response to transitions (*OFF* or *ON*) of an input signal and opens and/or closes contacts when a predetermined count is reached. Counters are internal to the processor and are not real-world devices.

CPU: An abbreviation for **C**entral **P**rocessing **U**nit. It is used interchangeably with **PROCESSOR**.

CRC: Short for **C**yclic **R**edundancy **C**heck, a common technique for detecting data transmission errors. Transmitted messages are divided into predetermined lengths that are divided by a fixed divisor. According to the calculation, the remainder number is appended onto and sent with the message. When the message is received, the computer or PLC recalculates the remainder and compares it to the transmitted remainder. If the numbers do not match, an error is detected.

DATA COMPARE: The process of comparing data in PLC memory for equality.

DATA MANIPULATION: The process of altering and/or exchanging data between memory words.

DATA TRANSFER: The process of exchanging data between PLC memory words and/or areas.

DECREMENT: A term used with counters to indicate that the value of the counter has decreased. When a counter value goes from 4 to 3, it is said to have decremented by 1.

DEFAULT: The initial setting of a value, or the initial assignment of a file by the software. Default values may or may not be changeable, depending on the software and the PLC manufacturer.

DERIVATIVE TIME: In the case of a process controller, the change in the output by an amount proportional to the rate of change of the process variable.

DIGITAL: The representation of numerical quantities by means of discrete numbers. It is possible to express in binary digital form all information stored, transferred, or processed by dual-state conditions; for example, *ON/OFF*, open/closed, etc.

DINT: A data type that stores a 32-bit signed integer value.

DISCRETE INPUT: An input that is either *ON* or *OFF*. Examples of discrete inputs are limit switches, push buttons, float switches, etc.

DISCRETE INPUT MODULE: A module that converts signals from real-world input devices to logic-level signals for use by the processor.

DISCRETE OUTPUT: An output that is either *ON* or *OFF*. Examples of discrete outputs are solenoids, motor starter coils, pilot lights, etc.

DISCRETE OUTPUT MODULE: A module that converts the logic levels of the processor to an output signal to control a real-world outputs.

DOWNLOAD: The process of taking a PLC project saved on a computer and loading it into the PLC controller memory.

DUPLEX: A means of two-way data communication. Also see **FULL DUPLEX** and **HALF DUPLEX**.

EEPROM or E2PROM: A memory chip that can be programmed using a standard programming device, and can be erased when the proper signal is applied to the erase pin. The initials stand for **E**lectrically **E**rasable **P**rogrammable **R**ead **O**nly **M**emory.

ELECTRICAL NOISE: Noise, or voltage spikes, that is generated whenever inductive loads such as relays, solenoids, motor starters, and motors are operated by "hard contacts" such as push buttons, selector switches, and relay contacts.

ELEMENT: A program instruction (normally open contact, timer, counter, etc.) displayed on a monitor. An element is also an addressable unit of data that is a sub-unit of a larger unit.

EMI: Electromagnetic induction. The term used to describe electrical noise.

ENABLED: A term used to indicate that a function or operation has been activated.

EVEN PARITY: The condition that occurs when the sum of a string of binary digits, 1s and 0s, is an even number. Parity is used for error checking.

EXAMINE OFF: An EXAMINE OFF PLC instruction is a true precondition if its addressed bit is *OFF* (0). It is false if the bit is *ON*, or 1.

EXAMINE ON: An EXAMINE ON instruction is a true precondition if its addressed bit is *ON* (1). It is false if the bit is *OFF*, or 0.

EXPRESSION: An expression is part of a complete assignment or construct statement and evaluates to a number or to a true or false state when using structured text programming.

FALSE: When relating to PLC instructions, an *OFF* state or condition.

FAULT: Any malfunction that interferes with normal operation.

FIELD WIRING: The electrical wires connecting the field devices (push buttons, pilot lights, motor starters, etc.) to the PLC I/O modules. Also considered all electrical wiring external to a PLC enclosure that houses the PLC and I/O modules.

FIFO: First-In First-Out. A reference to the way that information is stored and removed from an array or file.

FILE: A group of words, usually consecutive, that is used to store information.

FLASH MEMORY: Non-volatile computer memory storage medium that can be electrically erased and reprogrammed.

FORCE: A mode of operation or instruction that allows the operator (as opposed to the processor) to control the state of an input or output device.

FORCE OFF FUNCTION: A feature that allows the user to de-energize any input or output by means of the programmer, independent of the PLC program.

FORCE ON FUNCTION: A feature that allows the user to energize any input or output by means of the programmer, independent of the PLC program.

FRAME: A packet of transmitted information.

FULL DUPLEX (FDX): A mode of communications in which data may be simultaneously transmitted and received by both ends (sender/receiver).

GROUND: A conducting connection, intentional or accidental, between an electric circuit or equipment chassis and the earth ground.

GROUND POTENTIAL: Zero voltage potential with respect to earth ground.

HALF DUPLEX (HDX): A mode of data transmission capable of communicating in two directions, but in only one direction at a time.

HARD CONTACTS: Any type of physical switch contacts contrasted with electronic switching devices, such as triacs and transistors.

HARD DRIVE: A storage system that consists of an inflexible (hard) disk, as opposed to a portable memory device, that is used to store files, directories, software programs, etc.

HARDWARE: The mechanical, electrical, and electronic devices that constitute a PLC and its application.

HARDWIRED: Devices interconnected through physical wiring.

HEAT SINK: Heat sinks work on convection and are used to dissipate the heat generated by electronic devices.

HEXADECIMAL: The numbering system that represents all possible statuses of 4 bits with 16 unique digits (0–9 then A–F).

HOLDING REGISTER: A register or file that holds a value or values for comparison or for use in a user program.

HUMAN MACHINE INTERFACE (HMI): A computer-type device that provides the ability of a human to monitor and/or control a machine or process. Most HMIs are standard computers running special graphical interface software designed to communicate to one or more PLC controllers and/or other intelligent devices.

IEEE: An acronym for **I**nstitute of **E**lectrical and **E**lectronics **E**ngineers.

INCREMENT: A term used with counters to indicate that the value has increased. When a counter has counted up from 3 to 4, it is said to have incremented by 1.

INPUT DEVICES: Devices such as limit switches, pressure switches, push buttons, etc., that supply data to a PLC. These real-world inputs are of two types: those with common returns, and those with individual returns (referred to as isolated inputs). Other inputs include analog devices and digital encoders.

INSTRUCTION: A command or order that causes a PLC to perform a single prescribed operation.

INT: A data type that stores a 16-bit integer value.

INTEGRAL GAIN: In the case of a process controller, the change in the output by an amount proportional to the error and the duration of the error.

INTERFACING: Interconnection of a PLC with its input and output devices and data terminals through various modules and cables. Interface modules convert PLC logic levels into external signal levels, and vice versa.

INTERPOSING RELAY: A relay that is added to a PLC circuit to handle current values larger than can be handled by one terminal of an output module.

INTERRUPTIBLE: Interruptible refers to a timer that can be interrupted but still retain its accumulated time.

I/O: An abbreviation for **I**nput/**O**utput. For example, a group of input modules and output modules would be referred to as I/O modules.

I/O MODULE: The printed circuit assembly that interfaces between the user devices and the PLC.

I/O RACK: A chassis that contains I/O modules.

I/O SECTION: Interfaces the different signals from real-world devices and sensors to signals the CPU can use.

JUMPER: A short length of conductor used to make a connection between terminals, around a break in a circuit, or around a device.

KEYED: A term used with PLCs to indicate that a keying device has been installed to prevent I/O modules from being installed in the wrong slot.

KILO-: A prefix used with units of measurement to designate quantities 1000 times as great (as in kilowatt). The exception to K having a value of 1000 is when referring to computer memory. Computer memory is counted using the binary numbering system and one kilo (or K) of memory is 1024, not 1000 ($2^{10} = 1024$).

LADDER DIAGRAM: A complete control scheme normally drawn as a series of contacts and coils arranged between two vertical supply lines so that the horizontal lines of contacts appear similar to rungs of a ladder. A ladder diagram is normally the reference document used by the operator when entering the control program. (See **CONTACT SYMBOLOGY DIAGRAM.**)

LADDER DIAGRAM PROGRAMMING: A method of writing a user's PLC program in a format similar to a relay ladder diagram.

LANGUAGE: A set of symbols and rules for representing and communicating information (data) among people, or between people and machines.

LATCH: A device that continues to store the state of the input signal after the signal is removed. The input state is stored until the latch is reset.

LATCH INSTRUCTION: A PLC instruction that causes an output to stay *ON*, regardless of how briefly the instruction is enabled. (It can only be turned *OFF* by an **UNLATCH INSTRUCTION** in a separate rung.)

LATCHING RELAY: A relay constructed so that it maintains a given position by mechanical means until released mechanically or electrically.

LEAST SIGNIFICANT DIGIT (LSD): The digit that represents the smallest value. In the number 102, the 2 is the least significant number, or digit.

LED: Acronym for **L**ight **E**mitting **D**iode.

LIFO: Last-In First-Out. A reference to the way information is stored or removed from an array or file.

LIMIT SWITCH: A switch that is actuated by some part or motion of a machine or equipment to alter the electrical circuit associated with it.

LIQUID CRYSTAL DISPLAY (LCD): A reflective visual readout. Because its segments are displayed only by reflected light, it has extremely low power consumption, as contrasted with an LED display, which emits light.

LOAD: 1. The power delivered to a machine or apparatus. 2. A device intentionally placed in a circuit or connected to a machine or apparatus to absorb power and convert it into the desired useful form. 3. To place data (e.g., a ladder diagram) into the processor's memory.

LOCAL AREA NETWORK (LAN): A computer or PLC network that spans a relatively small area. Most LANs are confined to a single building or group of buildings. However, one LAN can be connected to other LANs over any distance via fiber optic cables, telephone lines, and radio waves. A system of LANs connected in this way is called a wide area network (WAN).

LOGIC LEVEL: The voltage magnitude associated with signal pulses representing ones and zeros (1s and 0s) in binary computation.

MAIN ROUTINE: The first routine to execute and be used to call (execute) other routines.

MANIPULATION: The process of controlling bits or words within a program to obtain the required program outcomes.

MASK: Bits in a word that are used to prevent other bits in a different word from being used. If there is a 1 in the bit location of the mask, the corresponding bit in the output word is enabled and can be turned *ON* and *OFF*. If the mask bit is set to 0, the corresponding bit in the output word is disabled, and does not allow the output to be turned *ON*, even if the program called for the bit to be turned *ON*.

MEMORY: The section of the PLC that stores the user program and other data. The storage may be either temporary or semi-permanent.

MENU: A display on the computer or PLC monitor that offers options or gives the operator choices to select from.

METROPOLITAN AREA NETWORK (MAN): A data network designed for a town or city. In terms of geographic area, MANs are larger than local area networks (LANs), but smaller than wide area networks (WANs). MANs are usually characterized by very high speed connections using fiber optic cable or other digital media.

MIDDLE DIGIT (MD): The middle digit of a three-digit number.

MILLIAMPERE (mA): One thousandth of an ampere or 0.001 ampere.

MILLISECOND (ms): One thousandth of a second or 0.001 second.

MICRO PLC: A scaled-down version of a standard PLC with small I/O capability.

MNEMONIC: A shorthand notation used with PLC instructions, such as OTE, the mnemonic for Output Energized; BST, for Branch Start; and so forth.

MODE: A selected method of operation (for example, *run, test,* or *program*).

MODULE: An interchangeable "plug-in" item containing electronic components that may be combined with other interchangeable items to form a complete unit.

MOST SIGNIFICANT DIGIT (MSD): The digit representing the greatest value. In the number 102, the 1 is the most significant number, or digit.

MOTOR CONTROLLER: A device, or group of devices, that serves to govern, in a predetermined manner, the electrical power delivered to a motor.

NAND LOGIC: NAND logic is a combination of a NOT gate and an AND gate. By placing the invert, or NOT, symbol at the output of the AND gate the output can only be true when one or both of the inputs are false, or set to 0.

NESTING: A programming technique that has a "branch-within-a-branch." Depending on the PLC manufacturer, nesting may or may not be allowed.

NETWORK: A group of connected logic elements used to perform a specific function. A network can range from one element to a complete matrix of elements, plus coil(s) as desired by the user. The size and configuration of the matrix or rungs varies with PLC manufacturers.

NETWORK ACCESS CONTROL: The means used to control access to a data network. Common network access control methods include Producer/Consumer, Token Passing, and CSMA/CD.

NETWORK MEDIA: In computer and PLC networks, media refers to the cables linking processors together. There are many different types of transmission media, the most popular being twisted-pair wire, coaxial cable, and fiber optic cable.

NODE: A common connection point between two or more contacts or elements in a circuit. In communication networks, a processing location. A node can be a PLC, a computer, or some other device, such as a printer. Every node has a unique network address, sometimes called a Node address, Data Link Control (DLC) address, or Media Access Control (MAC) address.

NOISE: Extraneous signals; any disturbance that causes interference with the desired signal or operation.

NONVOLATILE MEMORY: A memory that is designed to retain its information even though its power supply is turned off.

NOR LOGIC: NOR logic is the combination of a NOT gate and an OR gate. The NOR gate will only be logically true when both inputs, NOT and OR, are false, or set to 0.

NOT LOGIC: The NOT gate, often referred to as the inverter, will have only one input lead and one output lead. If the input is *OFF*, or set to 0, then the output will be *ON*, or set to 1. If the input is *ON*, or set to 1, then the output will be *OFF*, or set to 0.

OCTAL NUMBERING SYSTEM: A numbering system that uses a base of 8. Only the digits 0 through 7 are used.

ODD PARITY: The condition that occurs when the sum of 1s and 0s in a binary word is an odd number. Parity is used for error-checking.

OFF-DELAY TIMER: 1. In relay panel application, a device in which the timing period is initiated upon de-energization of its coil. 2. In a PLC, an instruction that starts the delay whenever the timer rung goes false.

OFF-LINE PROGRAMMING: A method of programming that is done in programming software as compared to programming on the device (PLC).

OFFSET: To offset is to equalize or compensate. In the case of linear interpolation formulas, the amount to be added +/− to the final calculated value.

ON-DELAY TIMER: 1. In relay panel applications, a device in which the timing period is initiated upon energization of its coil. 2. In a PLC, an instruction that starts the delay whenever the timer rung goes true.

ON-LINE OPERATION: Operations in which the PLC is directly controlling the machine or process.

ON-LINE PROGRAMMING: A method of programming by which rungs in the program may be inserted, changed, or deleted while the processor is running and controlling the process equipment.

OPERAND: 1. Either of the two numbers used in a basic computation to produce an answer. For example, in the computation $2 \times 3 = 6$, 2 and 3 are the operands. 2. Data required for the operation of a special function.

OPTICAL FIBER: A technology that uses glass (or plastic) threads (fibers) to transmit data. A fiber optic cable consists of a bundle of glass threads, each of which is capable of transmitting messages modulated onto light waves.

OPTICALLY COUPLED/OPTICAL ISOLATION: The use of a light-emitting diode and a photo transistor to communicate a signal or state to the processor. Optical coupling is used in input and output modules to isolate the logic level signal for line voltage sources.

OR LOGIC: Logic that has a "parallel" relationship. When two devices are in parallel, if either device one OR device two is true, the logic passes.

OUTPUT CIRCUIT: An output module point, real-world device (e.g., motor starter, digital readout, solenoid, etc.), and its associated wiring. The output module's function is to convert processor signal levels to field voltage levels necessary to control the real-world devices.

OUTPUT DEVICES: Devices such as solenoids, motor starters, etc. that receive data (control) from the PLC.

OUTPUT SIGNAL: A signal provided by the processor to the real-world output devices that controls their status (*ON* or *OFF*).

OVERLOAD: A load greater than that which a device is designed to handle.

PACKET: A piece of a message transmitted over a packet type network. One of the key features of a packet is that it contains the destination address in addition to the data.

PARITY BIT: An additional bit added to a binary word to make the sum of the number of 1s in a word always even or odd.

PASSWORD: A word used to gain access to a program or process. A password serves the same function as a hardware key, except that it is not a piece of hardware, but rather a word that when entered at the keyboard, gains access for the user to the program.

PC: Abbreviation for **P**ersonal **C**omputer (also used as an abbreviation for **P**rogrammable **C**ontroller). To avoid the confusion caused by using the same abbreviations for two different types of systems, the programmable logic controller is now most often referred to as a **PLC**.

PEER-TO-PEER: A type of network in which each station has equivalent capabilities and responsibilities. This differs from client/server or master/slave architectures, in which some stations are dedicated to serving the others. Peer-to-peer networks are often called token passing networks. Peer-to-peer networks are generally simpler, often found in PLC networks, but lack performance under heavy loads.

PILOT DEVICE: A device used in a circuit that performs a control function only. Pilot devices are limited to 10 amps of current-carrying capacity, and are to be used in control circuits only. They are not designed to control the power and current required by the operating equipment.

PLC: See **PROGRAMMABLE LOGIC CONTROLLER**.

POWER SUPPLY: Supplies the DC power for the CPU and for the I/O section. The voltage is typically +5 volts. The power supply can be internal with the processor, rack-mounted, or externally mounted as a separate unit. A separate power supply is required if DC voltage is required for the actual input and/or output devices.

PRIORITY: Order of importance.

PROCESS VARIABLE (PV): Refers to the physical measurement of a process component such as temperature, level, pressure, flow, speed, etc. The physical measurement is converted into an electrical or pneumatic signal that varies proportionally with a change in the measured quantity. Devices such as transducers and transmitters are used to convert the physical measurement into a corresponding electrical or pneumatic signal.

PROCESSOR/PROCESSOR UNIT: The part of the PLC that performs logic-solving, program storage, and special functions within a PLC system. It scans all the inputs and outputs in a predetermined order. The processor monitors the status of the inputs and outputs in response to the user-programmed instructions, and energizes or de-energizes outputs as a result of the logical comparisons made through these instructions.

PRODUCER/CONSUMER: Producer/Consumer is a communications method in which all nodes on a network can simultaneously access the same data from a single source (Producer). The data is produced only once, regardless of the number of nodes, and if a node needs the data, it consumes it, otherwise the data is ignored.

PROGRAM: A sequence of instructions executed by the processor to control a machine or process. Also, a set of related routines and tags found in the Logix 5000 controllers.

PROGRAMMABLE LOGIC CONTROLLER (PLC): A solid-state control system that has a user-programmable memory for storage of instructions to implement specific functions such as I/O control logic, timing, counting, arithmetic, and data manipulation. A PLC consists of the processor, input/output interface, memory, and programming device that typically uses relay-equivalent symbols. The PLC is purposely designed as an industrial control system to perform functions equivalent to a relay panel or a wired solid-state logic control system.

PROGRAMMER: A device that is needed to enter, modify, and troubleshoot the PLC program, and check the condition of the processor. The programmer may be hand-held or a personal computer.

PROGRAM SCAN TIME: The time required for the processor to execute all instructions in the program one time.

PROPORTIONAL GAIN: In the case of a process controller, it is the change in the output by an amount proportional to the error between the set point and process variable.

PROTOCOL: A defined means of establishing criteria for receiving and transmitting data through communication channels.

RACK: A PLC chassis that contains modules. Some PLC manufacturers, like Allen-Bradley, use the term *rack* to indicate a given number of I/O points rather than to identify a specific piece of hardware. In the Allen-Bradley scheme, a chassis could contain a number of racks. With most other manufacturers, however, rack and chassis are used interchangeably, and mean the hardware that holds the various modules, power supplies, etc.

RADIO AREA NETWORK (RAN): A communication network that uses radio waves as the medium that interconnects the computers or PLCs to form a network. Most RANs are used in the control field to interconnect PLCs where no other means is possible or practical. RANs can quite often be found in the water and wastewater industries.

RAM: Acronym for **R**andom **A**ccess **M**emory. Random access memory is a type that can be read from (accessed) or written into by the user.

RANDOM ACCESS: See **RAM**.

RATED VOLTAGE: The maximum voltage at which an electrical component can operate for extended periods without damage or undue degradation.

READ/WRITE MEMORY: A memory into which data can be placed (*write* mode) or accessed (*read* mode). The *write* mode destroys previous data; the *read* mode does not alter stored data.

REAL: A data type that stores a 32-bit IEEE floating-point value.

REGISTER: A word or group of words used to store numerical values.

REPORT: A display of data, or a printout, containing data and/or information that is useful to the user or operator. Reports can include operator messages, part records, production lists, etc. Reports are normally stored in a memory area separate from the user's program.

REPORT GENERATION: The printing or displaying of user-formatted application data by means of a programming device. Report generation can be initiated by means of either the user's program or a programming device keyboard.

RETENTIVE: To retain a value or time.

RETENTIVE OUTPUT: An output that remains in its last state (*ON* or *OFF*), depending on which of its two program rungs (one containing a **LATCH INSTRUCTION**, the other an **UNLATCH INSTRUCTION**) was the last to be true. The retentive output remains in its last state when both rungs are false. It also remains in its last state if power is removed from, then restored to, the PLC.

RETENTIVE TIMER: A PLC instruction that accumulates the amount of time, whether continuous or not, when the preconditions of its rung are true, and which controls one or more outputs after the total accumulated time is equal to the preset time. When the rung is false, the accumulated time is retained. Moreover, if the outputs have been energized, they remain *ON*. Additionally, the accumulated time and energized outputs are retained if power is removed from, then restored to, the PLC.

RING TOPOLOGY: All devices are connected in the shape of a closed loop, so that each device is connected directly to two other devices, one on either side of it. Ring topologies are relatively expensive and difficult to install, but they offer high bandwidth and can span large distances.

ROM: Acronym for **R**ead **O**nly **M**emory. A read only memory is a solid-state digital storage memory whose contents cannot be altered by the user.

ROUTINE: A set of logic instructions in a single programming language in a Logix 5000 controller and similar to a program file.

RS-232C: An Electronic Industries Association (EIA) standard for data transfer and communication.

RTD: Resistance Temperature Detector, or temperature probe, that can be connected to a special PLC input module to indicate temperature reading.

RUNG: A grouping of PLC instructions that control one output or storage bit. Some PLCs can have multiple outputs on the same rung. A rung is also referred to as a network.

SCAN: The time required for a program to make one complete scan through memory and update the status of all inputs and outputs.

SCHEMATIC: A diagram of a circuit in which symbols illustrate circuit components.

SCR: An acronym for **S**ilicon **C**ontrolled **R**ectifier. The SCR is used to convert AC current to DC current.

SEQUENCER: A controller that operates an application through a fixed sequence of events.

SERIAL COMMUNICATION: A type of communication or information transfer within a PLC whereby the bits are handled sequentially rather than simultaneously as they are in parallel communications. Serial operation is slower than parallel operation for equivalent clock rate. However, only one channel is required for serial operation.

SETPOINT (SP): The desired operating point that the process system is to operate at. The setpoint value typically has the same range as the process variable being measured and is determined by the operator of the system.

SHIELDING: The practice of confining the electrical field around a conductor to the primary insulation of the cable by putting a conducting layer around the cable insulation.

SIGNED BIT: A way to indicate that a number is negative or positive.

SINGLE PRECISION ARITHMETIC: Only one word is used to store the results of math or arithmetic operations. Single precision arithmetic is limited to a maximum value of 999. (See **DOUBLE PRECISION ARITHMETIC.**)

SINT: A data type that stores an 8-bit signed integer value.

SLOPE (RATE): An inclined or slanting direction downward or upward from the horizontal, as in the case of an *xy* graph. It is the degree of the incline or amount of change.

SOFTWARE: The manufacturer's program that controls the operation of a PLC.

SOLID-STATE: Circuitry designed using only integrated circuits, transistor, diodes, etc.; no electromechanical devices such as relays are utilized. High reliability is obtained with solid-state logic.

SOLID-STATE DEVICES (SEMICONDUCTORS): Electronic components that control electron flow through solid materials (e.g., transistors, diodes, integrated circuits, etc.).

STAR TOPOLOGY: All devices are connected to a central hub. Star networks are relatively easy to install and manage, but bottlenecks can occur because all data must pass through the hub. Today switches are used in place of hubs, all but eliminating bottlenecks.

STATE: The logic condition, 1 or 0, in PLC memory, or at a circuit's input or output.

STEP: A step represents a major function of your process and contains actions. Steps can be found in sequential function chart programming.

STORAGE: Synonymous with **MEMORY**.

STORAGE MEMORY: That part of the memory that stores the status of the input and output devices, numeric values for timers and counters, numeric values for arithmetic functions, status of internal relays, and information stored in holding and storage registers.

SURGE: A transient variation in the current and/or voltage at a point in the circuit.

SWITCHING: The action of turning a device *ON* and *OFF*.

SYMBOLIC NAME: A user designation for an application I/O device (e.g., S-1, LS-4, or SOL-7).

SYNCHRONOUS SCAN: The updating of I/O and execution of program logic in a prescribed order.

SYNCHRONOUS SHIFT REGISTER: An instruction that shifts information one bit at a time within a word or from one word to another.

TAG: A named area of memory where data is stored in the controller. Tags are the basic mechanism for allocating memory in the Logic 5000 controllers.

TASK: A scheduling mechanism for executing a program in the Logix 5000 controllers.

THUMBWHEEL SWITCH: A rotating numeric switch used to input numeric information to a controller.

TIMER: In relay panel hardware, an electromechanical device that can be wired and preset to control the operating interval of other devices. In a PLC, a timer is internal to the **PROCESSOR**, meaning that it does not exist in the real world, but can be controlled by a user-programmed instruction. A timer instruction has greater accuracy and timing range than a hardware timer.

TOGGLE SWITCH: A panel-mounted switch normally used for *ON* or *OFF* switching.

TOPOLOGY: The shape of a local area network (LAN) or other communications system. Topologies are either physical or logical. There are four principal topologies used in LANs: bus, star, ring, and tree.

TRANSFORMER COUPLING: One method of isolating I/O devices from the controller.

TRANSITION: A transition is the physical conditions that must occur or change in order to go to the next step in a sequential function chart programming.

TREE: A command that is used to display the organization of all the directories, subdirectories, and files on a given disk or hard drive.

TREE TOPOLOGY: A tree topology combines characteristics of linear bus and star topologies. It consists of groups of star-configured workstations connected to a linear bus backbone cable.

TRIAC: A solid-state component capable of switching alternating current.

TRUE: As related to a PLC instruction, an enabling logic state or *ON* condition. (See **FALSE**.)

TRUTH TABLE: A matrix that shows all the possible states (*ON* or *OFF*) of a single input device or combination of input devices, and the corresponding state (*ON* or *OFF*) of the output device(s).

TTL: Abbreviation for **T**ransistor–**T**ransistor **L**ogic, a family of integrated circuit logic. (Usually 5 volts is high, or 1, and 0 volts is low, or 0.)

TUTORIALS: Text included in software that provides the user with helpful information on how the software works or how to use functions of the software. Typically can be accessed by using the "Help" option.

TWISTED-PAIR CABLE: A type of cable that consists of two independently insulated wires twisted around one another. The use of two wires twisted together helps to reduce crosstalk and electromagnetic induction. Twisted-pair cable is the least expensive type of local area network (LAN) cable. Most computer and some PLC networks contain twisted-pair cabling at some point along the network.

2s COMPLEMENT: A convention for binary representation of negative and positive decimal numbers.

UNCONDITIONAL: A term applied to an output (or other instruction) that is always true.

UNLATCH INSTRUCTION: A PLC instruction that causes an output to unlatch, or turn *OFF*, regardless of how briefly the instruction is enabled. (It can only be turned back *ON* by a **LATCH INSTRUCTION** in a separate rung.)

UPDATING: A term used to indicate that the processor has scanned and checked the status of all input and output devices. After the status of the inputs and outputs is known, the data table is updated to reflect the current status.

UPLOAD: The process of saving a PLC project in controller memory to a computer file.

UPWARD COMPATIBILITY: The ability of a new version of software to support previous editions of the same software. If version 6.6 of a particular software allows documents and files from previous versions (5.2 and 6.0) to be read, the software is said to have upward compatibility.

USER MEMORY: The portion of memory that is set aside for the storage of the user program (i.e., ladder diagrams, program messages, etc.).

VALUE: 1. A number that represents a computed or assigned quantity. 2. A number contained in a register or file word.

VOLATILE MEMORY: A memory that loses its information if the power is removed.

WATCHDOG TIMER: A timer that is used within the PLC processor to verify that the program scan has been completed correctly in the allotted amount of time. If there is a program error, the scan is not completed in the prescribed amount of time, and the watchdog timer times out and indicates that there is a problem with the circuit.

WIDE AREA NETWORK (WAN): A computer network that spans a relatively large geographical area. Typically, a WAN consists of two or more local area networks (LANs). Computers connected to a wide area network are often connected through public networks, such as the telephone system. They can also be connected through leased lines or satellites. The largest WAN in existence is the Internet.

WORD: A grouping, or a number of bits, in a sequence that is treated as a unit.

WRITING OVER: A term used to indicate that information will be replaced (or the existing information will be written over) by new information.

XOR LOGIC: The XOR logic gate will only turn the output *ON* when either input A or B is *ON*—but not both *ON*.

Index